Spencer Christian's
Geography Book

Spencer Christian's Geography Book

by Spencer Christian

St. Martin's Griffin
New York

DEDICATION:

To my parents, gradeschool teachers, church and community leaders, and other wise and concerned adults who offered me guidance and encouragement when I was a child. These are the people who fueled my imagination and inspired my curiosity about the world around me. They taught me that knowledge and understanding are two important keys to freedom and empowerment.

Designed by Robert Engle Design; Robert Engle and Lorenz Skeeter

Production Coordination by John P. Holms

Library of Congress Cataloging-in-Publication Data

Christian, Spencer.
   [Geography Book]
   Spencer Christian's geography book / Spencer Christian.
     p.   cm.
   "A Thomas Dunne book."
   ISBN 0-312-13183-6 (pbk.)
   1. Geography — Popular works. I. Title. II. Title: Geography
book.
G116. C48   1995
910 — dc20   95-3761
                                      CIP

First St. Martin's Griffin Edition: September 1995
10 9 8 7 6 5 4 3 2 1

ACKNOWLEDGMENTS

Tom Biracree, whose tireless efforts provided me with the tremendous volumes of information which were necessary for the completion of this book.

The staff and management of *Good Morning Amerca*, for providing me the professional experiences which have greatly increased my knowledge and understanding of geography.

Tony Seidl, whose extensive experience and contacts in the publishing business made the publication of this book possible.

TABLE OF CONTENTS

INTRODUCTION

As weatherman, reporter, and ambassador of good will for *Good Morning America* I have had the very rich experience of traveling all over the United States, as well as across a good part of the world. So, it should not be surprising that I have acquired a considerable knowledge of geography; and along with that has come a keen understanding of the history, culture, and economy of the various states and regions of this country. This experience has not only been personally enriching and professionally helpful, it has deepened my appreciation for what it means to be an American.

My interest in geography began when I was a young child growing up in the rural central Virginia town of Charles City, about 25 miles southeast of Richmond. My family could not afford to travel—my father was a laborer, my mother a homemaker, and they worked very hard to provide just the simple comforts of home, food,

and decent clothing for my younger brother and me. Both of my parents had grown up in large families during the Great Depression, and they had faced all of the harsh indignities and denial of opportunity which were a part of being black in the South in those days. The greatest distance from home we had ever traveled during the first 13 years of my life was to Washington, D.C., which about 100 miles away. But, despite our economic deprivation, my parents instilled in me a great thirst for knowledge and curiosity about that vast world beyond Charles City.

I had learned to read by the age

of four, and, by the time I reached the 3rd or 4th grade, I had already developed a fascination with geography and knew that I wanted to see the world. Understanding that my parents could not afford to take us on exotic journeys, I traveled vicariously through whatever literature was available to me. In particular, I remember looking forward excitedly each week to the arrival of *The Weekly Reader*, a publication for elementary school students, which took me to far away places that I actually felt I had visited after reading about them. Whether my literary journeys were to other parts of the U. S., or to distant countries in other parts of the world, I would always go to the map, or globe, and find my destinations. Some of my classmates thought I was a little strange; I thought they were missing the boat. After, all, I could tell them about places from Key West to Seattle, from Angola to Antarctica.

When I finally did begin to travel—literally, as opposed to literarily—I discovered not only that my impressions of my travel destinations had been largely accurate, but that all of my reading had greatly enhanced my appreciation for and understanding of the places I visited. Even to my own amazement, I sometimes found myself "knowing my way around" upon my first visit to a U. S. city, simply because I was so familiar with major attractions, landmarks, history, and local customs. Can you imagine what a great feeling it is to go almost anywhere in America for the first time, and not feel like a stranger?

The main reason I am writing this book is to inspire others to take a greater interest in geography. In recent years seven surveys have shown that an alarming number of Americans (adults as well as children) cannot even locate their home states on a map. The level of geographic illiteracy climbs even higher, when people are asked to simply locate neighboring countries like Canada and Mexico. This is shameful!

But, there is a great deal more that I hope readers will get from this book than a knowledge of where boundary lines are drawn and what the various state capitals are. I want people to understand that geography is the mother of dozens of other academic disciplines, from history to social sciences to geology to economics to anthropology and many more (even meteorology—How could I describe the movement of weather systems each day on *Good Morning America* without thorough knowledge of geography?). In fact, the word "geography" means "the study of the earth and its features and the distribution of life on the earth", and that includes the real life human experiences and struggles which have resulted in the separations between two nations or two states. For example, a simple glance at theboundary line between Virginia and West Virginia may not mean much to you until you learn that they were once one state. Then, as the country moved toward the Civil War, the same conflicts that divided the nation divided the state.

So, you see, geography is about people as well as places. Geography tells us why French culture is so predominant in New Orleans; why Scandinavian immigrants settled in Minnesota; why great numbers of blacks migrated from the rural South to the urban North. It explains how we make a living, where we send our children to school, how we worshlp, and how we select our political leaders. In other words, geography touches on virtually every aspect of our lives. Geography has become so vital to our understanding of national and international affairs that, to my great delight, Good Morning America added a geography reporter, Prof. Harm de Blij, to its on-air staff several years ago. Harm's enlightening and insightful reports have informed and entertained me, and have even inspired me in the writing of this book.

In this book, I will take you from the origins of planet earth (according to accepted scientific findings) to the exploration, settlement, and growth of our nation, to America's natural wonders to detailed descriptions of all 50 states and the major U. S. territories. And, because you will encounter quite a few maps in this book, I begin with a section on how to read maps. I invite you to join me on what I believe is a fascinating journey—one which I began as a young child, and which I hope never ends. It is a journey that requires no budget, burns no fuel, and leaves no litter. Shall we begin?

—Spencer Christian

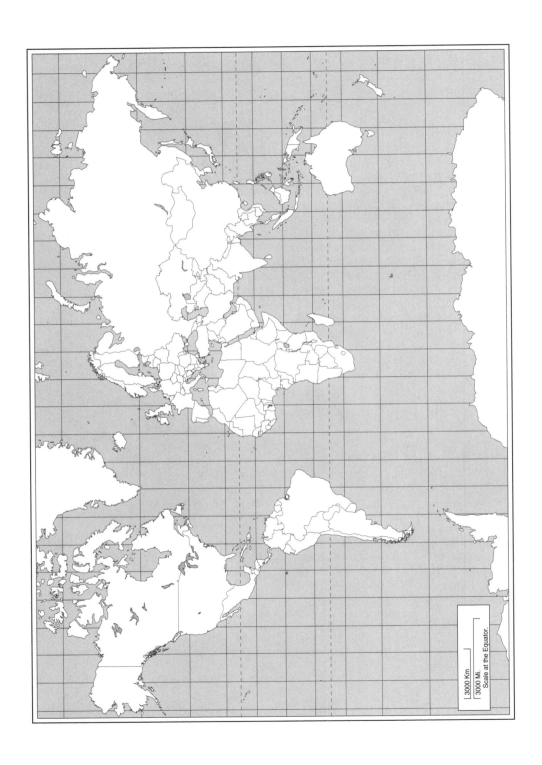

3000 Km

3000 Mi.

Scale at the Equator.

SECTION ONE

Worlds At Your Fingertips—
Everything You Should Know About Maps

You may be interested to know that less than half of the of the time I spend preparing my weather segments for "Good Morning America" is devoted to forecasting the weather. Our staff meteorologist and I receive extremely detailed computer models and forecasts from both the U.S. Weather Service and a private weather forecasting company. Most of the time, these models and forecasts are in general agreement, and editing them into a forecast is relatively routine (This is not the case when hurricanes, tornadoes, or other violent weather appears on the map!).

The really difficult task we face every day is presenting the weather forecasts and other information in a format that is easy for you viewers to grasp in the couple minutes I'm on camera. Our solution almost always involves maps. We can use maps in many different ways—imposed on satellite photos to show cloud cover, with lines to represent the movement of storm systems and fronts, with different colors to show temperature differences, with shades of colors that represent the intensity of precipitation, and so on. I can show maps of the world, of the U.S., of regions of the country, or of specific states.

After more than twenty years as a weather reporter, I've become quite a fan of maps. It would be remiss of me to do a geography book without introducing maps to you.

THE HISTORY OF MAPS

No doubt the first maps were directions to hunting grounds, berry patches, or other important sites drawn in dirt with a stick. How far back these simple maps were used, we'll never know, because they weren't mean to be preserved. Human nature being what it is, it's understandable that the earliest existing maps that have been found were land surveys cut into clay tiles by the Babylonian tax collectors about 2300 B.C. More extensive regional maps, drawn on silk and dating from the 2d centuryB.C., have been found in China. One of the most interesting types of primitive map is the cane chart constructed by the Marshall Islanders in the South Pacific Ocean. This chart is made of a gridwork of cane fibers arranged to show the location of islands. The art of mapmaking was advanced in both the Mayan and Inca civilizations, and the Inca as early as the 12th century AD made maps of the lands they conquered.

The first map to represent the known world is believed to have been made in the 6th century B.C. by the Greek philosopher Anaximander. It was circular and showed the known lands of the world grouped around the Aegean Sea at the center and surrounded by the ocean. One of the most famous maps of classical times was drawn by the Greek geographer Eratosthenes about 200B.C.. It represented the known world from England on the northwest to the mouth of the Ganges River on the east and to Libya on the south. This map was the first to be supplied with transverse parallel lines to show equal latitudes.. About 150 AD. the Egyptian scholar Ptolemy published his geography containing maps of the world. Following the fall of the Roman Empire, European mapmaking all but ceased; such maps as were made were usually drawn by monks, who often portrayed the earth inaccurately. Arabian seamen, however, made and used highly accurate charts during this same period. The Arabian geographer al-Idrisi made a map of the world in 1154. Beginning approximately in the 13th century, Mediterranean navigators prepared accurate charts of that sea, usually without meridians or parallels but provided with lines to show the bearings between important ports. These maps are usually called portolano or portolan charts. In the 15th century, editions of Ptolemy's maps were printed in Europe; for the next several hundred years these maps exerted great influence on European cartographers.

A map produced in 1507 by Martin Waldseemüller, a German cartographer, probably was the first to apply the name America to the newly discovered transatlantic lands. The map, printed in 12 separate sheets, was also the first to clearly separate North and South America from Asia. In 1570 Abraham Ortelius, a Flemish mapmaker, published the first modern atlas, Orbis Terrarum. It contained 70 maps. During the 16th century many other cartographers produced maps that incorporated the ever-increasing information brought back by navigators and explorers. It is Gerardus Mercator, however, who stands as the greatest cartographer of the age of discovery; the projection he devised for his world map proved invaluable to all future navigators.

The accuracy of later maps was greatly increased by more precise determinations of latitude and longitude and of the size and shape of the earth. The first maps to show compass variation were produced in the first half of the 17th century, and the first charts to show ocean currents were made about 1665. By the 18th century, the scientific principles of mapmaking were well established and the most notable inaccuracies in maps involved unexplored parts of the world.

By the late 18th century, as the initial force of world exploration subsided and as nationalism began to develop as a potent force, a number of European countries began to undertake detailed national topographic surveys. The complete topographic survey of France was issued in 1793; roughly square, it measured about 11 m (about 36 ft) on each side. Great Britain, Spain, Austria, Switzerland, and other countries followed suit. In the United States the Geological Survey was organized in 1879 for the purpose of making large-scale topographic maps of the entire country. In 1891 the International Geographical Congress proposed the mapping of the entire world on a scale of 1:1,000,000, a task that still remains to be completed. During the 20th century, mapmaking underwent a series of major technical innovations. Aerial photography was developed during World War I and used extensively during World War II in the making of maps. Beginning in 1966 with the launching of the satellite Pageos, and continuing in the 1970s with the three Landsat satellites, the U.S. has been engaged in a complete geodetic survey of the surface of the earth by means of high-resolution photographic equipment.

What Are the Different Types of Maps?

A map is a drawing of a geographic area done on a flat surface (if it's on a sphere, it's called a globe.). The information on a map depends its purpose. For example, roads, parks, or historical sites would just clutter up a weather map; on the other hand, a cold front and a few isobars won't help you find your way to Interstate 90. Among the common types of maps are;:

• Topographical maps show the natural features of an area, including rivers, lakes, mountains. They also frequently identify man-made or designated areas such as parks, campsites, and cities or towns.

• Nautical charts, which are used to navigate ships, not only show the surface of bodies of water and the shore, but also indicate depth of the water, the location of channels, hazards such as rocks or shoals, and the location of lighthouses and buoys.

• Aviation charts, like topographical maps, show the features of the land below. They also identify airways, the range of specific air traffic control towers, radio beacons, and airports.

• Political maps divide an area into artificially created units such as towns, counties, precincts, or legislative districts.

• Geologic maps show the geological structure of an area—for example, the type of rocks or soil and their age.

• Weather maps (my favorite, again) show the movement of weather systems, the distribution of rainfall or snowfall, or areas of equal temperature or air pressure.

• Social, scientific or economic maps are used for a great variety of purposes, from indicating the location of natural resources to the membership of various religious groups.

• Tourist maps show points of interest such as museums, historical houses, libraries, monuments, zoos, parks, etc.

• Satellite maps are outlines superimposed on actual photographs taken by cameras in orbit.

• Relief maps are three-dimensional models of terrain, with mountains, ridges and valleys sculpted out of clay or molded in plastic.

• Surveyor's maps show the exact dimensions of house lots, farms, or other pieces of property.

How to Read a Map

It takes a lot of skill to make a readable map, just like it does to produce a readable book (like this one!). But almost all maps have certain common elements. Systematically reading and understanding these elements will unlock the information contained on the map. These important elements are:

• The map title generally explains the primary purpose of map. For example, I just opened a book to a map of the United States entitled "Electoral Votes for President." The title tells me that the divisions must be states (identified by two letter abbreviations) and the number printed on each state must be the votes cast by that state in the Electoral College. All maps aren't this simple, but the title invariably provides a lot of information.

• Many maps have geographical grids that make it easier to locate places. The grid used on globes and national and international maps consists of lines of latitude and longitude. Lines of latitude show how far north or south of the Equator a specific location is. Latitude is measured in degrees north and south, with the Equator being zero degrees, the North Pole 90 degrees north, and the South Pole 90 degrees south. The lines of latitude are equal distance apart; Montreal, Canada, at 45 degrees north latitude, is halfway between the Equator and the North Pole. Each degrees of latitude is divided into 60 minutes.

Lines of longitude indicate how far east or west a location is from a line that runs from the North Pole to the South Pole. The problem was, for much of history, where that line should be drawn. Most countries used systems based on lines running through their capital cities, creating endless confusion. Finally, in 1884, all the nations of the world got together and decided to use a line running through England's Greenwich Observatory. So now lines of longitude and latitude run 180 degrees east and west of Greenwich, England. The line that represents 180 degrees is called the International Date Line (where Thursday becomes Friday). Memphis, Tennessee is at 90 degrees west latitude, or half way between Greenwich and the International Date line. Because the earth is curved, the lines of longitude are farthest apart at the Equator, and they all meet at the Poles. Each line of longitude is also divided into 60 minutes.

When we know a location's latitude and longitude, we can find it easily on a globe or world map. I find it interesting to follow lines of latitude around the globe—for example, we tend to think of Africa as being "south," but Cairo, Egypt is on the same latitude as Mobile, Alabama.

Obviously, latitude and longitude are too large to make a useful grid for a map of a state or city. So mapmakers generally create a grid, using letters (A-Z) and numbers in the other direction. A combination of a letter and a number (B-4) designates a box on the map, just like it does on a Bingo card.

• A map legend or key explains the symbols that are used on the map. On a weather map, for example, a cold front is designated by a line with triangles on one side, while a warm front is indicated by a line with half-circles on one side. Little x's indicate snow, short horizontal dashes mean ice, etc. A tourist map, on the other hand, could use an animal to mark a zoo and a book to mark a library.

• A map's scale is the relationship between the distance from one point to another on a map and the distance from one point to another on the earth's surface. For example, on most maps produced by the U.S. Geological Survey, one inch equals one mile. This scale is usually shown in the margin of the map as a line with divisions that indicate 5 miles, 10 miles, etc.

• Many maps indicate the varying heights, or relief, of the land they represent. In simple maps, hills or mountains may be indicated by little drawings; more accurate maps normally use contour lines to connect areas of equal elevation. The closer together the lines, the steeper the slopes. Finally, color shadings can also be used to indicate elevations.

Now, you can use these skills to interpret the following maps:

Agricultural Products

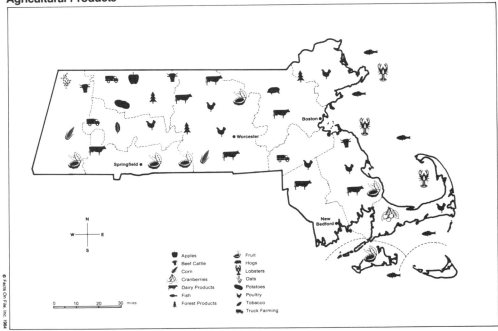

Apples
Beef Cattle
Corn
Cranberries
Dairy Products
Fish
Forest Products
Fruit
Hogs
Lobsters
Oats
Potatoes
Poultry
Tobacco
Truck Farming

Economic Map

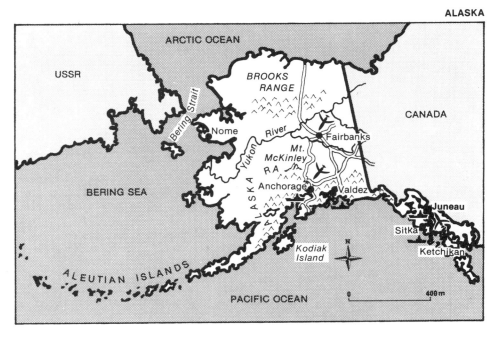

Topographical Map

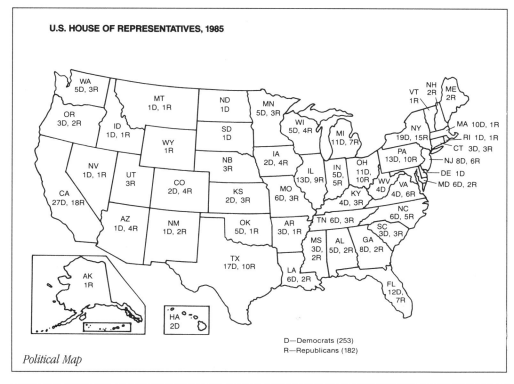

U.S. HOUSE OF REPRESENTATIVES, 1985

WA 5D, 3R
OR 3D, 2R
MT 1D, 1R
ID 1D, 1R
WY 1R
NV 1D, 1R
UT 3R
CA 27D, 18R
AZ 1D, 4R
NM 1D, 2R
ND 1D
SD 1D
NB 3R
CO 2D, 4R
KS 2D, 3R
OK 5D, 1R
TX 17D, 10R
MN 5D, 3R
WI 5D, 4R
IA 2D, 4R
MO 6D, 3R
AR 3D, 1R
LA 6D, 2R
MI 11D, 7R
IL 13D, 9R
IN 5D, 5R
OH 11D, 10R
KY 4D, 3R
TN 6D, 3R
MS 3D, 2R
AL 5D, 2R
GA 8D, 2R
WV 4D
VA 4D, 6R
NC 6D, 5R
SC 3D, 3R
FL 12D, 7R
NY 19D, 15R
PA 13D, 10R
VT 1R
NH 2R
ME 2R
MA 10D, 1R
RI 1D, 1R
CT 3D, 3R
NJ 8D, 6R
DE 1D
MD 6D, 2R
AK 1R
HA 2D

D—Democrats (253)
R—Republicans (182)

Political Map

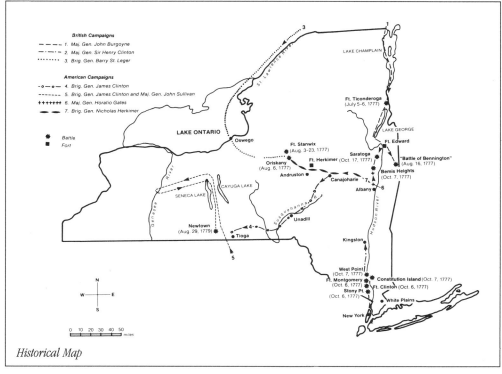

British Campaigns
- — — — 1. Maj. Gen. John Burgoyne
- —·—·— 2. Maj. Gen. Sir Henry Clinton
- ·········· 3. Brig. Gen. Barry St. Leger

American Campaigns
- -o-o- 4. Brig. Gen. James Clinton
- ------- 5. Brig. Gen. James Clinton and Maj. Gen. John Sullivan
- ++++++ 6. Maj. Gen. Horatio Gates
- ~~~~ 7. Brig. Gen. Nicholas Herkimer

✴ Battle
■ Fort

LAKE CHAMPLAIN
LAKE ONTARIO
LAKE GEORGE
St. Lawrence River
Genesee River
Susquehanna R.
Hudson River

Ft. Ticonderoga (July 5-6, 1777)
Ft. Edward
Oswego
Ft. Stanwix (Aug. 3-23, 1777)
Oriskany (Aug. 6, 1777)
Ft. Herkimer (Oct. 17, 1777)
Andruston
Saratoga
"Battle of Bennington" (Aug. 16, 1777)
Bemis Heights (Oct. 7, 1777)
Canajoharie
Albany
SENECA LAKE
CAYUGA LAKE
Unadill
Newtown (Aug. 29, 1779)
Tioga
Kingston
West Point (Oct. 7, 1777)
Ft. Montgomery (Oct. 6, 1777)
Constitution Island (Oct. 7, 1777)
Ft. Clinton (Oct. 6, 1777)
Stony Pt. (Oct. 6, 1777)
White Plains
New York

N
W — E
S

0 10 20 30 40 50 miles

Historical Map

Weather Map

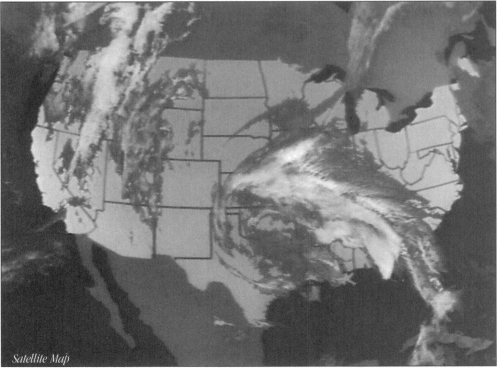

Satellite Map

SECTION TWO

The Landscape of America

In almost every field of endeavor, becoming better at what you do is a lot easier if you take the time to learn from the best people in your field. I've benefited enormously from working with many very talented people, including my present colleagues on "Good Morning America." But as a journalist whose job requires me to spend giant chunks of my time on the road, I have a special regard for Charles Kuralt, who recently retired from CBS. While it takes talent, experience, and hard work to cover headline news stories, there are many journalists who are up to the task. But Kuralt won the attention and affections of viewers all over the country through is unique ability to uncover fascinating stories in almost everyone of his thousands of stops along the backroads and in the small towns of America. By taking the time to look around him carefully and really listen to people, he was able, over decades, to fashion a mosaic of what life was really like in America.

When people ask me if I've grown tired of the constant travel after a decade of being GMA's Ambassador to America, I can honestly say that I enjoy it more and more every year. The reason is that I've become much better at looking and listening, and as a result, I find something fascinating at nearly every stop I make. Let me give you an example. There's a small town in Connecticut that's preserved a small tavern that dates back to the early 1700s. As a native Virginian and a history buff, I've visited thousands of Colonial buildings, and although this tavern was well taken care of, there was nothing really distinctive about it. When I walked inside, I must confess I was primarily concerned with how fast I could get back on the road and still be polite.

Then a wonderful woman, a volunteer tour guide, started talking. She explained that the tavern was a stop on the long carriage journey between New York and Boston in the days leading up to the Revolutionary War. She urged us to think of ourselves as weary travelers stepping out of a coach after a tedious, dusty few hours, and took us through what the experience of stopping at a tavern was really like. I found myself forgetting about time. Then she moved on to a day in 1777 when a large force of British troops approached the town on a march from interior Connecticut to the coast. Local farmers and merchants grabbed their muskets and joined a small company of American soldiers who fought a series of three skirmishes against the larger force as they marched down the main road. Casualties were heavy on both sides, and in the third confrontation, an American general was killed. As the British reached the town's main street, which I could see outside the tavern's window, they began torching houses. A handful of brave men stepped outside of the tavern and fired at the Redcoats, who returned fire—the musket balls are still embedded in the walls of the tavern. But the resistance discouraged the British from burning the tavern, and the bravery of the overmatched militia delayed the march long enough for reinforcements to defeat them on the coast.

When the guide's story was over, I looked at my watch and discovered two hours had passed. Instead of being

bored, I felt I had gained an insight into what those turbulent early days of the Revolutionary War had been like. It was yet another lesson in the value of looking and listening.

Of course, it's not only people and buildings that tell stories. This planet that we call home has been around for over 5 billion years. While we humans have been around for only a tiny fraction of that time, the story of those eons is not lost to us. Instead, the history is written on the face of the earth, in the rocks, mountains, hills, valleys, rivers, lakes, and other natural wonders that are all around us in this magical place we call America. I've learned, in my travels, that the geography of our country, the location of those mountains, plains, rivers and other features have played a critical role in the history of our land and the development of us as a people. Without a knowledge of geography, we can't possibly hope to understand who we are and where we are going.

Of course, appreciating the many stories that the world around us can tell requires a tour guide, a task that, with great humility, I'm taking on in this section of the book. I hope that I can measure up to the standards set by Charles Kuralt and that wonderful lady in Connecticut.

Meet Our Planet Earth

Anyone who's ever dealt with a real estate agent has heard that the three most important considerations when choosing a place to live are "location, location, location." Now, we human beings didn't get an opportunity to go galaxy-hopping with a cosmic broker to choose a place to evolve. But if we had, I doubt that we could have found another location as ideally suited for our species as our earth, a tiny but cozy speck in a huge and violent universe.

Before we begin to explore the geography of this very special planet, we'll take a little time to learn about where in the universe we live, how our planet was created, and what its structure is like.

WHAT ARE THE CHANCES THAT EARTH IS UNIQUE?

Next time you go to the beach, bend down and scoop up a handful of sand. That handful of sand probably contains about 10,000 grains. That seems like a pretty large number—until you learn that there are more stars in the universe than there are grains of sand on all the beaches on earth. Astronomers estimate that the universe contains at least 100 billion galaxies, each of which contains an average of 100 billion stars. Multiplying the two numbers together results in a string of zeros that can be summed up in one word—incomprehensibly huge.

Among these stars are hundreds of billions of stars just like our sun. It defies logic to suggest that planetary systems didn't form around a lot of these stars, or that none of these planets are capable of supporting life. Although we don't yet have a way to tell for sure, many scientists safely assume that there could be one million planets similar to earth in our galaxy alone.

Will we ever have contact with any other such planet? It's hard to predict, because the universe is so immense that we would have to cover vast distances merely to reach our nearest neighbors.

WHERE ARE WE?

The universe is mostly made up of emptiness: if you entered the universe anywhere at random, your chances of hitting a star or any other object would be about the same as your chances of shooting a bullet at the earth from outer space and hitting a single specified speck of dust. The distances between objects grow larger as we move farther away from our solar system.

Where in the solar system are we?

Our planet is bright and warm primarily because its orbit, in relation to the total solar system, virtually hugs the sun. Although Earth's orbit is elliptical, not circular, its distance from the sun averages about 93 million miles (actually, 92, 877,000). Our nearest neighbor is Venus, just 25 million miles away, while Mars and Mercury are a little over twice as far.

Beyond this tight ring of four rocky planets is a more vast solar system dominated by the four gas giants. Jupiter is 5 times farther from the sun than earth, Saturn is 10 times as far, Uranus is 20 times as far, and Neptune is 30 times as far. Pluto, a tiny, barren chunk of rock smaller than our moon, has a strange orbit that sometimes brings it closer to the sun than Neptune, and sometimes carries it 2 billion miles farther from the sun than Neptune.

The gravitational influence of the sun extends far beyond Pluto to approximately 100 times the distance of the earth from the sun. There is some evidence that in this vastness orbits a tenth planet that we are yet unable to detect.

BASIC FACTS ABOUT THE PLANETS

Planet	Distance from Sun (miles)	Diameter (miles)	Rotation on Axis (Earth time)	One Orbit of Sun (Earth time)
Mercury	36,000,000	3,032	59 days	88 days
Venus	67,000,000	7,519	243 days	225 days
EARTH	93,000,000	7,926	23 hr 56 min	365 days
Mars	141,000,000	4,194	24 hr 37 min.	687 days
Jupiter	484,000,000	88,736	9 hr 55 min.	11.9 years
Saturn	887,000,000	74,978	10 hr 40 min.	29.4 years
Uranus	1,784,000,000	32,193	16 hr 48 min.	84.0 years
Neptune	2,796,000,000	30,775	16 hr 11 min.	165 years
Pluto	3,666,000,000	1,423	6 days 9 hr	248 years

Where in the galaxy are we?

Our sun may be special to us, but it's only one of more than 100 billion stars that revolve around a central point. This group of stars, or galaxy, is called the Milky Way. The Milky Way is a spiral galaxy, with stars located on a gigantic flat disc that features spiral arms that stretch out from the center.

The sun is located on the outer edge of one of those arms about 26,000 light years from the center of the Milky Way, and it takes our friendly star about 250 million years to make one complete revolution. The diameter of the entire galaxy stretches about 100,000 light years from edge to edge.

When we peer up into the heavens on a clear night, we can see with our naked eyes about 8,000 stars, a tiny portion of our galactic relatives. Our nearest stellar neighbor, Alpha Centauri, is 4.3 light years away. If we launched a space ship that could travel at 6,000 miles per hour (the limits of our current technology), it would take that ship 40,000 years to reach our neighbor.

Where in the universe are we?

The Milky Way (along with the Andromeda galaxy) is one of the two largest of 20 galaxies that make up a small unit known as the Local Group. The Local Group, in turn, joins thousands of other galaxies in the Virgo Cluster. Throughout the universe, clusters join together to form super clusters, one of which, the Great Wall, stretches a distance of 500 million light years.

Sophisticated technology has allowed astronomers to detect galaxies that contain up to 13 times as many stars as the Milky Way. To date, the faintest galaxies we can detect are about 6 billion light years away, but astronomers theorize that the universe extends another 6 to 9 billion light years beyond that.

WHERE DID THE UNIVERSE COME FROM?

The current theory is that our universe began with an event called "the Big Bang." This theory supposes that everything began at a single point that had infinite mass and infinite density (please, don't ask me how this is possible). This "singularity" exploded, sending forth at great speed a huge amount of heat and matter that consisted almost exclusively of hydrogen and helium.

If you take a handful of sand and scatter it on the sidewalk, you'll see clumps of sand in certain areas rather than an exactly uniform distribution. Similarly, the Big Bang unevenly distributed the matter it gushed. This matter clumped into clouds, then condensed into the nuclear reactors we call stars. When the largest of these stars ran out of fuel to burn, they exploded. The force of the explosions converted some hydrogen and helium into heavier elements such as iron, carbon, silicon, oxygen and many others. The resulting dust became the building blocks of further stars.

To this date, the universe is still expanding. The shape of the universe is exactly like that of a balloon that is being gradually filled with more and more air. All matter that makes up the universe is on the surface of that balloon.

One of the first questions this theory brings up is what is going to happen to the "balloon." Is it going to keep expanding forever or is it going to "pop?" Physicists believe the answer lies in how much matter the universe contains. If enough matter is present, the expansion will eventually come to an end. Then the universe will begin to contract, eventually ending in the singularity with which it began. If sufficient matter isn't present, the universe will continue to expand forever.

The second big question is what's inside and outside the balloon? Some physicists, including the famous British scientist and author Stephen Hawkings, believe that there are "wormholes" in the fabric of the universe that are "shortcuts" to distant points or even pathways to other universes. Others have postulated that our entire universe may be a tiny entity in a bigger universe in the way that our sun is just a tiny entity in our universe.

How was our sun formed?

The universe is a giant star-making factory. The "raw material" is vast clouds of dust and gases (mostly hydrogen and helium) scattered by the explosions of dying giant stars known as supernova. About 4.6 billion years ago, a supernova explosion condensed a cloud of interstellar debris in our area of the Milky Way to the point where gravity began pulling in more and more material. As the cloud collapsed, friction between the densely packed atoms built up heat. When the temperature at the center of this mass reached 20,000,000 degrees Fahrenheit, the cloud became a nuclear reactor, converting hydrogen into helium in a process that gave off immense amounts of heat and light. This brand new nuclear reactor was a star we now call the sun.

SUN FACTS

Size: The sun has a diameter of 865,000 miles, 108 times that of earth. The sun is a typical mid-size star. The largest stars are 1,000 times larger and 1,000,000 million times brighter, while the smallest are .01 times the size and one-thousandth as bright.

Energy: Every second, the sun converts 600 million metric tons of hydrogen to helium, releasing the energy of 100 billion one megaton atomic bombs.

Temperature: The interior temperature of the sun is approximately 29,000,000 degrees Fahrenheit, while the surface is a comparatively cool 9300 degrees.

Sunspots: Huge dark splotches called sunspots, which appear and disappear in a 22 year cycle, are magnetic storms caused by convection patterns in the sun. They often disrupt communications on earth.

Future of the sun: The sun's hydrogen will last approximately another 4.5 billion years. After it's exhausted, the sun will begin converting helium. For half a billion years, the sun will grow dramatically into a red giant so huge that it may engulf the earth. Even if it doesn't actually swallow the earth, intense heat will boil away all of our planet's water and atmosphere, turning it into a barren rock. Then, its helium exhausted, the sun will shrink into a comparatively tiny white dwarf star, slowly cooling over billions of years.

How was Earth formed?

Star making is a sloppy process, and the brand new star was surrounded by a cocoon nebula, a flat ring of left-over particles of gas and dust. These particles frequently collided. Because space was so crowded, most of the particles moved so slowly that the collisions were gentle, and the particles clumped together like snowflakes in a blizzard. Over the course of just 100,000 years, space began to clear as some of these bodies grew to be the size of mountains.

These larger bodies, called planetesimals, were the embryos of the planets. As their number decreased, their speed increased, and a kind of celestial demolition derby occurred as they smashed into each other. By random chance, one grew larger than the others, and its gravity caused it to attract other planetesimals in a process called accretion. The bigger it got, the more rapidly it grew. When it reached a diameter of about 500 kilometers, gravitational forces molded it into a spherical shape. This object was the embryo that became Earth.

Man On The Move

If you've ever awakened in the morning feeling just as tired as you were the night before, I'll give you a good excuse—the earth that we're all riding on is hurtling at great speed in all kinds of directions every moment of every day. Here's where we're going and how fast:

* *The earth rotates on its axis. The earth spins on its axis, completing one rotation in 23 hours, 56 minutes, 4.1 seconds. The speed of the rotation at the equator is 1,000 miles per hour, while at Portland, Oregon (45 degrees north latitude) the speed slows to 667 miles per hour.*
* *The earth orbits the sun. Every 365 and one-quarter days, the earth travels the 583,400,000 mile path around the sun at an impressive 66,000 miles per hour.*
* *Our entire solar system is making a gigantic revolution around the center of the Milky Way. Right now, our sun and the planets are hurtling toward the constellation Hercules at 45,000 miles per hour.*
* *The Milky Way is moving through space toward the constellation Leo at an astounding 1.35 million miles per hour.*

Almost leaves you exhausted enough to head back to bed.

Our area of the solar system had so much debris that the Earth reached half its present size in about one million years—almost overnight, in cosmic terms. For millions of more years, the Earth was subjected to an incredible rain of asteroids that fell with an impact rate of one billion times the rate at which meteorites enter our atmosphere today. At least once a month it was struck by a planetesimal so large that dust and debris from the collision blotted out all sunlight. The heat generated by the impacts was so intense that the entire surface of the planet was a sea of molten lava.

The earth reached about 98% of its present size in the first 100 million years. The newly formed planets had swept up so much material that the impact rate gradually fell off over the next billion years.

WHAT IS THE STRUCTURE OF THE FINISHED PRODUCT?

The great science fiction writer Jules Verne may have anticipated modern submarines in his book Twenty Thousand Leagues Under the Sea, but another of his fictional adventures, Journey to the Center of the Earth will never be duplicated in real life. For when you understand the structure of our planet, you'll see why we humans should be content to live exactly where we do.

The mass of matter that makes up our cozy planet is divided into five basic parts:

The atmosphere of gases that surrounds earth.

If a time machine existed that could magically transport you back to the surface of our planet 3.5 billion years ago, the trip would be fatal. The reason: the atmosphere had little or no breathable oxygen. The baby earth was enveloped primarily by carbon dioxide, sulfur dioxide, nitrogen and water vapor spewed forth by gigantic prehistoric volcanoes—an atmosphere very similar to the present atmosphere of Venus.

Gradually, carbon dioxide from the atmosphere dissolved in the oceans. By some magical process that we are still striving to fully understand, the carbon molecules became the building blocks of life. The oceans eventually became covered with algae that, like today's plant life, took in carbon dioxide and gave off oxygen. It took 1.5 billion years before oxygen made up 1% of the atmosphere. Only 570 million years ago did the oxygen level become rich enough to support oxygen-breathing marine life, and only 400 million years ago did enough exist to support land animals.

Our current atmosphere consists of 78% nitrogen, 21% oxygen, and a mixture of argon, carbon dioxide, water vapor and other trace elements making up the last 1%. The atmosphere rises 70 miles from the earth's surface, but half its mass is in the lowest 3.5 miles.

The hydrosphere, the water covering much of our planet.

The very young earth was rocky and barren, because it had no atmosphere to shield the sun's rays and prevent water vapor from escaping into space. But as we've seen, volcanic activity spewed forth water made from oxygen and hydrogen trapped in the earth's interior. Added to this supply was water melted from icy comets thrown into earth's orbit by the formation of Jupiter, Saturn, and the other outer planets.

Today, water covers almost three-quarters of the earth's surface. The average depth of the world's oceans is over 12,000 feet.

The lithosphere, or the surface crust.

The young earth was a very hot place that acted much like a modern blast furnace: when rock turns molten, the denser materials fall to the bottom and the lighter materials rise to the top.

The least dense rock, which consisted primarily of oxygen and silicon compounds, formed a crust over the entire surface of our planet that is thinner, in relative terms, than the skin on an apple. This crust, or lithosphere, varies in thickness from as little as five miles under the oceans to as much as 80 miles under the world's tallest mountain chains.

The mantle, which lies below the lithosphere

The top of the 1800 mile thick mantle is a layer of plastic-like rock called the asthenospere—for comparison, think of hot taffy or molasses. This rock is made malleable by heat rising from the interior of the earth in a process called convection. To visualize how convection works, boil some water in a pot on your stove. You'll notice that currents churn the water. Hot water rises to the surface, cools, then drops back to the bottom as newly heated water rises. Drop a cork into the boiling water. It's moved around by the currents in the water. While the amount of movement of the earth's crust isn't as dramatic, it's caused by the same process.

Underneath the asthenosphere, the mantle is made of solid rock that is denser than the crust but less dense than the core. The mantle is punctured by vents that allow molten rock, or lava, to rise towards the surface.

The core, the center of the planet

The earth's core, which is about 2180 miles thick, is made up primarily of iron with a smattering of nickel. Because the temperature of the core is about 12,000 degrees Fahrenheit—hotter than the surface of the sun—the 1380 mile thick outer core is molten liquid. However, the intense pressures at the very center of our planet produce a 800 mile thick inner core that is solid despite the intense heat.

Our old friend, convection, is also at work in the outer core. The movement of molten rock acts like a generator that produces an electromagnetic field. That field is the reason that compasses always point north.

HOW BIG IS EARTH?

Distances are relative—how far one place is from another depends a lot on exactly what kind of transportation you have available to get from one place to another. North America seemed enormous to pioneers who needed months to cross the continent in covered wagons; on the supersonic Concorde jet, it's a short hop.

But by any standards, earth is a rather large place to us humans. Our planet is not quite a perfect sphere, bulging about 13 miles at the equator. That bulge means that if you took a giant tape measure and stretched it around the earth at the equator you'd get a circumference of 24,902.4 miles, 42.2 miles longer than a similar measurement taken around the north and south poles. The diameter of the earth at the equator is 7926.42 miles, a little over 26 miles longer than the length of a giant arrow stuck through both poles.

The surface area of our planet is 196,930,000 square miles, 70.8% of which is covered by water. Encircling the land area is an atmosphere of gases than extends about 70 miles into space.

WHERE WE LIVE

We reside on the lithosphere, or crust, of our planet. To most of us, nothing seems more solid than the earth beneath our feet. Those of us who own property have a deed that fixes the exact location of our piece of rock and gives us and our descendants the right to live in that location in perpetuity.

As we learn a little about the structure of the earth, we realize that it isn't just a chunk of rock hurtling through space, but a dynamic, living organism. And the land on which we live isn't perpetually frozen in place, but is more like a piece of wood floating in a bathtub. In the last two decades, scientists have learned that the thin layer of light rock upon which we live is divided into about twelve plates that "float" on a sea of moving rock below. Although the movement seems insignificantly slow to us—the North American plate moves about .4" per year—over hundreds of millions of years that same continent has drifted as far north as the north pole and as far south as the equator.

In the next chapter, we'll explore the fascinating history of the continent on which we live.

HEROES OF GEOGRAPHY:

THE EARLY ASTRONOMERS

Imagine getting up in front of an audience made up of America's leading scientists, politicians and religious leaders to announce that twenty years of meticulous scientific research had led you to the inescapable conclusion that the world was flat. No doubt you'd nearly go deaf from the laughter and derision before they led you off to a mental institution.

Now, let me mentally place you in a time machine and whisk you about 400 years into the past in front of an equally prestigious audience—but this time you're announcing that you've come to the conclusion that the earth and the other planets orbit the sun. The reaction—not only laughter and derision, but a healthy dose of anger. The reason is that for all of mankind's recorded history the belief that the earth was the stationary center of the universe had been an unchallenged tenet not only of science, but of religion, as well.

NICOLAUS COPERNICUS (1473-1543), in his treatise "On The Revolutions of the Celestial Sphere," was the first astronomer in 2,000 years to state that the sun, not the earth, was the center of the solar system and that the earth made one full rotation on its axis every 24 hours. Even though Copernicus's theory was supported by observations of the movement of the planets and sun, it was embraced by only a handful of scientists.

TYCHO BRAHE (1546-1601) began the era of precise measurement of the movement of heavenly bodies by painstakingly observing the skies every night for twenty years. Although he never gave up the belief that the earth was the center of the universe, he did state that the other planets revolved around the sun. He calculated the length of the year at 365 days, 5 hours, 48 minutes, and 45 seconds, leading to the adoption of our modern calendar.

GALILEO (1564-1642) built the very first telescope. By gazing through it, he discovered that the moon's surface was pock-marked by mountains and craters, that the Milky Way was made up of stars, that giant sunspots appeared and disappeared on the surface of the sun, and that Jupiter had at least four moons. Galileo's observations made him such an influential and vocal proponent of Copernicus that he was tried for heresy by the Catholic Church.

JOHANNES KEPLER (1571-1630), an assistant to Tycho Brahe, used his data to develop three laws of planetary motion that confirmed the theory of Copernicus. Among other things, these laws stated that the orbits of the planets were elliptical and that the orbit speed of the planets increased as the distance to the sun decreased.

SIR ISAAC NEWTON (1643-1727) went beyond Kessler's laws to state 3 universal laws of motion that fit the behavior of objects on earth as well as the planets and other objects in the heavens. He also identified gravitation as the universal force that governed the structure of the solar system.

SIR EDMUND HALEY (1656-1742) encouraged Newton to write and publish Philosophiae Naturalis Principia Mathematica, in which he presented the concept of gravity and his 3 laws of motion. Haley, for 18 years Britain's Royal Astronomer, mathematically demonstrated that comets were part of the solar system, moving in elliptical orbits around the sun. When the large body now known as Halley's Comet returned exactly as Halley had predicted in 1756, it put the final nail in the coffin of the centuries old belief in an earth-centered universe.

Why are there seashells in Montana?

The Incredible History of
Our North American Continent

One of the most fascinating things about life is the tremendous role chance plays in determining our future. No better example exists than the way in which I became a weatherman. Although many people assume I had years of scientific training before I stepped before the cameras, the truth is that my goal was to become a network news anchor-

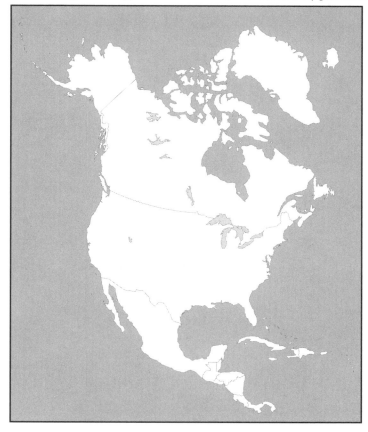

man. After I finished school, I climbed on the first rung of that career ladder by winning a job as a general assignment news reporter for a Richmond, Virginia television station. About a year later, the station's long-time weatherman suddenly resigned. The news director asked to fill in for two weeks until a permanent replacement could be hired.

Because I loved what I had been doing, I was more irritated than honored at my selection. However, I threw myself into the temporary assignment with enthusiasm, and I received a very warm response from viewers. After some verbal persuasion (and a hefty raise), I agreed to becomea a weatherman, and my career path was permanently and profoundly altered.

Another—and in terms of history, a much more important—example of the impact of chance is the story of Christopher Columbus. In the fifteenth century, the European

appetite for spices, silks, and jewels from Asia meant that the cargo in the hold of just one ship could bring fame and fortune. Unfortunately, the voyage to the Far East around the Cape of Good Hope was very long and extremely dangerous. Columbus finally received royal support for his "crazy" obsession with reaching Asia by sailing west because the cost was so small in relation to the possible payoff that it was worth the risk.

Of course, Columbus didn't reach Asia. But he did open the door to European exploration and settlement of North America. The Europeans didn't find many spices, silks, or significant amounts of jewels and gold. But instead, they discovered a sparsely populated land with a moderate climate, millions of acres of arable soil, networks of navigable rivers and lakes, vast expanses of forests, and huge reserves of minerals. This continent's natural bounty soon nurtured the richest nation on earth.

NORTH AMERICA IS A VERY SPECIAL PLACE

The land mass we call North America wasn't just sitting still for eons on the other side of the Atlantic Ocean, waiting for Columbus and his cohorts. Rather, it's the fortunate survivor of violent and dynamic forces that have moved and shaped it over a period of four billion years. Our home continent has drifted as far north as the North Pole and as far south as the Southern hemisphere. It has crashed into other continents, raising mountain chains higher and more vast than the Himalayas that have since eroded to nothing. It has been cratered by giant meteorites, sundered by massive earthquakes, punctured by volcanoes, and scraped and carved by glaciers while watching twenty oceans come and go. The story of how this nearly perfect place for modern man came to be is one of the most fascinating tales in all of geography.

What Are The Continents and How Did They Get Their Names?

Earth has seven continents. These seven are listed below in order of size, along with the way each got its name:

1 Asia, the largest continent, covers 17,297,000 square miles, 30% of the earth's land. The word "asia" in Greek means "region of the rising sun." The Greeks apparently applied the term only to the area we call Asia Minor, but it gradually became used for the entire continent.

2 Africa covers 11,708,000 square miles and contains 52 countries, the most of any continent. The continent was named by the Romans and was probably derived from the Latin word "apricus," meaning "sunny," which in turn comes from the Greek word "aphrike," meaning "without cold."

3 & 4 North America (9,406,000 square miles) and South America (6,883,000 square miles) got their common name from a decision made in 1507 by a German geographer and mapmaker named Martin Waldseemuller. In drawing up a map of the barely known "fourth part" of the world (after Europe, Asia, and Africa), Waldseemuller proposed naming it "America" after the Italian explorer Amerigo Vespucci, who had made four voyages across the Atlantic. The explorer's name in Latin was Americus Vespuccus, but Waldseemuller changed the masculine ending "us" to the feminine ending "a" because the Latin words Europa and Asia were feminine.

5 Antarctica, the "peopleless continent," consists of 5,405,000 square miles of land buried under a permanent ice cap. "Arctic" comes from the Latin word "arcticus," which means "bear." The constellation Ursa Major, the Great Bear, pointed the way to the North Star. So "arctic" came to mean "north," as in Arctic Ocean. The "ant" in Antarctica is short for "anti," or "opposite." "Antarctica" is the opposite of the Arctic.

6 Europe, which covers 3,835,000 square miles, is the most densely populated continent at 174 people per square mile. Scholars believe that the word "europa" meant "mainland" in ancient Greek. They believe the Greeks used the term to refer to the extensive but unexplored lands that stretched far off to the north.

7 Australia, at 2,967,000 square miles, is the smallest continent but largest island in the world. From the Middle Ages on, geographers speculated about the existence of a "terris australis incognita," or "unknown southern land." The island was first explored most extensively by the Dutch, who called it "New Holland." But after it was settled by the English, the old world word for "south" was revived as the continent's name.

HOW AND WHEN WERE THE CONTINENTS FORMED?

If you were a cosmic vacationer about four billion years ago, the baby planet earth would not have been a wise destination. Earth was a violent and chaotic place way back then, with the constant bombardment of meteors from space producing such heat that the surface was often turned into a sea of molten lava. Later, in a period geologists call "The Big Burp," gigantic volcanoes spewed forth massive quantities of water vapor that eventually condensed and fell as rain, covering our planet with one huge ocean. Because the surface was formed and reformed so many times, no geological record of this first half billion or so years remains.

However, a little less than 4 billion years ago, areas of the crust that were thicker because of volcanic activity and convection currents in the mantle began to emerge from the waves. These small kernels of continents, called "cratons," pinwheeled around the surface, adding to their area when they crashed into one another like bumper cars in a carnival ride. About 5% of the current continental surface was formed by 3.5 billion years ago. Today, three areas of ancient land, called "shields," still remain. Half the current land mass was created by the end of the Archaen era 2.5 billion years ago, and almost all existed 500 million years ago. Although the creation of new land has virtually come to a halt, the map of the Earth's surface is far from static. Since the first years of their existence, continents have never stopped moving.

WHY DO THE CONTINENTS DRIFT?

The theory that continents move around the surface of our planet was suggested by a piece of evidence that's obvious to elementary students who glance at a world map—if you cut out North America, South America, Africa, and Europe the pieces fit together nearly perfectly. Way back in 1915, a German meteorologist named Alfred Wegener (we weather people tend to be Renaissance men) determined that rock formations in Africa and South America were identical in age, structure, and fossil content. In other words, not only did the pieces of the puzzle fit, but the picture on the pieces was the same. That's why Wegener published a book arguing that the continents are continually on the move.

For nearly half a century, few scientists accepted Wegener's theory. The reason—no one could think of a force powerful enough to move something as massive as a continent. But in the 1960s we discovered that the heat of the earth's core—which is as hot as the surface of the sun—generates currents in the massive layer of molten rock that surrounds it, just like the hot burner of your stove produces currents in boiling water. Molten rock, or magma, moves upward to the quasi-plastic layer of mantle just below the crust, then horizontally along it, then plunges back down toward the core again when it cools.

The lighter rock that forms the earth's crust is divided like a jigsaw puzzle into about a dozen gigantic pieces. The currents in the mantle drag along the plates that "float" on them. These puzzle pieces of crust, which include the continents as well as parts of the sea floor, have come to be called "plates," and the study of their movements is known as "plate tectonics."

WHAT HAPPENS WHEN PLATES MEET?

To us mortals, the movement of the earth's plates continents is incredibly slow—an inch or less per year, slower than the growth rate of your fingernails. North Americans will travel a grand total of six feet during your lifetime. But over the immense time spans with which we measure the history of the earth, the continents have covered enormous distances. And the effects of the interaction between the plates is equally dramatic.

Geologists have found out that three different kinds of activity take place at the edges, or "margins," of the plates:

- Constructive margins: In the middle of the Atlantic Ocean is a range of enormous underwater mountains that runs almost all the way from the North Pole to the South Pole. If you could drain the water from the ocean and peer down at these mountains from a satellite, you'd be astounded to see that there is a gash right down the middle of the entire range, as if they'd been cut by a gigantic knife. This gash is called the Atlantic rift, which is the margin between plates that are moving apart—the North and South American plates on one side and the Euroasian and African plates on the other. Through this rift, hot rock seeps up from the earth's mantle and forms new crust on the ocean floor. That's why Iceland, which sits right on top of the rift, is a land of volcanoes and geysers so hot that their waters heat the island's homes. Another huge rift girdles Antarctica, bisects the Indian Ocean, then curls around through East Africa.
- Destructive margins: Since the area of our planet isn't increasing, new crust being created in some portions of the globe must mean that old crust is being destroyed or distorted in others.

Destruction takes place when an ocean plate crashes into a continental plate, which has been happening for the last 60 million years on America's west coast The rocks that form the crust of the Pacific plate are heavier than the rocks of the North American continent. So the Pacific plate is being pulled under the continental plate and melted by the heat of the mantle. The continental plate has been lifted up, forming the Rocky Mountains and the other western ranges. The stresses produced by the clash of these huge plates has and will result in lots of volcanic activity and earthquakes all along the margin.

Distortion takes place when two continental plates crash into each other. For twenty million years, India has been smashing into Asia. The crust has no place to go but up, forming the Himalayas, the highest mountain range in the world.

* Conservative margins: Finally, two plates can slide past each other with no new crust being created and none being destroyed. This is taking place in some parts of the Pacific right now.

WHAT IS THE HISTORY OF THE NORTH AMERICAN CONTINENT?

Since we don't have a time machine that allows us to take satellite photos of the ancient earth, geologists have to rely on subtle clues provided by rocks and fossils to decipher the past location of land masses and plates. Because older rocks are less common and because life didn't emerge from the seas until the last half billion years, plotting the earlier movement of continents is much more difficult. So we have to look at the movement of the land mass we call North America through a lens that shows a picture that sharpens considerably as we move closer to the present day.

4000 MILLION TO 2500 MILLION YEARS AGO

The core of North America is a Y-shaped landmass called the Laurentian Shield, which extends from about the present Great Lakes region northward, splitting into a Y around the present Hudson Bay. Like the other two remaining shields in Africa and Australia, it was formed about 3.9 billion years ago when a chain of volcanoes gushed forth massive amounts of lava. Over the course of 1.5 billion years, the volcanoes eroded to a virtually flat plain, and the sediment from that erosion built up land upon its flanks. By 2.5 billion years ago, this shield had merged with five or six other cratons (mini-continents) to emerge as one of the large true continents.

2.5 BILLION TO 600 MILLION YEARS AGO

If you find a time machine to take you back to North America two billion years ago, bring your long underwear. Our baby continent was moving far north as all the large continents began to drift toward each other. About 1.9 billion years ago, the Baltic plate (which consisted of what is now northern Europe) smashed into North America, producing a huge mountain range in what is now northern Canada that has since largely eroded.

After this collision, we believe that all the continents were joined together in a long, narrow land mass. At one extreme of this huge supercontinent was a very frigid "baby" North America, which straddled the North Pole. This land mass stretched southward all the way to the equator.

1.4 BILLION TO 800 MILLION YEARS AGO

This supercontinent broke up into four or five smaller continents. The Baltic plate rifted from North America along a line that stretches from Western New York down through Kentucky and Tennessee and curves westward through Texas. Three hundred million years later, the Baltic plate reversed course and smashed into North America again, raising a continent-long mountain chain before moving off a second time. During this entire time range, all of the continents were moving southward, and North America began to thaw out.

800 MILLION TO 400 MILLION YEARS AGO

If you thawed out from your trip to North America two billion years ago, you might want to set the dial for 700 million years in the past. North America had tipped on its side, and what is now our east coast faced directly south. The equator bisected our continent, and the southern shore was tropical beach front lapped by the waters of what was called the Iapetus Ocean (in Greek mythology, Iapetus was the father of Atlas, for whom the Atlantic Ocean was named). Much of the inland area of the continent was covered by a shallow sea—the only way to get from Missouri to Nevada was by boat.

The land masses which are now Africa, South America, Florida, Spain, Central Europe, Arabia, India, Australia, and Antarctica were fused into a giant continent called Gondwana, which straddled the South Pole. Much of the land was covered by an ice cap, which is why massive glacial deposits are found today in the Sahara Desert and the tropical regions of both Africa and South America.

North America had joined with what is now Greenland and parts of Asia into a supercontinent called "Laurasia." About 570 million years ago, the Iapetus Ocean began to close. What is now New England was crumpled, forming-what are now the Taconic, Berkshire, and Green Mountains. An island called Avalonia broke up and pieces fused onto the New England coast.

400 MILLION TO 200 MILLION YEARS AGO

The face of the earth began to change dramatically as the immense Gondwana moved northward and the huge ice cap covering it began to melt. Sea levels rose, flooding low lying land all over the globe. This flooding widened the inland sea that covered much of the central United States. Swamps and marshes covered Pennsylvania and other eastern states, and the remains of the dense vegetation that grew in these areas were compressed into the vast seams of coal that we have mined in the last two hundred years. Vegetation in seas covering what is now the Middle East decayed, producing most of the world's supply of oil.

Inevitably, Gondwana and Laurasia collided. The first impact, around 380 million years go, raised the northern Appalachian Mountains; 90 million years later, the Iapetus Ocean closed completely, forming the southern Appalachians. All the major land masses on earth were then fused into a single continent called Pangaea, which was surrounded by the Panthalassa Ocean.

By this time, dinosaurs had evolved and were beginning to flourish. With no oceans to bar migration, these giant reptiles and all other life forms spread everywhere, which is why fossils of dinosaurs are found on every continent today. What is now the east coast of North America was completely landlocked—it was only a short stroll from what is now Boston to North Africa.

During the entire period, Pangaea was moving northward and rotating counterclockwise (gradually tilting North America upright again). About 245 million years ago, the east coast of North America finally returned to the northern side of the equator. The climate became less tropical and the continent's inland seas dried up, leaving vast salt deposits and, in the western U.S., huge salt lakes.

200 MILLION TO 65 MILLION YEARS AGO

One hundred million years after its formation, Pangaea began to break up. The continents were "unzipped" from bottom to top, with first South America and then Africa separating from North America, forming the Atlantic Ocean. Ocean levels rose, flooding the central U.S. once again. Volcanic activity along the rift area on the east coast produced huge lava flows. The west coast of North America expanded as a number of large islands collided with and stuck to the continent. In particular, Alaska is constructed largely of many different chunks of land.

The end of this era was definitively marked by what's called the "Great Extinction." In a breathtakingly short period of time, every species of dinosaurs as well as every other type of animal that weighed 50 pounds or more disappeared. The most credible explanation for this mass destruction was the impact of a giant meteorite up to ten miles in diameter. Scientists believe that this killer visitor hit the earth off the eastern coast of Mexico. A mile-high tidal wave battered coasts world-wide and firestorms generated by the heat of impact ignited much of the world's forests. Dust from the impact was so thick that the earth's surface was as dark as night for several months. The planet cooled dramatically, vegetation died, and all but the smallest and most adaptable creatures disappeared.

65 MILLION YEARS AGO TO THE PRESENT

After the dust eventually settled, North America continued to rotate toward it's present position as it moved westward. The Atlantic widened nearly 1,000 miles since the extinction of the dinosaurs. The Appalachian Mountains, once a range as towering as the Himalayas, continued to erode. On the west coast of the continent, the collision between the North American and Pacific plates raised the Rocky Mountains, which become the continent's dominant mountain chain.

About 4 million years ago, the climate in the Northern Hemisphere began to cool after a few hundred million years of warmer weather. Glaciers began to grow in central Canada and slowly move south. For most of the last two million years, the glaciers have expanded and retreated in roughly 100,000 year cycles, several times covering the northern third of the United States. The scraping and melting of the giant ice caps dramatically changed the face of the land they covered. The Great Lakes and the tens of thousands of lakes in Minnesota and Canada were created and filled by

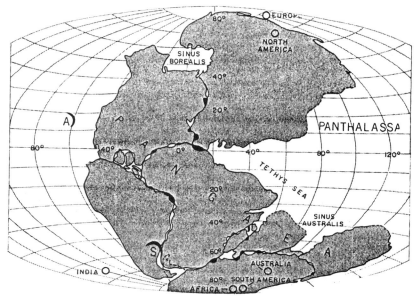

Map of earth 200 million years ago.

Meet North America

The North American continent today is entirely in the Northern Hemisphere, stretching from just north of the equator at the tip of Central America to the Arctic Ocean in the far north. But the vastness of the continent, as proven in the information below, may come as a surprise:

▲ 71% of all the land is currently in the Northern Hemisphere. Land covers 42% of the Northern Hemisphere, compared with just 17% of the Southern Hemisphere.

▲ North America, the third largest continent, covers 9,406,000 square miles, about 16% of the total land on earth. Our continent is a little more than half the size of Asia, about 20% smaller than Africa, and about 25% larger than South America.

▲ North America spans eight time zones, one-third of the distance around the globe. The time difference between easternmost and westernmost Alaska (3 hours, 50 minutes) is greater than time difference between New York and Los Angeles.

▲ The population of North America is about 430 million people, which represents just 7% of the world's 5.4 billion population. Asia, the most populous continent, has 3.1 billion people.

▲ A parallel line running through the southernmost point of North America would run through Nigeria, pass south of the entire Arabian peninsula, touch the southern tip of India, and run right through Ho Chi Minh City (the former Saigon), Viet Nam.

▲ North America is more than 4,000 miles wide at its northernmost border, but only 31 miles wide at its southernmost border.

▲ Canada contains 50% of the world's lakes, and North America as a whole has nearly three-quarters of the world's lakes. These lakes were carved out by glaciers from the last ice age and filled by melted waters.

▲ North America is not directly north of South America. Atlanta, Georgia is further west than the entire South American continent.

the glaciers. The last wave of glaciers, which melted just 20,000 years ago, left New England strewn with rocks, pock-marked ice-carved valleys and hills, and marked by piles of debris that include Long Island.

What Will Happen To North America In The Next 50 Million Years?

We'll spend the rest of this book on the wonderful continent that resulted from this incredible 4 billion year odyssey. But before we do, let's look into the crystal ball. Over the next 50 million years, North America will continue to move away from Europe and Africa, widening the Atlantic and shrinking the Pacific. California west of the San Andreas fault as well as Baja California will separate from the mainland, becoming an island. The land link between Alaska and the Soviet Union could once again emerge from the sea.

On the map of the world, East Africa may break off along the east African rift. The rest of the continent will move northward, eventually swallowing the Mediterranean Sea. Australia will move northward and mountains will continue to be pushed upward along the India-Tibet border.

Over the course of 300 or 400 million years, as water from the oceans began to evaporate into the atmosphere, some of the thickest areas of crust crested above the surface to form microcontinents. As these microcontinents drifted, they encountered islands formed by volcanoes or by debris piled up by the impacts of giant meteors. About 5% of the current continental surface was formed by 3.5 billion years ago, about half by 2.5 billion years ago, and almost all by 500 million years ago. Today, the continents include about 57.7 million square miles, which represents 29.3% of the earth's surface area.

Where Is The North American Plate Today?

The North American plate includes North America, Central America, the islands of the Caribbean, and Greenland. This plate extends roughly mid-way across the Atlantic Ocean, to the Atlantic rift, where new crust is currently being formed. Iceland lies right on the separation line, which is the reason that island is so many active volcanoes.

On the other side, the North American plate collides with the Pacific plate north from Mexico to southern Alaska, then west along the Aleutian trench. The margin between the North American plate and the Nazca plate off the western coast of Central America is also destructive.

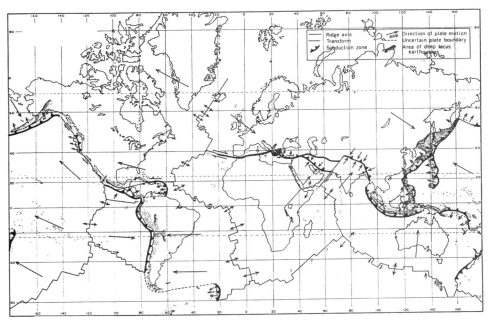

Map of earth's plates.

Meet America's Real Movers And Shakers:

Earthquakes, Volcanoes, And Other Forces That Shape The Land

We've just seen how the powerful energy trapped inside the earth drove our continent all around the globe, then parked it in a nearly ideal latitude and longitude. But continental drift is far from the only agent that has and does shape the wonderful land we call home. Other forces are at play that may not be powerful enough to propel Cleveland to the north pole, but are more than strong enough to have an incredible impact on our lives.

Though the power of television (my favorite medium), nearly all of us have been witnesses to such events as the 1980 eruption of Mount St. Helens, the 1989 San Francisco earthquake, the 1994 Los Angeles earthquake, and the 1993 flooding in the Midwest. While the focus of television was naturally on human suffering and loss of property, these acts of nature also changed the face of the land. To us, the specific effects of these changes don't seem so great— 400 feet lost from the height of Mount St. Helens, a few million cubic feet of topsoil washed away in the Midwest. But thousands of these events taking place over hundreds of thousands of years have had a profound impact. Before we talk about the richness of landforms that make up the geography of the United States, we'll spend a little time talking about the fearsome and fascinating forces of nature.

INTERNAL AND EXTERNAL FORCES ARE AT PLAY

There are two different types of forces that change the face of the land:
- Forces generated inside the earth from convection currents and plate movements, such as earthquakes and volcanoes.
- Forces generated on the surface and above, such as erosion caused by ice, wind, water, and weathering, meteorites impacting from space, and constructions projects of us humans.

VOLCANOES

"Flames were then spurting straight up into the air, now and then waving to one side or the other for a moment and then again leaping suddenly higher up. There was a constant muffled roar. It was like the largest oil refinery in the world burning up on the mountain top. There was a tremendous explosion about 7:45, soon after we got in. The mountain was blown to pieces. There was no warning. The side of the volcano was ripped out, and there was hurled straight toward us a solid wall of flame. It sounded like thousands of cannon. The wave of fire was on us and over us like a lightning flash. It was like a hurricane of fire. The fire rolled in mass straight down upon St. Pierre and the shipping. The town vanished before our eyes and the air grew stifling hot and we were in the thick of it."

In this vivid language, a crew member of a British ship described the 1902 volcanic eruption of Mount Pelée on the Caribbean island of Martinique that killed 30,000 people, completely destroyed the port city of Saint Pierre, and sank 17 ships moored in the harbor. Ironically, the sole survivor on land was a convicted murderer awaiting execution in a dungeon deep below the ground. Above ground, not a single stick of wood remained unburned. It is no wonder that until this century, most of mankind believed that volcanoes were gateways to a hellish world of fire that existed below the surface of the world. The word itself comes from the island of Vulcano, north of Sicily, which the Romans believed to be the entrance to the netherworld and lair of Vulcan, the blacksmith god.

Those ancient beliefs were true in one respect: an inferno does rage not far below the Earth that we know. Located 3 to 100 miles beneath the crust are huge reservoirs of liquid rock called "magma," which ranges in temperature from 1300 to 2500 degrees Fahrenheit. Sometimes the Earth's skin cracks, forming a chimney or vent from the magma reservoir through solid rock to the surface. We call these chimneys "volcanoes,", and the fiery magma that bursts forth from them provides one of nature's most awesome displays.

Our name for the magma that reaches the surface is lava. Lava is a molten stew that hardens into several different kinds of rock, from pumice (which is so light that it floats) to basalt, which is so hard that it can't be eroded by water or wind. Lava is red at temperatures of 2000 degrees Fahrenheit or greater, but it turns gray or black as it cools. The average lava flow moves at 5 or 6 miles per hour, but lava has been observed traveling more than 50 miles per hour down the slopes of Mount Vesuvius.

Volcanoes are pipelines to the center of the Earth. Through these pipelines have flowed all of the earth's gold, silver, diamonds, copper, tin, tungsten, and many other valuable minerals. Mount Erebus, an active volcano in Antarctica, spews forth crystals of pure gold, although these crystals are dispersed too widely to be collected at the present time. Volcanic ash and debris have also added rich nutrients such as phosphate, calcium, potash, and boron to the soil. Because most volcanic rock is porous, the plains and plateaus it has created have excellent drainage which is necessary for agriculture.

Hardened lava has also created some of the most impressive natural wonders in the United States. The Palisades, a majestic cliff that borders the Hudson River for 50 miles north of New York City, is made of horizontal layers of hardened lava called "sill." The giant rectangular pillars of rock (such as Ship Rock) that dot the desert in Arizona and New Mexico are plugs of basalt that hardened in the throats of ancient volcanoes that have since been eroded by wind and water.

THERE ARE THREE BASIC TYPES OF VOLCANOES:

- Composite cones are steep mountains created by successive layers of volcanic ash and lava. The highest volcano in the world is Cerro Ojos del Salado, which rises 23,292 feet in Chile's Andes Mountains. Many other famous volcanoes are composite cones, including Mount Vesuvius, Mount Kilimanjaro in Kenya, Mount Fujiyama in Japan, Mount Hood in Oregon, and Mount Rainier in Washington.
- Shield volcanoes, named because their shape resembles an ancient warrior's shield, are more gentle slopes created by less violent but more frequent eruptions of lava. The Hawaiian Islands are a composite of shield volcanoes. Mauna Loa on the island of Hawaii, one of the two active volcanoes in the Islands, is the largest volcano in the world. The base of the mountain is 15,000 feet below sea level and it extends another 13,679 feet into the sky, making a total height of nearly 29,000 feet. Thousands of lava flows averaging ten feet thick have made Mauna Loa a dome 60 miles long and 30 miles wide.
- Fissure volcanoes are not mountains, but cracks in the earth from which lava quietly flows and floods up to thousands of square miles. Many plateaus, including the Columbia River Plateau and the Snake River Plateau in the western United States, were created by fissure volcanoes. In modern times, the approximately 100 volcanoes in Iceland are the only active fissure volcanoes in the world. The heat from these volcanoes melts the glaciers in parts of that country, creating pools hot enough for bathing underneath the ice. Residents of Iceland tap this hot water to heat their homes. Pools of magma trapped just below the surface create three other natural phenomena:
- Hot springs occur when steam from a reservoir of molten rock rises, condensing into water and at the same time heating existing ground water. The minerals and chemicals contained in the condensed steam are the reason for the therapeutic value attributed to bathing in hot springs. The springs at Baden-Baden on the edge of Germany's Black Forest have been used for healing wounds and injuries for perhaps as long as 10,000 years.
- Geysers are rare phenomena that result from the formation of very narrow cracks that extend down into the earth above a pool of magma. For a time, hot water rising up the crack is stopped by the weight of cool ground water seeping down from above. Periodically, the ground water is heated to the boiling point and the geyser erupts in a spectacular display of steam and hot water. In a handful of geysers, such as Yellowstone Park's Old Faithful, the timing of the eruptions is very regular. More commonly, eruptions can be days, even weeks apart. Geysers only occur in the U.S., Iceland, New Zealand and Chile.
- Fumeroles are cracks or conduits in the earth through which hot steam steadily flows. In several countries, the steam in these conduits, which is called geothermal energy, has been tapped to generate electric power.

WHAT WERE THE MOST DESTRUCTIVE VOLCANIC ERUPTIONS?

Mount St. Helens—Like majestic Mount Rainier and the other peaks in Washington's Cascade Mountains, Mount St. Helens was formed by volcanic activity. But St. Helens had last erupted in 1857, and the event was long forgotten by the hikers who walked its serene trails and fisherman who cast into the idyllic Shadow Lake at its foot. Then, at 8:32 a.m. on May 18, 1980, Mt. St. Helens exploded with a force 500 times that of the atomic bomb that fell on Hiroshima. Thousands of acres of timber were flattened by the shock wave, 400 feet of the rim vanished, and a cloud of ash, gas and debris rose 12 miles upward into the stratosphere, darkening the skies for thousands of square miles. Shadow Lake vanished, its waters boiled away almost instantly by the 2000 degree Fahrenheit temperature of the lava that spilled out of the crater. All the life in the surrounding 70 square miles was destroyed, including 60 people. Mount St. Helens was not only the most memorable, but the most destructive American volcanic eruption.

Krakatoa—Krakatoa was a remote island that rose 2,700 feet up from the Indonesian waters between Java and Sumatra. On August 27, 1883, at about 10 a.m., this island was shattered by perhaps the most colossal explosion ever documented by man. The eruption of Krakatoa was heard as far as India and Australia, 3,000 miles away. The blast threw 5 cubic miles of rock and soil 50 miles into the atmosphere, blackening the sky for hundreds of miles. The shock wave, so powerful that it traveled around the entire world seven times before it died out, created a 120 foot high tidal wave that killed 36,000 people on Java and Sumatra. When the reservoir of molten rock was exhausted, Krakatoa collapsed into a vast empty magma chamber. What was once a towering island was now a four mile wide, 900 foot deep crater in the ocean floor.

Vesuvius—On August 24 in the year 79 AD., more than 20,000 inhabitants of the cities of Pompeii and Herculaneum were going about their daily chores on the slopes of beautiful but peaceful Mt. Vesuvius. Suddenly, the volcano that had been quiet for centuries exploded with such force that the toxic gases released suffocated those 20,000 people almost instantly. A 15 to 25 foot deep layer of falling ash and hot volcanic mud buried the cities so quickly that the death agony on the faces of many victims was vividly captured in the hardening mud. Since the Romans did not have the heavy equipment needed to dig out the buried city, it was left to modern archaeologists to discover how totally devastating the eruption had been.

After the violence of 79 AD., Vesuvius was relatively quiet for 1,500 years. Then, on December 16, 1631, the mountain exploded again, destroying 15 towns and killing 4,000 people. Smaller scale eruptions occurred periodically over the next 300 years. In the early days of World War II, the British discussed dropping bombs into the volcano to try to trigger a massive eruption as a weapon against the Axis forces. But the plan, which was dismissed as useless by most experts, was never implemented. Although Vesuvius has been quiet since World War II, the potential for further eruption is such that Vesuvius today is classified as the only active volcano on the European continent.

Tambora—Tambora, a giant volcano on the island of Sumbawa, west of Java, exploded in April, 1815 in a roar heard 1,500 miles away. Blocks of rock weighing several pounds were thrown distances up to 25 miles, and 3,000 to 4,000 feet of the mountain were hurled into the sky. For three days there was total darkness for a distance of 300 miles. Estimates are that 12,000 people were killed directly by falling stones and ash, and 80,000 people later starved to death because millions of domesticated animals died and fields were destroyed.

However, the devastation caused by Tambora was not limited to Indonesia. The dust thrown into the atmosphere caused world wide global cooling for more than a year. 1816 was called the "year without a summer" in the United States, with hard frosts and snow recorded every month in New York State, including July and August.

Santorini—In 1500 or 1400 B.C., the huge volcanic Mediterranean island of Santorini erupted, belching forth a volume of lava and gas that scientists estimate was 5 times greater than that generated by Krakatoa. The ash covered many islands in the Aegean Sea, flattening crops and killing livestock. Sea waves swamped hundreds of cities and settlements, and the sky was darkened for many days.

Historians now believe that the eruption of Santorini ended the glorious Minoan civilization. Many also believe that this island that disappeared below the sea was the legendary lost continent of Atlantis that the Greek philosopher Plato wrote about one thousand years later.

EARTHQUAKES

To get an idea of where the power of an earthquake comes from, picture a bow and arrow. As you slowly pull on the string, stress builds up in the bow. When you suddenly release the string, the energy is transferred to the arrow, sending it toward the target at great speed.

Earthquakes are a product of the great strain produced in rocks in certain areas of the earth, primarily those in which two of our planet's huge plates are colliding. Eventually, the strain causes rock to fracture or shift, sending forth waves of energy in all directions (picture a stone landing in the water). The rock can shift up and down as well as horizontally. Movement along fault lines associated with earthquakes raised several mountain ranges in the Great Basin area, such as Utah's Wasatch Range, and they piled up sediment on the eastern base of the Rocky Mountains.

Major earthquakes are most devastating when they occur in or near major population centers. The strongest earthquake ever recorded in North America (9.2 on the Richter Scale) was fortunately centered in wilderness 80 miles east of Anchorage, Alaska when it struck in March, 1994. Even though it was felt 8500 miles away, it resulted in only 117 deaths. By comparison, the 1989 San Francisco earthquake that was seen live by tens of millions of World Series viewers was 1,000 times less powerful, but destroyed 100,000 buildings and killed 67 people. The 1994 Los Angeles earthquake was perhaps the costliest natural disaster in U.S. history.

About three-quarters of all the seismic (from the Greek word "seismos," meaning "earthquake") energy released every year occurs around the same "Ring of Fire" that includes most of the world's volcanoes. The boundaries between the Pacific plate and the surrounding plates are marked by fractures in the rocks called "faults." For example, California's famous San Andreas Fault occurs where the coastal part of the state is moving northwest as it scrapes against the North American plate. If some celestial mechanic could lubricate these faults so that the rock would slide smoothly, no earthquakes would ever occur. But in reality the rocks on each side frequently catch, producing various degrees of stress. The larger the built-up strain, the larger the earthquake.

While most earthquakes occur along plate boundaries, continental movement can produce stresses that build up in the middle of plates. No area of the U.S. is totally earthquake safe—a series of quakes around New Madrid, Missouri in 1811 and 1812 were among the most severe ever recorded in North America. Because construction requirements aren't as rigid elsewhere as they are on the West Coast, a powerful quake could cause many times the loss of life and property damage in New York or Chicago as an equivalent quake could in Los Angeles. However, every significant American earthquake of this century has occurred in California or Alaska, and these two areas will continue to suffer major earthquakes for tens of thousands of years.

About 25% of all earthquakes are not triggered by plate movements. Lava rising toward the surface before a volcanic eruption can also produce small earthquakes—in Hawaii, as many as 1,000 have occurred before a major eruption. We humans can also induce small earthquakes by filling reservoirs, detonating large bombs, or pumping fluids into wells.

Geologists estimate that several million earthquakes occur each year, but the vast majority of them are too small to measure. Approximately 50,000 are strong enough to be felt (3 to 4 on the Richter scale). About 800 earthquakes measure 5 to 6 on the Richter scale, an intensity strong enough to produce damage in urban areas. About a dozen major earthquakes (7 to 8) and one superquake (8-9) have resulted in an annual average of 30,000 deaths worldwide over the last three decades.

THE FORCES OF EROSION

The quiet forces of erosion, which re-shape the world built by the violent processes of continental drift, volcanoes, and earthquakes, are almost undetectable because they work over periods measured in thousands or ten thousands of years. The perfect example is the Grand Canyon, dug out of the plateau by the Colorado River at the rate of one foot every ten thousand years. Yet the final result is vivid testimony to the ultimate power of erosion, the force that levels even the most monumental structures of nature and man. The instruments of this force are water, ice, wind, and weathering.

WATER

Volcanoes and earthquakes can unleash the power of hundreds or millions of nuclear bombs. But if you ask any scientist what the most powerful landshaping force is, he or she would reply, "Water." While a single drop of water is the softest and gentlest of objects, the collective impact of rain, rivers, waves, tides, and currents levels giant mountain ranges, reshapes the outlines of continents, and reforms the topography of nearly every parcel of land.

WHAT WERE THE WORLD'S MOST DESTRUCTIVE EARTQUAKES?

Perhaps no U.S. city has been as completely devastated as was San Francisco after the earthquake of April 18, 1906. The quake, which measured an estimated 7.9 on the Richter scale, produced such intense vibrations that the soil in some parts of the city acquired the properties of a liquid, literally swallowing hundreds of buildings. But more destructive than the actual quake were the fires that followed. Undermanned and poorly equipped city forces couldn't begin to cope with an inferno that consumed almost all of the central business district and many residential districts—an area of more than 4 square miles. More than 500 people were listed as dead or missing, and in today's dollars, the property loss would be measured in the billions.

MEASURING EARTHQUAKES

If you watched me report from earthquake-devastated Los Angeles in 1994, you probably heard me report that the quake measured 6.9 on the Richter Scale. Now, finally, I have a chance to explain what that means.

Earthquakes generate seismic waves, a kind of energy we can measure just like we measure light, radio signals, brain activity, or any other type of waves. Seismic waves are measured on an instrument called a seismograph, which was invented early in this century and which came into general use about 1935. About fifteen years later, a scientist named Charles Richter developed a scale that allowed comparison between the amounts of energy released by different earthquakes. Unfortunately, at least for simplicity's sake, Richter made his scale logarithmic (a logarithm is the exponent that indicates how many times a number should be multiplied by itself. In the case of the Richter Scale the exponent is 10). As a result, an earthquake that measures 3.0 released 10 times as much energy as one that measures 2.0, while a 4.0 quake releases 100 times as much energy. The highest Richter scale reading ever assigned was the recently revised 9.2 given to the 1964 Alaska quake.

If you're still a bit confused by the Richter Scale, you might be interested in the more user-friendly Mercalli Scale. This system places earthquakes in categories Roman numeral I - XII based on the actual effects the quake has on people and their surroundings.

Following is what is officially called the Modified Mercalli Intensity Scale:

I. Not felt except by a very few under special circumstances.

II. Felt only by a few persons at rest, especially on the upperfloors of buildings. Delicately suspended objects may swing.

III. Felt quite noticeably by persons indoors, especially on the upper floors of buildings. Many people do not recognize it as an earthquake. Standing motor cars may rock slightly. Vibrations like that of a passing truck.

IV. Felt indoors by many, outdoors by a few during the day. At night, some awakened. Dishes, windows, doors disturbed; walls make cracking sound. Sensation like that of a heavy truck striking the building. Standing motor cars rocked noticeably.

V. Felt by nearly everyone; many awakened. Some dishes, windows broken. Unstable objects overturned. Pendulum clocks may stop.

VI. Felt by all; many frightened. Some heavy furniture moved; some cases of falling plaster. Damage slight.

VII. Damage negligible in buildings of good design and structure; slight to moderate in ordinary well-built structures; considerable damage in poorly built or badly designed structures; some chimneys broken.

VIII. Damage slight in specially designed structures; considerable damage in ordinary substantial buildings with partial collapse. Damage great in poorly built structures. Fall of chimneys, factory stacks, monuments, columns, walls. Heavy furniture overturned.

IX. Damage considerable in specially designed structures; well-designed frame structures thrown out of plumb. Damage great in substantial buildings, with partial collapse. Buildings shifted off foundations.

X. Some well-built wooden structures destroyed; most masonry and frame structures destroyed with foundations. Rails bent.

XI. Few if any structures remain standing. Bridges destroyed. Rails bent greatly.

XII. Damage total. Lines of sight and level distorted. Objects thrown into the air.

Next time you hear about the Richter Scale measurement of an earthquake, use the following chart to related it to the Mercalli Scale:

Richter Scale	Mercalli Scale
0 to 2.9	I, II
3.0 to 3.9	III, IV, V
4.0 to 5.9	VI, VII
6.0 to 6.9	VIII, IX
7.0 to 7.9	X
8.0 +	XI, XII

Rain produces both chemical and mechanical erosion. You may be surprised to learn that all precipitation is actually acid rain—falling water picks up carbon dioxide from the atmosphere to form carbonic acid (the acid rain that has become an environmental hazard refers to precipitation that has a much more corrosive effect because of higher levels of carbon dioxide in the air that result from pollution). Rain can chemically alter even the hardest rock, turning granite to clay and basalt to iron oxide.

All you have to do is take one look at the Grand Canyon to understand that rivers produce the most dramatic effects of any erosive force. They carry away stones and many other kinds of material loosened by other forms of erosion. This material in turn acts sort of like steel wool, scouring river banks and beds to loosen more material which is carried downstream and deposited in the ocean or other large body of water into which the river flows. The amount of material carried by a river is staggering. Every year, the Mississippi River transports 400 million cubic yards of silt to the Gulf of Mexico. More incredibly, every 7,000 to 9,000 years, the Mississippi and its tributaries wash away enough soil to lower the elevation of the entire 26 state drainage basin by 12 inches.

If you've watched me cover any of the major hurricanes of the last few years, you've seen dramatic footage of the impact waves and tides can have on beaches and beachfront properties. Yet for all the volume of ocean water and the power of the tides, the erosion that they produce tends to be very flat and shallow. Oceans sculpt the shape of the coastline, but rivers have a much more profound effect on the topography of the land.

ICE

Sheets of ice cover 10% of the land surface of the world and about 12% of the oceans. But as recently as 18,000 years ago, massive glaciers had crawled southward from the arctic until they covered 30% of all the land area on earth. When they finally retreated 10,000 years ago, they left a profoundly changed landscape behind.

Glaciers form when snow doesn't melt in the summer. In the areas in which glaciers form, the snow can accumulate very quickly—in rare cases, quickly enough to cover, freeze and perfectly preserve a creature as large as a woolly mammoth.

As more snow falls, the snow below is compressed into ice pellets called "firn." In 10 to 20 years, depending on the annual snowfall, the firn is compressed into solid ice. When this ice begins to move like a giant frozen river, it is called a glacier.

How can something as large as a glacier actually move? Although the tops of glaciers are rigid, a combination of pressure from the ice mass above and heat rising from the earth's core makes the lower ice plastic enough to flow. Because of gravity, mountain glaciers can move as much as 100 feet per day. On flatter land, where the movement results only from the pressure of ice above, the flow can slow to a few inches per day.

The rate of flow often varies from one part of a glacier to another. The resulting stresses can cause deep fissures in the glacier that are called "crevasses".

Glaciers change the land over which they pass in several different ways. The freeze-thaw cycle at the top of glaciers causes rock to split and crack. As the glacier moves, it pushes a mass of debris ahead of it and plucks stones from the valley's sides. Loose rock embedded in the glacier also rubs abrasively against the valley sides, loosening more rock. The result tends to sharpen mountain peaks and carve out steep, U-shaped valleys called "cirques." When the glaciers from the last ice age melted, sea water flooded such U-shaped valleys on the coasts of Norway and Alaska to form fjords. Glaciers also carve out hollows in the ground they pass over.

Friction between the ground and the glacier produces a steady flow of water under the glacier. This emerges as "meltwater," strong streams that can carve the land like rivers. Meltwater also carries fine sediment ground by the glacier, which it deposits in its path.

When glaciers melt, they leave debris behind. Masses of debris which have built up at the head and sides of glaciers are left behind as a land formation called a "moraine." Long Island is a moraine left behind by the last ice age. The glaciers also leave behind scattered rocks and boulders that had been caught up in the ice and carried many miles. Melting water fills hollows carved by the glaciers, producing lakes.

SNOW

If you don't think that snow is an erosive force, you've never seen an avalanche. An avalanche is a sudden cascade of snow moving from one height to another. While most of the hundreds of thousands of avalanches that occur each year go unnoticed in uninhabited mountain regions, some are devastating to man. During World War I, 60,000 soldiers were killed by avalanches in the Alps, more than all the deaths produced by weapons of war combined.

Avalanches result from the same build up of snow that leads to the formation of glaciers. Layers of snow become packed into a tight, cohesive mass. Underneath the snow, pressure or a rise in temperature produces a layer of lubricating water or ice. On flat or gently sloping terrain, such build ups may move very slowly. But on steeply sloping terrain, they can suddenly give way, producing avalanches that pack the energy of hurricanes or tornadoes. In some cases, the mass of snow is balanced so precariously that the weight of a skier, a clap of thunder, or a gun shot can set off an avalanche.

An estimated 150 people are killed in avalanches each year. The worst avalanche disaster in U.S. history occurred on March 1, 1910, when three passenger trains were snow-bound in a pass in the Cascade Mountains of Washington. An avalanche swept many of the cars off the track and down the mountain, killing more than 100 people.

WIND

Strong winds rocking trees and swaying bushes can be an awesome sight, and a tree landing on a house, automobile, or power line can greatly inconvenience our lives. But in terms of changing our environment, landscapes covered with vegetation suffer few erosive effects from the gales. Rather, wind plays a major role in shaping areas covered by sparse vegetation or none at all. In these landscapes, wind can carve stone, level mountains, scoop out hollows for lakes, and create awesome seas of sand.

The force of the wind works in three basic ways:

- First, wind picks up loose particles from the ground and carries them away, a process called "deflation." Deflation is part of the process in which marginal land becomes desert. When rainfall drops below a certain level, crops won't grow. Wind then attacks the bare soil, lifting it up and blowing it away. In a very short span of time, the nutrient-rich top soil that traps rain water and feeds vegetation is gone.

  Deflation turned millions of acres of land in the western Great Plains into the "Dust Bowl" in the 1930s, and some areas have yet to recover. Millions of acres have similarly been lost in southern Australia. To stop the loss and recover arable land, the Australian government plans to plant 1 billion trees by the year 2000 to slow down the wind.

- Second, wind-blown particles "sand-blast" rock in a process called "abrasion." In the deserts of the western U.S., the wind has carved mushroom-shaped rocks called "rock pedestals" and smooth, flattened rocks called "ventifacts."

The wind can also carve ridges in an area where soft rock alternates with harder rock. Eventually, large areas are worn down to broad stretches of resistant flat rock called "plateaus."

- Third, wind-blown particles erode as they rub against each other in a process called "attrition." Eventually, the particles are reduced to dust, which is often carried up high into the air and carried great distances by the prevailing winds.

All wind-blown particles eventually come to rest. The most extensive features created by wind-blown material are sand dunes. Most dunes are formed when sand is blown against a crop of rock, vegetation, or other natural elevation. If wind blows constantly in the same direction, the dunes are commonly found in long, narrow ridges perpendicular to the wind direction. If the sand supply is limited, the dunes can form a horse-shoe shape. Winds that frequently change direction can produce dunes in complex patterns such as stars.

Prevailing winds can also deposit layers of dust in certain areas which is cemented together by the weight of additional material piled above. Geologists call this material "loess," the German word for "loose." In the United States, it is commonly called "adobe," which is found in the desert Southwest and a wide swath of Texas, Oklahoma, Arkansas, and Missouri.

WEATHERING

We are accustomed to seeing the effects of weathering on wood that's been exposed to the elements. But weathering also has a long term effect on rocks in almost every climate of the world. Weathering most often weakens rocks, making them more vulnerable to wind, water, and ice.

There are two different types of weathering processes:

- "Mechanical" weathering breaks up rock without altering the materials of which the rock is made. This type of weathering is most prevalent in temperate or cold climates. Rocks can be weakened and broken during thermal weathering, which takes place when there is a large daily temperature change that freezes rocks, then heats them. The same process produces cracks in cement, which is man-made rock.

Other agents of mechanical weathering are frost, water that seeps into cracks and freezes, plant roots, and even the action of burrowing animals.

- Chemical weathering occurs when chemicals in air and rainwater eat away at the rock. Rocks can be oxidized by oxygen in the air, turning to rust. Carbon dioxide in the air mixes with rain water to form carbonic acid, which dissolves some rock. Many other chemicals introduced to the air and water as industrial waste have greatly increased the amount of chemical weathering that takes place in many parts of the world.

In some cases, chemical and mechanical weathering work together. For example, chemical weathering can dissolve salts from rocks that seep into cracks as a liquid, then harden and expand as they turn into crystals.

WHEN AND WHY DID THE GREAT ICE AGES OCCUR?

For about 90% of the last 3.6 billion years, the earth has been virtually ice-free (except for glaciers high in the mountains.) However, for reasons we don't completely understand, ice ages lasting a few million years have occurred periodically—in the last billion years, ice ages have occurred about every 150 million years.

The earth's first ice age occurred about 2.3 billion years ago, when rapidly moving glaciers appeared to cover most of the land. Scientists speculate that this ice age resulted when the proliferation of algae on the oceans took so much carbon dioxide from the atmosphere that the greenhouse effect was drastically reduced. Because the sun shone with less energy than it does today, the earth cooled and an ice age resulted. Eventually, the sun warmed and the ice melted.

Another great ice age, a period of time called "the longest winter" occurred about 950 million years ago. Other periods of widespread glacial expansions began about 600 million years ago, 435 million years ago, and 300 million years ago. There was no ice age 150 million years ago, perhaps because all the continents were joined together in a giant continent that straddled the equator.

The most recent ice age, which began about 2.5 million years ago, produces our mental pictures of fur-clad cave men hunting mammoths. It began after a gradual decline in the average temperature of earth over a period of 60 million year ago. During this long ice age, glaciers ebbed and flowed in cycles a few hundred thousand years apart. The most recent retreat of the ice sheets took place just 10,000 years ago. At their peak, glaciers covered about one-third of Europe and about one-quarter of North America.

In the United States, the periods of glaciation are named for the states that now exist at the glaciers' farthest southern penetration. From earliest to latest, these periods are the Nebraskan age, the Kansan, the Illinoian, and the Wisconsin.

The current belief is that ice ages result from a combination of factors, including slight variations in the earth's orbit around the sun, changes in ocean currents resulting from continental drift, changes in climate caused by continental uplift, and changes in the carbon dioxide content of the atmosphere. More importantly, we have no answer to the big question: is this latest ice age truly over, or will the glaciers eventually move southward again.

WHERE ARE THE ICE SHEETS TODAY?

Antarctica is completely covered in a massive ice sheet as much as 14,000 feet thick. This ice sheet is so heavy that much of the continent has been pressed down to below sea level. The other great ice sheet covers most of the island of Greenland in the north. Glaciers are found in Iceland, in the Alps, and in other high mountain ranges.

In the U.S., Alaska has 2,900 square miles of glaciers. Some of these have become tourist attractions in the summer. In the continental U.S., a few glaciers exist in the mountains of the northwest.

WHAT ARE ICEBERGS?

Icebergs are formed when chunks break off from the ice sheets covering Antarctica and Greenland, a process known as "calving." Because about four-fifths of the mass of these icebergs is under the water, they are a hazard to shipping. Antarctica calves the largest icebergs. In 1956, a 12,000 square mile iceberg broke loose from that continent, an area larger than Belgium.

MAP 5-7   PRESENT DAY ACTIVE GLACIERS IN THE UNITED STATES

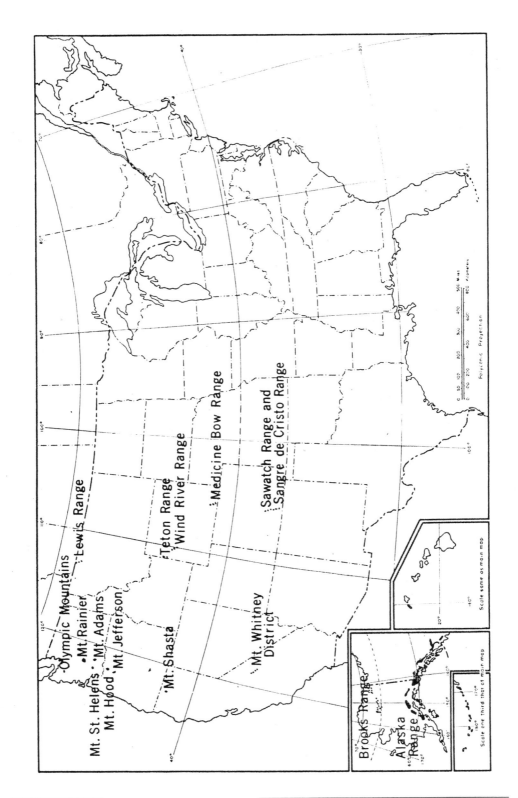

Mountains, Plains, Plateaus, and Basins

American Landforms

My job as "Good Morning America's" Ambassador to America limits the time and energy I have available for foreign travel—a vacation for me is a week when I don't have to rush to the airport for a long flight. But in a way, my countless visits to every part of this vast country of ours qualifies me as a seasoned world traveler. The reason is that within the borders of the United States you can find examples of nearly every terrain and land form that exists on earth. The state of California alone has six major types of terrain—you can go from snow-capped peaks to sun-drenched beaches, from barren desert to the world's most fertile agricultural valley, from grassland to forest.

Now that we've talked about all the different forces that shape our land, it's time to explore the mountains, plateaus, plains, and other land forms they've created for our enjoyment and wonder.

MOUNTAINS

A mountain is a portion of the earth's surface that rises above the surrounding area. Much of the earth's most majestic terrain consists of towering mountain ranges like the Rocky Mountains. Such soaring peaks are so awe-inspiring it seems to us that they are as permanent as the sun, moon, and stars. The truth is, however, that mountains are as temporary a phenomenon over the long life span of earth as buildings are over the life span of mankind. No sooner are they created out of the violence of the movement of the great plates that cover the earth or of erupting volcanoes than such forces of nature as wind, water, and ice begin the longer process of erosion that will eventually grind them down to sea level.

Mountains are and always have been very important to mankind. Mountain ranges serve as barriers to transportation. They have a profound affect upon the weather. Many mountainous areas are rich sources of minerals. And, finally, mountains are sources of recreation ranging from hiking to skiing to mountain climbing.

Mountains Are Like Icebergs

Continents float on the earth's mantle like icebergs on the oceans. Just as ice is only slightly less dense than the water on which it floats, the rock that makes up the crust is only slightly less dense than the mantle. That means that more than 80% of any land mass is below the surface.

The higher the mountain, the deeper the base extends into the earth. Because rock begins to melt about 43 miles below the surface, mountains can rise to a maximum height of about 6 miles, or about 31,000 feet. The highest mountain in the world today, Mount Everest, rises 29,108 feet above sea level.

When mountains erode, the rock below the surface rises in proportion to the decrease in height. When a mountain is eroded all the way to sea level, a process that can take 100 million years or more, the rock seen at the surface was originally 9 to 15 miles below the surface.

How Are Mountains Built ?

We're often told that we're living in the age of technology, and we're inordinately proud of the machines that determine where and how we live. But from time to time it's wise to remind ourselves about how puny our collective power is compared with that of Mother Nature. If you have any doubts, compare the Rocky Mountains to the Sears Tower, and think about the immense force necessary to lift that endless expanse of snow-covered peaks out of a flat plain.

Mountain building, or "orogenesis," occurs over millions of years. There are five different types of processes that can make mountains.

1. The world's greatest mountain ranges result from the collision of two of the plates that float on the earth's surface. Because these collisions take place across such a broad area, major mountain chains can stretch thousands of miles, the tremendous pressures involved raise the tallest of all peaks. The Rocky Mountains began to rise 65 million years ago when the North American plate collided with the Pacific plate, and the movement of the South American plate means the huge Andes Mountains are still active. The Himalayas, which include the 48 tallest mountains in the world, are still being raised up by the collision of India with Tibet. The Alaska Range, North America's youngest and tallest, is a result of plate movement which crashed more than 50 smaller pieces of crust called terreins into the continent. Undersea mountains can be raised when two ocean plates meet. The Mid-Atlantic range includes peaks topping 10,000 feet.

   When the tops of the mountains break the surface, they form island arcs like the Aleutians.
2. Movement of great sheets of rock along fault lines can cause one sheet to be lifted up. For example, California's tall Sierra Nevada Mountains, which include Mt. Whitney (the tallest peak in the contiguous 48 states), consist of one gigantic slab of rock that has been tilted upward towards the east.
3. Volcanic activity can also build mountains. Most of this activity takes place along with the other orogenesis that occurs when two plates collide. For example, the Cascade Mountains are a combination of volcanic peaks such as Mt. Rainier, Mt. Hood, and Mount St. Helens that erupted while the land was being forced upward by plate collisions.

   Volcanoes erupting in the oceans can build mountains in a relatively short period of time (from a geologic standpoint, that is). The Hawaiian Islands were formed when a stream of hot magma punched through the thin oceanic crust moving over it. Thousands and thousands of eruptions piled up layer after layer of hardened lava until the rock protruded above the surface to form islands. Mauna Loa, one of two active volcanic peaks in the islands, stretches 13,679 feet above the surface. But adding the 15,000 feet below the surface makes it the largest mountain in the world. The entire mountain forms a dome 60 miles wide and 30 miles long.

   Isolated volcanic mountains can occur on land. Mt. Paracutin in Mexico erupted from a farmer's field and grew to 1,500 feet in less than two years.
4. Magma can well up from the depths of the earth in the middle of plates as well as at plate margins. The resulting pressure can force the crust to bend upward like an inflating dome, raising mountains at the surface. New York's Adirondack Mountains are still rising due to pressure from below. The rock pushed to the surface is well over 1 billion years old, some of the most ancient on the continent.
5. Erosional mountains are blocks of harder rock that remain standing after softer rock has been eroded by wind and water. The best known erosional mountain in the U.S. is New Hampshire's Mount Monadnock, which rises 1,800 feet. As a result, other such structures are known as "monadnocks." The Ozark Mountains of Missouri and Northern Arkansas are elevations carved by erosion from the Ozark Plateau.

Even Mt. Everest Won't Last Forever

The process of erosion begins as soon as a mountain emerges, but it can last hundreds of millions of years. Because rock rises from below as rock above erodes, leveling a mountain that was originally 3 miles high requires the removal of 12 miles of rock. The remnants of higher mountain ranges exist for an incomprehensibly long period of time.

For example, the Appalachian Mountains were a mighty chain formed when Africa and North America collided 350 million years ago—perhaps as huge as the Himalayan Mountains. They had worn down considerably in size by the time the two continents separated 180 million years ago. Water running off the still existing chain eroded the mountains nearest the ocean more quickly. That's why the remaining stubs of the original Appalachian Mountains are found inland today.

What happens to the sediment eroded from these mountains? It washes down into the ocean to become the building blocks of mountain chains which will be formed millions of years in the future.

Mountains Effect Climate and Weather

The areas on opposite sides of mountain ranges usually have very different climates. The most dramatic effects are those of the great mountain ranges that run east-west, such as the Himalayas and the Alps. Italy, Greece, and the south of France are roughly on the same latitude as New York and Chicago, but they have much milder winter weather

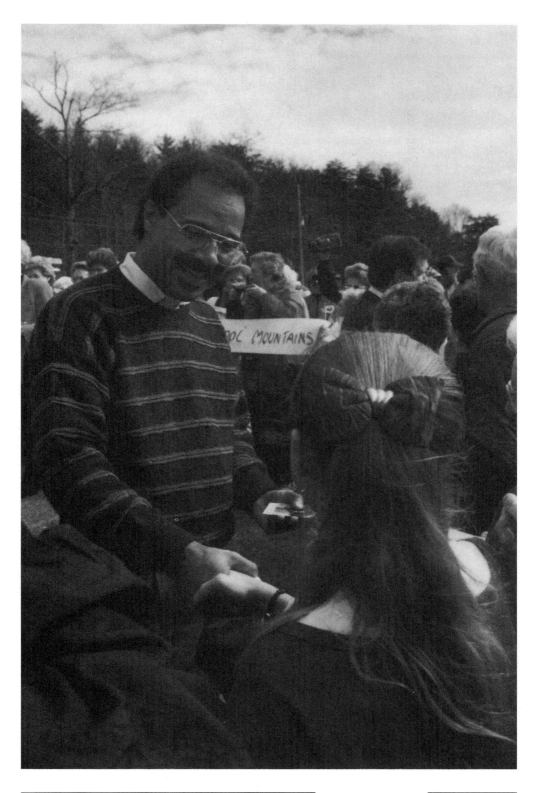

because the Alps block the cold air sweeping down from the far north. Similarly, the blocking provided by the Himalayas give India a tropical climate.

There is also a major difference in climates on either side of the major U.S. mountain ranges that run north-south in the western U.S., where the prevailing winds sweep in from the Pacific Ocean. Because the ocean's temperature doesn't vary widely, such cities as San Francisco, Seattle, and Portland have a rather even temperature range year around. The moisture-laden winds also produce frequent clouds and a lot of precipitation. The Pacific Northwest is a region I have visited many times, and I continue to be amazed by the dramatic differences in the micro-climates of the area. What a difference a mountain range makes.

The downwind side, on the other hand, has a much greater temperature range and is drier and sunnier. For example, Tacoma, Washington, on the west side of the Cascades, averages 37 inches of rain per year; Yakima, 100 miles away on the eastern side, averages just 7 inches of rain per year.

Every mountain, large or small, has an effect on the prevailing winds in its area. In the morning, mountain tops are warmed first, so the wind blows uphill from the cooler valleys. In the evenings, the valleys retain heat longer, so the wind direction is reversed.

WHAT IS A PLATEAU?

A plateau is a large flat or sloping land mass that is significantly elevated above the surrounding land. Plateaus are often punctuated by mountain ranges or gouged by erosion into canyons or valleys.

Among the most notable American plateaus are:

- The Columbia Plateau, the area between the Rocky Mountains and the Cascade Mountains in southern Washington, eastern Oregon, and western Idaho, consists primarily of vast ancient lava beds.
- The Colorado Plateau covers southern Utah, southwestern Colorado, northwestern New Mexico, and northern Arizona. It's cut by many canyons and gorges, including the Grand Canyon.
- The Great Basin is a triangular shaped area (broadest at the northern end) that separates the Rocky Mountains from the coastal ranges such as the Sierra Nevadas. This region includes Utah's Great Salt Lake, the Mojave Desert, and Death Valley. Wetter eras long ago, it was largely covered by two vast lakes, Lake Bonneville (of which the Great Salt Lake is a remnant) and Lake Lahort.
- The Great Plains are a vast plateau between the central lowlands and the Rocky Mountains that runs from the Canadian border all the way into Texas.
- The Cumberland Plateau runs from southern Virginia through eastern Tennessee into northern Georgia and Alabama. This plateau is an eroded remnant of the once majestic Appalachian Mountain chain, and several mountain ranges rise from it.
- The Ozark Plateau, which runs from southern Missouri and northern Arkansas into eastern Kansas and Oklahoma, is an elevated area of harder rock that resisted erosion better than the surrounding central lowlands.

Jutting from plateaus in the American West and Southwest are areas of level land jutting from rock formations with steep rock cliffs on all sides. These mini-plateaus are called mesas. Similar formations that are smaller in size are called buttes. One of the most amazing froups of mesas and buttes I've ever seen can be found in Monument Valley, in northern Arizona

PLAINS

A plain is a large area of flat or rolling land that is relatively treeless. The United States has two main plain areas. The Atlantic and Gulf Coastal Plains include the area between the Appalachian Mountains and the Atlantic Ocean and the Gulf of Mexico. This land, an extension of the continental shelf (a submerged part of the North American plate that extends miles into the Atlantic Ocean), was formed by sediment washing down from the eroding mountains.

Our country's vast central plain that stretches from the Appalachian Mountains to the Rocky Mountains has two parts. The central lowlands were frequently covered by shallow inland seas during the last few hundred million years. These seas were eventually filled by sediment and decayed organic material, which created very fertile soil. To the west are the higher Great Plains, an area that receives less rainfall and has less fertile soil.

OTHER LAND FORMS CREATED BY EROSION

Now that we know the basic "lay of the land" in the U.S., we can take a moment to talk about the other common shapes that have been sculpted from the land. Before you know it, you'll be an expert, impressing your friends and family as you point out the window and exclaim, "Now that's a real butte!"
- Valleys are depressions in the land that are almost always carved by rivers. The water tends to cut more quickly and deeply in areas that are drier and have less vegetation. All valleys widen and become more U-shaped as they get older.
- Canyons are valleys with steep, rocky walls. They are made by faster-flowing rivers, generally in plateaus which are rising at the same time.
- Gorges are small, narrow canyons or parts of canyons.
- Hills are rounded elevations of land that are smaller than mountains and almost always created by some force of erosion.
- Moraines are large deposits of earth and stone deposited when glaciers melt. Long Island is a moraine.
- Mesas are isolated, flat-topped rock formations that have steep rocky cliffs. They occur primarily in sparse, arid areas of the Southwestern United States where they have been carved from larger plateaus.
- Buttes are small mesas.
- Caves are underground chambers formed primarily in limestone by the action of water and acidic compounds of water, carbon dioxide, and organic matter. Water eats away at the limestone until it works its way down to the water table. Fluctuations in the water table over long periods of time produce multi-level cave systems like Kentucky's Mammoth Cave. Caves can also be formed by lava, ice, wind, and sea water.
- Marshlands are treeless expanses where the water table is at, just above, or just below the surface. Marshes can be covered with fresh water or salt water. Florida's Everglades are America's largest area of marshland.
- Deltas are areas of land built up from soil or silt dumped when a river empties into a lake or ocean.

MAKE YOURSELF AT HOME WITH THE RANGE

If you want to talk about mountains and sound like an expert, you've got to know the lingo. Here are the definitions of the terms used to talk about groups of mountains:

- A mountain range is a ridge containing several peaks or a number of ridges containing peaks of similar size, area, and shape. The Bitteroot Mountains of Montana are a range.
- A mountain system is group of closely related ranges. The Northern Rockies are a system that includes the Bitteroot Mountains.
- A mountain chain is a series of systems. The Rocky Mountain chain includes the Southern, Central and Northern Rockies.
- A cordillera is a complex of ranges, systems, and chains. Collectively, the mountains of the western United States form a cordillera.

HOW HIGH DOES A MOUNTAIN HAVE TO BE TO BE SNOW-CAPPED?

The temperature drops a dramatic 1 degree Fahrenheit for every 300 feet of elevation above sea level. That's why even at the equator mountains have permanent snow caps above 18,000 feet. The snow line is 3,000 feet lower in the Alps, and much lower in the Canadian Rockies and in Alaska.

Vegetation on the slopes of mountains changes with the temperature. In New Mexico, the southernmost part of the Rockies, trees grow to about 11,000 feet, then give way to arctic-like tundra before the permanent snow cap begins. The tree line drops to 6,000 feet at the Canadian-US. border and to just 2,500 feet in Alaska.

THE TALLEST MOUNTAINS IN THE U.S.?

The United States contains three huge chains of mountains, each formed at a different period in time. Because mountain ranges are formed by destructive plate margins, all of the 93 U.S. mountains over 14,000 feet exist where the North American plate has collided with the Pacific plate. The Rocky Mountains, one of the world's great highland areas, were raised between 65 million and 30 million years ago; relatively young by geological standards. Today, the Rockies stretch 3,200 miles from New Mexico through Canada to Alaska, their width varying from 100 to 400 miles. The Rockies, with steep slopes and deep valleys, create some of the most spectacular scenery in the world.

Beginning 25 million years ago, volcanic activity and captured terreins pushed up the newest chain of mountains beginning with the Alaska Range, extending down through the Coastal Mountains of Canada through the Cascades of the Northwest to the Sierra Madres of California and Nevada. The newest mountains are the tallest, and Mount McKinley, at 20,320 feet, is the highest point in the U.S.(but just the 93rd highest mountain in the world). The seventeen next highest peaks are all in Alaska. In the continental United States, the tallest mountain is California's Mount Whitney (14,494 feet), which is part of the coastal chain.

In the eastern part of the country, no new mountain ranges have been formed for hundreds of millions of years because the North American plate extends halfway across the Atlantic Ocean to a constructive margin. Although ranges like Vermont's Green Mountains, New Hampshire's White Mountains, New York's Adirondack Mountains, and North Carolina's Blue Ridge Mountains are famous, they are worn down stubs of the ancient Appalachian chain that stretches from Georgia to northern Maine. Most of the southern mountains in this chain are worn and rounded, as their names indicate (e.g., Clingman's Dome) because their patterns of erosion. Comparing the heights of America's tallest mountains show a striking difference between western and eastern mountain chains as shown by the chart below.

WESTERN UNITED STATES			EASTERN UNITED STATES		
Colorado	Mount Elbert	14,433 feet	Maine	Mount Katahdin	5,267 feet
Washington	Mount Ranier	14,411 feet	New Hampshire	Mount Washington	6,288 feet
Wyoming	Gannett Peak	13,804 feet	Vermont	Mount Mansfield	4,393 feet
Utah	Kings Peak	13,528 feet	New York	Mount Marcy	5,344 feet
New Mexico	Wheeler Peak	13,161 feet	Pennsylvania	Mount Davis	3,213 feet
Nevada	Boundary Peak	13,146 feet	West Virginia	Spruce Knob	4,861 feet
Montana	Granite Peak	12,799 feet	Virginia	Mount Rogers	5,729 feet
Idaho	Borah Peak	12,662 feet	North Carolina	Mount Mitchell	6,684 feet
Arizona	Humphrey's Peak	12,633 feet	Tennessee	Clingman's Dome	6,643 feet
Oregon	Mount Hood	11,239 feet	Kentucky	Black Mountain	4,139 feet
			Georgia	Brasstown Bald	4,784 feet

I grew up in Virginia, not far from the majestic Blue Ridge and Shenandoah mountains. But after visiting every other major mountain range in the continental U.S., I have grown totally impartial. Each range has its own spectacular beauty and character, and I would urge you to visit as many as you can. However, if you are looking for just one natural wonder in a mountain range to touch the very depths of your soul, go to the southern Rockies and see the Grand Canyon. It is hauntingly spectacular.

The Wellsprings of Life

The story of oceans, seas, lakes, rivers and other parts of the water cycle

If an interstellar roving reporter from "Good Morning Galaxy" were to drop in to our solar system and catalog its celestial bodies, no doubt he or she would describe earth as the "water planet." Almost three-quarters of the planet's surface (71%), is covered in water that gives earth its distinctive blue appearance from space. Early in its existence, our world was completely covered by one immense ocean, and in these waters life was miraculously created. Although we take the oceans for granted, they play a crucial role in maintaining a climate and environment that allows the continued existence of life.

We all understand that life on earth couldn't exist without fresh water (and a few other things, like vintage Bordeaux). But because we're so used to water gushing out at the turn of a tap, we may not understand how precarious a resource it is. Even though the total volume of earth's water is 326 million cubic miles (at more than 26 billion gallons per cubic mile, an incomprehensible amount), 97% of that is in the oceans. Another 2.1% is locked up in the ice caps, leaving less than 1% for all the people, animals, and plants in the world. To put it further in perspective, if all the water on earth were represented by 25 gallons, all life would have to exist on just one teaspoon.

Because we depend so greatly on both the oceans and fresh water resources, the geography of water is a vital subject in the United States as well as around the world. Saving water is a vital step in saving the earth.

WHERE DID OUR WATER COME FROM?

Because of the intense direct bombardment of the sun, when the young earth was three-quarters its present size, it had less water than the next outer planet, Mars. However, the intense heat in the earth's core released water trapped in rocks as they melted, and formed more by combining two hydrogen atoms with one atom of oxygen. After the magma seas hardened into a crust, numerous volcanoes spewed forth gases that consisted of carbon dioxide and 60% to 70% water vapor. Because the sun at that time shone with only 75% of its current intensity, the average temperature of the earth was only about 52 degrees Fahrenheit. In these cool temperatures, this water vapor condensed easily and fell as rain, forming the first lakes and oceans. The clouds generated by volcanoes were so thick that it may have taken 200 million years of torrential rains to condense enough moisture out of the sky to allow the sun to shine through. At the same time, ice-rich planetismals thrown toward the sun by the formation of the huge outer planets frequently struck earth and melted, adding to the water supply. By that time, about 4.2 billion years ago, the fully-formed earth was completely covered by one gigantic ocean.

Where is our water today?

The amount of water in our biosphere has neither increased nor decreased in billions of years. Today, the vast majority of all water, 97.2%, is salt water in the world's oceans. The distribution of our fresh water is:

	Volume (cubic miles)	% of total
All freshwater lakes	30,000	.009
All rivers	300	.0001
Antarctic icecap	6,300,000	1.9
Arctic icecap & glaciers	680,000	.21
Water in atmosphere	3,100	.001
Ground water (within 1/2 mile of surface)	1,000,000	.31
Deep ground water	1,000,000	.31
Total fresh water	10,013,400	2.7391

Fresh water in lakes and rivers would soon disappear if it weren't replenished by rain or snow. At any given time, there's only enough water in the atmosphere to drop 1" of rain over all the earth's surface, just a ten to twelve day supply. Fortunately, an immense amount of water is constantly evaporating from the oceans—104 million billion gallons, about one-quarter of the oceans' total volume, evaporates every year. Most of this water falls back into the ocean. But some is blown over land, falling as rain or snow to replenish fresh water supplies. It replaces water that makes its way down streams or rivers into the ocean.

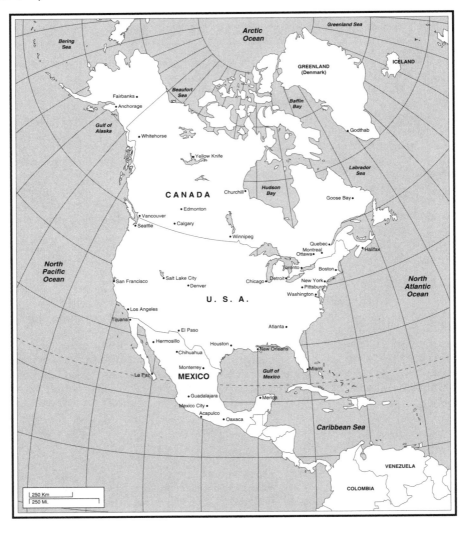

THE OCEANS

With the exception of some interior parts of Antarctica, vast tracts of the world's oceans are the last unexplored parts of our planet. The United States is bordered by three of the four oceans, the Pacific, Atlantic, and Arctic which play a critical and complex role in every aspect of our daily lives from the weather to our recreation.

What's the difference between an ocean and a sea?

"Ocean" is a term used for the immense and interconnected body of salt water that covers about 71% of the world's surface. Geographers divide this body into 3 to 5 separate oceans. A "sea" is a large body of salt water significantly smaller than an ocean. Nearly everyone accepts the Pacific, Atlantic, and Indian Oceans, while there is disagreement about whether the Arctic Ocean and the Antarctic Oceans are oceans or seas. In this book, we refer to the Arctic Ocean, which has a very clear line of demarcation between it and the Atlantic, as a fourth ocean.

The International Hydrographic Organization identifies 66 bodies of water as seas, gulfs, bays, straits, channels or passages. Their definitions are:
- Gulf: Part of an ocean or sea that extends inland. For example, the Gulf of Mexico.
- Bay: Part of an ocean or sea that extends inland, but is smaller than a gulf. For example, San Francisco Bay.
- Strait: A narrow body of water that connects two larger bodies of water. For example, the Straits of Magellan off the tip of South America which connects the Atlantic and Pacific Oceans.
- Channel: A narrow body of water that connect two larger bodies of water, but is smaller than a strait. For example, the English Channel.

Many of these bodies of water are subdivided into many smaller bodies. For example, the Mediterranean Sea is divided into the Straits of Gibraltar, the Balearic Sea, Ligurian Sea, Tyrrhenian Sea, Ionian Sea, Adriatic Sea, and Aegean Sea. The Gulf of Mexico has five bays on the U.S. side, including Galveston Bay and Tampa Bay.

What was the first ocean like?

All the water on the young earth was fresh water. Our oceans became salty only after absorbing salt and other minerals washed off land masses over billions of years. Because the moon was only half its present distance from the earth 4.2 billion years go, the waters of the oceans were pulled to and fro by huge tidal forces. These forces, along with awesome storms, quickly eroded early land masses formed by volcanic activity or the impacts of meteors.

Most scientists believe that the ocean itself was boiled away by the impact of giant meteors several times during its first half billion years of existence. Only after the intensity of the bombardment dropped dramatically about 3.8 billion years ago did the ocean become more stable.

What is sea water like today?

Sea water, on the average, is 96% pure water and 4% dissolved minerals that were washed from the land masses and dissolved in the seas over billions of years. Chlorine and sodium, which combine to form sodium chloride, or table salt, make up 85% of chemicals. This concentration doesn't sound like much, but if you were to let one cubic foot of sea water evaporate, you'd be left with 2.2 pounds of salt.

The saltiest water is found in the Red Sea and the Persian Gulf, where high temperatures lead to more evaporation and low levels of rainfall cut down on the replenishment of fresh water. The least salty water is found near the poles because of fresh water melted from the ice caps.

WHAT ARE THE LARGEST SEAS?

The Arabian Sea (1,492,000 square miles) and the South China Sea (1,331,000) are the largest seas. The Caribbean Sea, which borders on the Gulf of Mexico, is the third largest at 1,063,000 square miles.

By far the most important smaller body of salt water bordering the United States is the Gulf of Mexico, which covers 596,000 square miles. Four of the six busiest ports in the United States— New Orleans (2), Houston (3), Baton Rouge (5) and Corpus Christi (6)—are on the Gulf.

What is the geography of the ocean floor?

The earth's crust below the oceans is younger and thinner than the crust below the land masses. The reason is that new crust is created and old crust pulled down, or "subducted," into the mantle underneath the oceans. The oldest rocks found on the ocean floor date back just 200 million years, and the thickness of the crust averages six miles.

Much of the ocean floor is the relatively flat abyssal plain ranging from two to three miles below the surface. But there are also the following significant features:

- Continental Shelves consist of gently sloping rims that are the true edges of continents. These shelves are shallow (no more than 450 feet deep) and they range in width from 30 to 200 miles. They spawn the richest variety of ocean life.
- Continental Slopes are more steeply dropping surfaces that bridge the continental shelves and the plains that form the bulk of the ocean floor.
- Continental Rises are gentle upslopes formed by masses of sediment that come to rest at the base of continental slopes. This area can range from 3 to 100 miles wide.
- Seamounts are underwater volcanoes that rise more than 3300 feet above their surroundings. Flatter seamounts that were once volcanic islands but are now submerged are called "guyots."
- Spreading ridges are underwater mountains formed when magma wells up between two separating plates. The Mid-Atlantic Ridge is a huge line of underwater peaks that surface as Iceland and Ascension Island.
- Trenches are the deepest parts of the ocean. They occur where one plate bends down under a less dense plate, destroying old crust to compensate for the new crust being created in spreading ridges. The Mariana Trench in the Pacific Ocean is 35,840 feet deep, the lowest part of the ocean floor.

Most of the ocean floor consists of rock covered by a mixture of mud, sand and other sediments washed from land along with the remains of dead ocean organisms.

How far does sunlight penetrate the ocean waters?

Sunlight penetrates only about 2,000 feet, less than half a mile. Below that level, the ocean is completely dark. The absence of photosynthesis, however, does not mean the absence of life. In the depths of the oceans are micro-organisms that draw their energy from sulfurous hot springs rather than the sun. These organisms may be the ancestors of the oldest surviving living things, bacteria which survived deep in the ooze at the bottom of ancient oceans vaporized by giant meteor impacts more than 4 billion years ago. A variety of strange blind sea creatures feed on these micro-organisms.

How do the waters of the oceans move?

The waters of the ocean are far from still, and their movement plays important roles in shaping the land and in moderating earth's climate. The basic types of movements are:

- Waves. Waves are undulations primarily set in motion by the wind. In open waters, the waves pass through the water without moving it forward - the particles of water move in a circular motion up and down. But in shallow water, the wave crests grow closer until they pile up, making the waves higher and tipping them over. This causes a forward movement of water that ends when the waves break on the shore.

  The height of the waves depends on the strength of the wind and the distance over the water the waves have traveled. Waves driven by hurricanes can be particularly devastating, especially when the storm strikes at high tide.

  The most destructive waves of all are "tsunamis," which are triggered by underwater earthquakes. They occur most commonly in the Pacific, because of the many active earthquake zones on its margins. A tsunami can travel at speeds of 435 miles an hour or more, which means one generated off the southern coast of Alaska will reach Australia in less than a day. Be cause waves don't move water forward in mid-ocean, ships at sea aren't affected. However, shallow waters of such islands as Hawaii or Japan can build waves to a devastating 30 feet or higher.

- Tides. Tides are twice daily movements of billion of tons of water caused primarily by the gravitational pull of the moon and the sun. Because the moon exerts twice the gravitational pull of the sun, its movements govern the rhythm of the tides. Since the moon takes about 24 hours, 50 minutes to orbit the earth, the time of high and low tides advances a little less than an hour each day.

  The strength of the tides varies according to the relative positions of the moon and the sun. When the moon is either new or full, it is in a straight line with the earth and sun. This produces "spring" tides, high high tides and low low tides. In between, at the time of quarter moons, the sun is at right angles to the moon and earth. This provides the least gravitational pull and the lowest variation between high and low tides, a condition call a "neap" tide.

On the average, tides only change the height of the water two or three feet. But in certain shallow, narrow bays and inlets, the force of the tide is compressed, leading to much higher variations. An example is the Bay of Fundy, which separates New Brunswick from Nova Scotia. Tides in this narrow bay reach a height of 60 feet or more.

- Currents. Ocean currents, which have a profound effect on the world's climate, are of two types. The warm surface currents are caused by a combination of wind and the rotation of the earth. These tend to flow northeast and southwest from the equator. Colder, denser, more saline water of the polar regions sinks to form deep ocean currents that flow toward the equator.

How have the oceans sculpted the coasts of the United States

According to the National Oceanic Service, the U.S. has 12,283 miles of coastline:
- 2,069 miles on the Atlantic Ocean
- 1,631 miles on the Gulf of Mexico
- 7,623 miles on the Pacific Ocean (5,580 of it in Alaska)
- 1,060 miles on the Arctic Ocean

WHAT ARE THE FOUR OCEANS THAT EXIST TODAY?

The four oceans are:

1 The Pacific, which at 64,000,000 square miles is larger than all the land masses combined. The Pacific has the deepest average depth (13,215 feet) and the deepest point (Mariana Trench, 35,840 feet). The Pacific is the oldest ocean, which evolved from Panthalassa, the ocean that surrounded the supercontinent Pangaea more than 200 million years ago. The ocean floor makes up the Pacific plate, which is being pressed by the American plate, the Eurasian plate, and the Indo-Australian plate. These destructive margins spawn not only earthquakes, but a gigantic circle of active volcanoes known as the "Ring of Fire." The crust at the edges of the Pacific plate is diving under the continental plates and being melted in the mantle. As a result, the Pacific Ocean will shrink over the next tens of millions of years.

2 The Atlantic, the second largest ocean at 31,815,000 square miles, is also the youngest, born about 135 million years ago when North America began to separate from Eurasia. The South Atlantic opened up about 60 million years later when South America separated from Africa.

Because the American, Eurasian, and African plates are continuing to separate the middle of the Atlantic, the ocean is widening at the rate of about an inch per year. New crust is being created along the vast Mid-Atlantic Ridge, producing volcanic activity in Iceland. This ridge is part of the vast Mid-Oceanic Ridge which winds its way 45,000 miles through the world's oceans.

But because the edges of the Atlantic are part of the continental plates, its shores are geologically tranquil and its waters teem with ocean life.

3 The Indian Ocean, which covers 25,300,000 square miles, was created about 180 million years ago when India separated from Africa, Antarctic, and Australia. The Indian Ocean is continuing to expand at the expense of the Pacific Ocean.

4 The Arctic Ocean, at 5,440,000 square miles, is by far the smallest of the four oceans. It was born about 180 million years ago when North America and Eurasia separated and began rotating to form a circle around the North Pole. The Arctic Ocean is now almost completely circled by land.

The Polar ocean is crossed by three major underwater mountain ranges, some peaks of which reach above the surface to form islands. Much of the Arctic Ocean is covered by a permanent ice cap that includes the North Pole. This ice cap coves about 7% of the total ocean area of the Northern Hemisphere.

When all the shorelines of offshore islands, bays, sounds, rivers, and creeks touched by tidal waters is added in, the total soars to 88,633 miles.

These shores seem to serve as a magnet for Americans—in the last half century, the percentage of the U.S. population living within 50 miles of the coast has doubled from 25% to 50%. Many more people flock to the shores for vacations. So we all have an interest in the effects of the seas on that land.

The oceans change the shoreline in three general ways:

A. Destroying the land: The sea is probably eroding the land where cliffs or rocks rim the shore. This erosion takes place through a combination of three actions:

- As waves slam against the rock, air is forced into cracks and it expands, breaking off pieces.
- Small stones and bits of rock are hurled against the rock by waves, breaking off additional pieces.
- Rocks rubbing against each other breaks them down eventually to smaller rocks, then sand.

Because most cliffs and rocky shorelines are made up of a combination of softer and harder rocks, the action of the sea can produce sea caves, platforms, arches in rocks, and other similar features. This same action can cut out entire bays, leaving headlands that are first attached to the mainland, then cut off as one or more small islands.

B. Submerging the land. About 10,000 years ago, during the last Ice Age, so much water was locked up in glaciers that sea level was about 330 feet lower than it is today. When the glaciers melted, much of the existing coastline was flooded, creating the following types of features:

- Longitudinal, or Pacific, coastlines are formed where low mountain ridges run parallel to the shoreline. Because the mountains of our West Coast run north and south along the shore, the coastline used to be marked by a north-southridge of foothills. When the sea level rose, the valleys of these foothills became sounds, like Washington's Puget Sound, and the ridges became narrow islands parallel to the new shoreline, like Santa Catalina, San Clemente and the other islands off the coast of Southern California.
- Fjords were valleys with sheer, high sides that were cut out by the glaciers before being flooded with water. Although Norway is famed for its fjords, equally awesome formations exist all along the coast of southern Alaska.
- Rias formed where U-shaped river valleys with rounded, gently sloping sides emptied into the ocean. When these flooded, they formed what is also called an "Atlantic" type coastline, with ridges of land running perpendicular to the shore.
- Estuaries are the broad, tidal plains of great rivers like the St. Lawrence, Hudson, and Chesapeake. They are funnel shaped at the mouth, flanked by mud-flats, and marked by low-lying land that emerges at low tide.

COULD LIFE HAVE COME FROM OUTER SPACE?

Scientists have discovered organic molecules and amino acids in meteorites that have struck the earth. It is possible that some life forms developed from these amino acids. However, the ocean produced a vastly larger quantity of the building blocks of life on its own, which suggests that most, if not all, of our life is home-grown.

- Deposits of glacial rocks partially submerged by water account for the small rocky islands of Boston Harbor and many other parts of the New England coastline.

C. Building up the land. The debris taken away from one shoreline is often deposited on another to form the following types of shoreline:

- Boulder beaches are narrow belts of rock and gravel deposited at the bases of cliffs. Many beaches of New England and the coast the northwest U.S. are of this type.
- Bay beaches are sandy crescents deposited between rocky headlands.
- Lowland beaches are broad, sloping sandy stretches often backed by sand dunes.
- Bars are offshore strips of sand paralleling the coastline. Bars include Fire Island off the coast of Long Island, the barrier islands of North Carolina, and such South Carolina resort locations as Hilton Head Island and Kiawah Island. Bars border much of the southeast and Gulf coasts of the U.S.

How do the oceans affect our climate and weather?

The oceans serve four vital functions critical to the widespread development of life on our planet:

A. The oceans moderate the temperature. The oceans are slow to take in heat and slow to release it. Without the oceans, the temperatures at the equator would be much hotter during the day and much colder at night. The more extreme latitudes would have much hotter summers and much colder winters, conditions not conducive to life. For example, inland Canada experiences summer temperature as high as 104 degrees Fahrenheit and winter temperatures as low as -50 degrees Fahrenheit. Vancouver, on the coast at the same latitude, has an average temperature range of 37 degrees to 63 degrees Fahrenheit.

B. The oceans help distribute warmer air toward the polar regions. The prevailing winds that blow warmer air from the equator toward the poles generate strong currents of warm water flowing in the same directions. For example, the Gulf Stream, a gigantic river of warm water more than 50 miles wide, flows from the Caribbean past Cape Hatteras and Cape Cod, then veers eastward toward the British Isles. The effects of this warm water are so pronounced that subtropical trees grow on the southern coast of Scotland. Overall, the distribution of warm air northward greatly moderates the world's climate. Without the distribution, the average temperature at the poles would be -180 degrees Fahrenheit instead of -80 degrees. Cities like New York and London would be ice-bound and arctic all winter long.

To compensate for warm surface currents moving north and south from the equator, cold water from the polar region sinks and forms deep-flowing currents that move toward the equator. Because these currents are rich in oxygen, they are filled with plankton and other microscopic life that nurture vast schools of fish when they meet warmer water.

C. Water evaporating from the oceans is blown over land, then falls as rain or snow. Although 90% of this water quickly evaporates, some of the rest is captured by plants and by the lakes, rivers, reservoirs, acquifers that serve as the source of water for man and for animal life.

D. The ocean absorbs carbon dioxide from the atmosphere. Algae and other plant life originally produced the levels of oxygen in the atmosphere that made it possible for living creatures to exist out of the water. The oceans also play a critical role in maintaining a healthy atmosphere today.

What was the role of the ocean in spawning life?

Carbon dioxide dissolves very easily in water, as any of us who drink carbonated beverages can testify. This has led to a large concentration of carbon atoms in the water. Carbon has two very special properties that were crucial to the development of life. First, carbon atoms have the ability to easily bond to each other and to atoms of hydrogen, oxygen, nitrogen and other elements. While water is a simple compound of two hydrogen atoms and an oxygen atom, carbon-based molecules can be very complex, including thousands, even millions of atoms.

Second , certain carbon-based molecules have the ability to split down the middle into two identical halves. This chemical property is not life, but it is the basis of life. That's why carbon-based molecules are called "organic molecules."

Scientists believe that soon after it developed, the ocean became an organic soup teeming with complex molecules. At that time, there were two special conditions in the earth's environment that contributed to the development of life, two that don't exist today. First, because the atmosphere contained no free oxygen, no ozone layer existed to filter out the sun's ultraviolet rays. Second, the lack of free oxygen in the water meant that carbon molecules bonded with each other. Both of these conditions encouraged a chemical reaction, possibly triggered by lightning striking the water, that caused organic molecules to join together in a chemical reaction and form amino acids. As early as 1953, this process was duplicated in research laboratories.

These amino acids, through a process we still don't understand, evolved into living cells. A cell, the smallest unit that still retains the properties of life, is a complex system of molecules that has the power to reproduce. We don't know exactly when these first cells were formed, but we do know that by about 3.5 billion years ago, organisms consisting of clumps of cells left microscopic fossils found in western Australia. It is from these clumps of cells, over the course of billions of years, that the diversity of life we see today evolved.

Most of this life, the part that we're familiar with, exists on land. But the oceans represent fully 95% of the area of our planet that is hospitable to life. And although man, animals, insects, and many plants live on land, our cellular structure is still water-based and we need water to survive.

LAKES

Lakes are among the most beautiful and most temporary of nature's creations. Most disappear in just a few thousand years, a fleeting period of time in the long history of the earth. But while they're around, lakes are important to us in many ways: primarily as sources of fresh water; as highways for transport of people, goods, and materials; and as an added bonus, great recreational resources for swimming, boating and fishing.

What is a lake?

A lake is a body of water completely surrounded by land. A lake can have salt water or fresh water, and it can be shallow or very deep. The largest lake in the world in surface area is the Caspian Sea (143,000 square miles), which is bordered by Russia, Kazakhstan, Turkmenistan, and Iran. The Romans called this body of water a "sea" because it had salt water. The deepest lake in the world is Lake Baikal in Siberia, the depths of which reach more than a mile (5,318 feet).

How are lakes formed?

Lakes are formed in a number of ways:

- Depressions scooped out by moving glaciers are the primary reason Canada contains more than half the world's lakes. The province of Quebec alone has more than 2 million lakes. Minnesota's more than 10,000 lakes, almost all produced by the glaciers, cover 4,854 square miles. Together, the northern U.S. states covered by glaciers in the last ice age have many more lakes than do the southern states that were unaffected by the ice.
- Movements of the earth's crust can also create lakes. The Caspian Sea was cut off from the Black Sea by an uplift of land. The Dead Sea in the Middle East formed in a rift valley caused by a crack in the earth's crust. Crustal warping was partially responsible for creating the Great Lakes.
- Extinct volcanic craters can fill with water to form lakes. Oregon's beautiful Crater Lake fills an ancient volcanic shell more than 1,000 feet deep. Another unusual feature of this lake is that no streams or springs feed it—its waters are replenished only by rain and melting snow.
- Depositing of rocks or sediment can dam rivers, creating lakes. Montana's Earthquake Lake was formed when a rock slide dammed a river. Sediments deposited by a river can form lakes in the delta, while sand bars and dunes can create shallow coastal lakes.
- Erosion by ice or wind can carve out hollows later filled to form lakes.

How do lakes disappear?

Most lakes gradually shrink over time as they are filled by mud and silt washed in by rivers. These disappearing lakes create much of our wetlands. Some become swamps, areas of land permanently wet, waterlogged, or flooded. Lakes that are covered by moss become bogs, which are soft, wet, and spongy. Marshes can be created when an area formerly covered by a lake is periodically flooded by a nearby river.

Other lakes are doomed when the climate becomes drier and they lose more water to evaporation than they receive from run-off, springs, and rivers. The Caspian Sea is shrinking rapidly for this reason. The Great Basin Lakes, which once stretched more than 400 miles across the Nevada, Idaho, and Oregon, disappeared completely when the amount of rainfall was dramatically lowered.

Finally, lakes can disappear as a consequence of other changes in the landscape. The bursting of an ice dam emptied Montana's huge prehistoric Lake Missoula in a flood ten times greater than the flow of all the rivers on earth combined. A river eroding a canyon wall caused ancient Lake Bonneville, the precursor of the Great Salt Lake, to lose one-third of its water in six weeks. Earthquakes and volcanoes can also empty lakes.

How are man-made lakes created?

The largest man-made lakes are created by building dams to restrict the flows of rivers. Dams are built for one or more of three reasons:

- To control flooding. Without a series of dams to hold back the water, Los Angeles would be swept by devastating floods each spring as the mountain snows melt.
- To collect water for consumption. Our towns and cities need a steady daily supply of water, but nature doesn't cooperate. So dams were built to store water from storms and melt-off so that it could be doled out to meet daily consumption.
- To generate electricity. All of the world's largest dams, all but a handful of which have been built in the last thirty years, harness the power of swift-flowing rivers to turn giant turbines that generate hydroelectric power. Although hydroelectric power only accounts for about 4% of the electricity consumed in the U.S., Washington's Grand Coulee Dam (completed in 1942) has the second largest capacity of any hydroelectric plant in the world.

Many smaller lakes have been constructed by diverting part or all of the flow of rivers into a lowland area. These lakes are created primarily for recreational purposes.

Where are the largest dams?

There are two ways to measure the largest dams—height and overall size. In terms of total size, Arizona's New Cornelia Tailings Dam, completed in 1972 with 274,015,000 yards of building materials, is the largest in the world in overall size. Also in the top ten are Montana's Fort Peck Dam (125,628,000 cubic yards) and South Dakota's Oahe Dam (92,000,000 cubic yards).

WHAT ARE THE GREAT LAKES?

The five Great Lakes rank among the fifteen largest lakes in the world. They are:

Superior	31,821 square miles
Huron	23,010 square miles
Michigan	22,400 square miles
Erie	9,940 square miles
Ontario	7,540 square miles

Superior is the second largest lake in the world, with Huron ranking fifth and Michigan ranking sixth. Michigan is the only one of the five complete surrounded by U.S. territory.

The Great Lakes are themselves the remnants of the giant prehistoric Lake Algonquin that covered most of the area before the last ice age. Much of the waters of Algonquin, one of several huge lakes that used to cover vast areas of North America, were drained away when the glaciers retreated.

HOW WERE THE FINGER LAKES FORMED?

Western New York State's Finger Lakes are long, narrow lakes formed when rocks deposited by the melting glacier dammed valleys. Their name results from an Indian belief that the lakes were formed when a mighty god put his handprint on the earth.

WHY IS THE GREAT SALT LAKE SALTY?

All "fresh" water contains salts—on the average, about one-third the level found in ocean water. The Great Salt Lake is a remnant of Lake Bonneville, 10 to 20 times its present size, that lost much of its water to evaporation. Evaporation concentrated the salts in the remaining water, turning it increasingly salty.

Part of the evaporated lake bed of the former lake became Utah's Bonneville Salt Flats, a vast flat expanse upon which so many land speed records have been set.

WHAT ARE THE OTHER LARGE U.S. LAKES?

Besides the Great Lakes and the Great Salt Lake, the U.S. contains 29 other lakes that cover 100 square miles or more. The largest are Florida's Lake Okeechobee (700 square miles), Louisiana's Lake Pontchartrain (625 square miles), Lake Champlain (435 square miles) on the New York-Vermont border, and Lake Tahoe (193 square miles) on the California-Nevada border.

WHAT'S SO UNUSUAL ABOUT LAKE OKEECHOBEE?

Lake Okeechobee is not drained by a river, but by the Florida Everglades, a permanent swamp that is really a vast slowly moving sheet of water. South Florida cities such as Palm Beach are built on a limestone ridge that surrounds this sheet of water.

The highest dam in the world is the Rogun Damin Vakhsh, Russia, which soars 1066 feet above the river bed. The tallest American dam is California's Oroville Dam (770 feet). Just behind is the famous Hoover Dam (726 feet), which was considered an engineering wonder when completed in 1934. Having visited Hoover Dam, I can attest not only to its practical value, but to its awesome appeal as a tourist attraction.

Where are the largest man-made lakes?

The size of the man-made lake created when a dam is completed depends not so much upon the size of the dam, but upon the volume of water carried by the blocked river and by the size of the terrain filled by the new lake. The largest man-made lake is Uganda's Owens Falls, which contains more than 204 million cubic meters of water. The most famous is the lake created by the damming of the Nile River by Egypt's Aswan Dam. This lake flooded a valley rich with archaeological treasures, many of which were moved in a massive effort by the world's scientific community.

The creation of artificial lakes is somewhat controversial in the scientific community. For example, the Aswan Dam generates enormous electric power and allows the Egyptian government to control the annual flooding. At the same time, the soil of the farms down river from the dam is no longer enriched by silt deposited by the flood waters. The loss of silt carried by the river may also effect the future course of the river and its mouth in unpredictable ways.

RIVERS

Oceans, seas, and lakes are basically bodies of water that stay in place; rivers, on the other hand, are water on the move. According to the dictionary, a river is a body of water that flows by gravity from higher ground to an outlet in a sea or lake. To me, however, that seems a much too simple a definition of natural phenomena that have had such a profound influence on the history of mankind. I prefer to think of rivers as the arteries of the massive bodies of land on which we live. Rivers provide us with water and food (fish), and by overflowing their banks they have created fertile soil on which we grow crops and graze animals. Equally important, they are highways for inexpensive movement of people and goods from one area to another.

Civilization began on the banks of rivers such as the Tigris, the Euphrates, and the Nile. And rivers were absolutely crucial to the exploration and settlement of America—for example, as you'll see later in this book, the Mississippi River figures prominently in the history of more than half of our states. Although we have developed lots of other means of transportation in the last century or two, we still depend on rivers in many ways. Let's learn a little more about them.

The anatomy of a river

What's the difference between a river, a stream, and a creek? Like beauty, it's in the eyes of the beholder, because streams and creeks are really small rivers. If people call it a river, it's a river.

The course of a river is the path it takes from its source to its outlet or mouth, which is usually another river, a lake, or the sea. A river basin or drainage basin is the area in which surface water or groundwater from springs flows into the river. A river system includes all the smaller rivers or streams that flow into a larger river.

The course of a river has three basic parts:

- The headwaters of the river are the area near its source, which is usually in mountains or other areas of higher ground. Rivers normally flow faster near their headwaters, so they tend to cut more narrow and steeper courses, and flooding is less common.
- In the middle of their courses, rivers tend to flow through broad, relatively flat valleys that have been cut over tens of thousands of years. In such valleys, rivers can take a serpentine course that can change over time. For example, a river often cuts a new straight path at a place where it has been U-shaped, making what is called an ox bow lake out of the isolated body of water left behind. Because the banks are less steep, rivers sometimes flood in the middle of their course. The soil and nutrients left behind when the river returns to its normal course is called an alluvial plain, and tends to be rich farmland. To prevent such flooding, many communities build artificially high banks called levees, or they build dams to hold back floodwaters in times of heavy rain or spring melting.
- Rivers flow most slowly near their mouths. They frequently overflow their low banks, creating an area called a floodplain. When they discharge into a sea, they deposit silt that forms new land called a delta.

Why rivers are important?

There are times when rivers can be nuisances — when they flood, or when the nearest bridge is fifteen miles away. But their usefulness far outweighs the trouble or inconvenience they cause. We need rivers for:

- Water. Rivers were the earliest source of drinking water for many of our major cities, and they still serve that function. But many people don't realize how valuable they are as a source of water for irrigation that has opened vast areas for agriculture. The water from just one dam on the Columbia River, the Grand Coulee Dam, irrigates 1.2 million acres.
- Transportation. Of the more than 150 American cities we profile in this book, more than 90% were founded on the banks of rivers . The reason—the river allowed them to receive and ship goods. Even today, long after the nation's railroad and interstate highway system has been completed, rivers (and artificial rivers called canals) are important to transportation. For example, most of vast quantities of grain grown on farms in the Midwest is transported on barges down the Mississippi, to the Port of Louisiana, which is our nation's busiest.
- Hydroelectric power. Dams take the energy of flowing water and convert it to electricity for industrial and home use. Energy from rivers has been especially crucial to the development of the western U.S., which has faster flowing rivers. Hydroelectric plants, which are non-polluting and don't generate hazardous waste contribute about 10% of our nation's electricity.
- Recreation. We boat and swim and fish in America's rivers, all of which are important leisure time activities. I don't know about you, but I also find it relaxing sometimes to just sit on the banks of a river and watch the waters flow by. There's something almost mystical about rivers—when I watch the Mississippi, I swear I can almost see Huck Finn and Jim float by on their raft. River watching is not only relaxing, its cheaper than therapy.

America's most important rivers

North America generally has lots of precipitation (thank goodness—if there weren't rain and snow, there wouldn't be much need for weathermen), so we have lots and lots of rivers. Some are among the greatest and most important in the world. Let's take a look at a few.

- The Mississippi River rises in Lake Itasca in northwestern Minnesota as a stream just 12 feet wide and 1.5 feet deep; by the time it reaches the Gulf of Mexico, this river has traveled 2,330 miles, received the waters of more than 250 tributaries that drain most of the area between the Rocky and Allegheny Mountains, 1,257,000 square miles. Since Robert Cavalier, sieur de La Salle, traveled the length of the river in 1682 and claimed the entire basin for France, the Mississippi has been our nation's most important waterway. The Mississippi is navigable from the Falls of St. Anthony to the Gulf of Mexico. A series of reservoirs near the head of navigation store release water when levels drop. An extensive series of levees has been built on nearly the entire length of the river to control flooding, but in 1993, the upper Mississippi overflowed anyway, causing $15 billion in damage.
- The Missouri River, a tributary of the Mississippi, stretches 2,466 miles to become America's longest river. The Missouri is formed by the junction of the Jefferson, Gallatin, and Madison Rivers at Three Forks in southwestern Montana, then flows north to skirt the Rocky Mountains before turning southeast. It is navigable from Fort Benton, Montana to its junction with the Mississippi 17 miles north of St. Louis. The Missouri drains an area of 529,000 square miles.
- The Columbia River begins in Columbia lake in southern British Columbia and flows straight north before doing an about face and heading south to Washington. This mighty river, which drains 260,000 miles during its 1245 course, is one of the fastest-flowing major rivers in the U.S. Because if its speed, it has cut many spectacular canyons and is ideal for damming. Just one dam along the Columbia, the Grand Coulee Dam, provides hydroelectric power for a wide area and irrigates 1.2 million acres of farmland.
- The Rio Grande begins in the San Juan Mountains of southern Colorado and flows 1885 miles to the Gulf of Mexico. For 1300 miles, it is the border between Texas and Mexico. The Rio Grande is too shallow for navigation, but it provides water for farming for millions of acres. In Mexico, the Rio Grande is called the Rio Bravo.
- The Ohio River is formed when the Allegheny and Monongahela Rivers meet near Pittsburgh's aptly named Three River Stadium. The Ohio forms the borders between Ohio and West Virginia, between Ohio and Kentucky, between Indiana and Kentucky, and between Illinois and Kentucky. Because it is navigable for its entire length, it has been vital to the development of areas around its 981 mile length. The Ohio joins the Mississippi at Cairo, Illinois.
- The Arkansas River begins more than 14,000 feet above see level in the Sawatch Mountains of the Colorado Rockies and plunges down rapidly until it reaches the plains of Kansas. As it flows southwestward through Kansas and into Oklahoma, its water level varies widely as it flows 1,450 miles to its junction with the Mississippi in southeastern Arkansas. Many dams have been built to regulate the flow. The completion of the Arkansas River Navigation System in the 1970s made the river navigable to Tulsa, Oklahoma.

NORTH AMERICA'S LONGEST RIVERS

River	Outlet	Length
Mackenzie	Arctic Ocean	2,635
Missouri	Mississippi	2,440
Mississippi	Gulf of Mexico	2,340
Yukon	Bering Sea	1,979
Rio Grande	Gulf of Mexico	1,885
Arkansas	Mississippi	1,459
Colorado	Mississippi	1,400
Red	Mississippi	1,290
Columbia	Pacific Ocean	1,245
Peace	Slave River	1,210
Snake	Columbia River	1,038
Churchill	Hudson Bay	1,000
Ohio	Mississippi	981
Pecos	Rio Grande River	926
Brazos	Gulf of Mexico	923
Canadian	Arkansas River	906

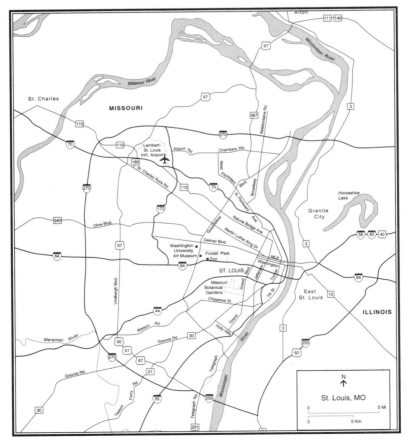

The Missouri and Mississippi River's meet at St. Louis.

A Cornucopia of Climates:

The Rich Diversity of U.S. Weather

Of all the climates in the world, the one we're all most concerned about is the climate where we live. If you look at weather as a series of theatrical events, the U.S. is the world's "Broadway", a stage for the world's most interesting weather. Mark Twain once described New England as a place with 134 different kinds of weather—all of which took place on the same day. Although the great humorist was exaggerating a little bit (as we great humorists tend to do), his comment does emphasize that our country is a jigsaw of climate types that display the richest possible diversity of weather events.

As I said before, the key elements of climate are temperature and precipitation. And the patterns in which these elements occur depend in turn on three basic factors:
- The prevailing winds and ocean currents
- The location of semi-permanent air masses and high and low pressure systems.
- Geography: the location of our continent, mountain ranges, oceans, lakes, and plains.

Previously, I've explained the prevailing winds, ocean currents, and the meandering of the jet stream. I've also described the semi-permanent air masses and highs and lows that generate our weather. But before we talk about climate, I have to spend a little time talking about the geography of our continent.

MEET NORTH AMERICA

Any of you who have endured wind chills of 40 degrees below zero when an Arctic high dips down from Canada into the States, or have experienced the terror of hurricanes storming in from the Caribbean have no doubt uttered a profound wish that weather stopped at our borders. But weather doesn't pay the slightest bit of attention to lines drawn on a map. To understand the geography of U.S. weather, we have to take a look at the entire North American continent and its surroundings.

The North American continent today is entirely in the Northern Hemisphere, stretching from just north of the equator at the tip of Central America to the Arctic Ocean. Because it stretches almost the entire distance from the equator to the North Pole, you might say that nature gave North America a lot of latitude (no doubt a lot more than you're likely to give me). You'll understand how far south our continent stretches when you visualize a parallel line running through the southernmost point of North America would run through Nigeria, pass south of the entire Arabian peninsula, touch the southern tip of India, and run right through Ho Chi Minh City (the former Saigon), Viet Nam.

Geography and our weather—a broad overview

I hope you remember the sequence of the wanderings and collisions that produced North America, because they have had a profound effect upon our weather. If the collision that formed the mountains across Canada had occurred 70 million years ago rather than 1.9 billion years ago, the tall peaks of this chain would have served as a barrier against the cold Arctic air masses. If the Appalachian Chain had formed most recently, the Northeastern and Mid-Atlantic states might have a Mediterranean climate with palm trees instead of pines.

But, of course, this didn't happen. So unless we figure out a way to attach a giant motor to North America and drive it someplace else, we're stuck with this basic configuration: a land mass sandwiched between two oceans with a large body of warm, moist tropical air to the southeast.

Pacific air moving on shore makes much of the West Coast the cloudiest and dampest part of the U.S. But all this moisture falls from the clouds when the air slams into the Rocky Mountains. The western half of the interior of the U.S. lies in the "rain shadow" of the mountains, making it the driest part of the country

The vast central plain of North America is a battleground for the year around clashes of warm moist air moving northward and cool dry air moving southward. In the Winter, cold air from northern Canada and the Arctic dominates. The north central part of the U.S. has very cold and relatively dry Winters; the southern states have less severe Winters and more precipitation. In the Summer, the warm air pushes northward, bringing thunderstorms and other violent weather with it. If the edge of the warm air pushes unusually far northward into Canada, record heat and drought can punish the central plains.

The eastern states from Florida to New England have substantial moisture year-around, thanks to storm systems moving northward up the coast from the Gulf of Mexico and the Caribbean and "northeasters" moving southwest from the Icelandic low. The Atlantic Ocean moderates the climate of New England and the Mid-Atlantic states, giving them less severe Winters than the northern plains.

Our seasons

Astronomically, our seasons in the Northern Hemisphere are defined by the orbit of the earth around the sun: Winter begins on December 21, when the sun is at its southernmost point; Summer on June 21, when the sun reaches its northernmost point; Spring and Fall begin on the two equinoxes, March 21 and September 21. Meteorologically, though, the seasons are defined by average temperatures. Winter is the coldest season of the year and by December 21, much of the northern U.S. is two or three weeks into the season. On the other hand, by the same definition, Winter is over by March 1, three weeks before astronomical Spring.

To understand the climate of the U.S., it's useful to take a look at the meteorological season across the country.

- Winter. Another common definition of the beginning of Winter is the date on which the average daily temperature drops below freezing. This can be as early as the first ten days of November for North Dakota and northwest Minnesota, the last ten days of November for New England, upstate New York and the upper midwest, and December 15 for the mid-Atlantic states and much of the midwest. The average daily temperature never drops below 32 degrees Fahrenheit in much of the south and the Pacific coast, so those regions don't have Winter (by this definition) at all.

  A third marker for the coming of Winter is the first snowfall. Flakes can be spotted in the Rocky Mountains in late September, late October or early November in the northern plains and mountains of New England, and by late November or early December in other areas where the average daily temperature falls below 32 degrees.

- Spring. This season marks the end of Winter. Average daily temperatures climb above 32 degrees. For most of the country that has Winter, this occurs between late February and the second week of March.

  Those people with a botanical bent define Spring as the time when the average high temperature reaches 50 degrees, the point at which plant life wakes from its Winter sleep. By this definition, Spring arrives about the first of February in the south and as late as May 1 in the northern U.S.

- Summer. If you define Summer as the hottest season of the year, it begins somewhere between late May and June 15. The peak Summer normally begins about a month after the first day of the season.

  Summer can also be defined as the season during which average daily temperatures top 68 degrees. This plateau is reached by May 1 in most of the south, by June 1 across the central U.S., and by June 15 in the northern U.S. However, some parts of the Pacific coast and some mountainous areas don't have any Summer by this definition.

- Autumn. If you consider autumn the three months in between the hottest and coldest quarters, then it begins in most of the U.S. between the last week in August and the first week in September. The exception is much of the California coast, where warm temperatures persist as late as September 30.

  If, on the other hand, you start Fall when the average daily temperatures drop below 68 degrees, Summer ends in late August in the northern states, in September in the central U.S., and well into October in the southern states.

The weather from sea to sea: regional climates of the continental U.S.

Geography, wind patterns, the location of semi-permanent air masses, highs, and lows, and the movement of the sun all combine to determine the climate where you live. Since most U.S. weather moves from west to east, we'll go in the same direction in explaining a little bit about regional climates.

- The Pacific Northwest. Washington, Oregon, the California coast from San Francisco northward, and parts of Idaho share a climate that features the smallest annual range of temperatures in the U.S. Because this area is dominated by Pacific Maritime air, it has mild Winters and only moderately warm temperatures. A mixture of clouds and fog make it the least sunny region of the U.S. Its climate is very similar to that of Great Britain and northwest Europe.

  The Pacific Northwest has precipitation all year, but Winter is the wettest season. In the Summer, the Pacific high expands, directing major storm systems northward; the U.S. coast is often foggy but seldom rainy. Seattle, Washington averages just 2.7 inches of precipitation in June, July, and August, less than half the average 5.6 inches that falls in December.

  The lack of temperature contrast means that the Pacific Northwest experiences very little violent weather—thunderstorms are infrequent and tornadoes very rare. A combination of prevailing westerly winds and the cool waters of the Pacific prevent tropical depressions from reaching this region.

- California. Except for the slice of northern coast, California has a Mediterranean climate featuring very dry Summers and wetter Winters. In the Summer, California bakes as the strong Pacific high keeps clouds away—Los Angeles averages only .3 inch of precipitation from May through September. The cool Pacific current that flows southward keeps Summer temperatures moderate along the coast, but the average daily high temperatures soar to 116 degrees inland in Death Valley.

  In the Winter, when the Pacific high shrinks, low pressure systems move in from the Pacific northwest, bringing heavy snowfalls to the mountains and rain to the coast and inland valleys. Winter temperatures remain mild.

  California experiences relatively few thunderstorms and very infrequent tornadoes. Tropical cyclones seldom reach California at hurricane strength. However, because the soil of this semi-arid area is so hard-packed, tropical storms and other systems that bring heavy rains can cause extensive flooding and damaging mud slides.

- The Southwest. Arizona, Nevada, and New Mexico are part of the great Western Cordillera, the mountain ranges and plateaus that stretch from the Mexican border into Canada. Because these states are dominated by persistent high pressure and because they are in the rain shadow of the California mountains to the west, annual precipitation is very low. The Southwestern states are the sunniest region of the country, and the low humidity makes the climate a healthy one for people suffering from respiratory distress.

  The yearly temperature range depends on altitude. Summer high temperatures average in the hundreds in Phoenix at an altitude of 1100 feet, but reach a maximum of 80 degrees at Santa Fe, which has an altitude of 7,000 feet.

  Lack of precipitation means little violent weather. The most dangerous weather conditions are flooding caused by heavy rainfall from the occasional tropical storm moving northward from Mexico.

- The Northern Rocky Mountains. The climate of Colorado, Utah, Montana, Wyoming, and most of Idaho is colder and wetter than the mountain states to the south. But because most of the moisture from Pacific air falls on the Cascades, annual precipitation averages less than 20 inches per year over most of the area, and some plateaus are actually deserts or semi-arid.

  The region is relatively sunny. Great temperature swings from day to day are relatively common—October temperatures in Cheyenne, Wyoming range from a record high of 85 degrees to a record low of 5 degrees below zero.

  Thunderstorms caused by the lifting of moist air as it hits the mountains are relatively frequent. Tornadoes, on the other hand, are relatively rare.

- The Great Plains and Northern Midwest. This vast region of the U.S. (which includes North Dakota, South Dakota, Minnesota, Wisconsin, Michigan, Kansas, Nebraska, Missouri, Illinois, Indiana, Ohio, Iowa, and Kentucky) has a true continental climate, featuring severe Winters and warm, humid Summers. In the Winter, when the weather is dominated by bitterly cold, dry air moving south from northern Canada and the Arctic. These cold fronts, often called "Alberta clippers," drop little precipitation—except when they pick up moisture crossing the Great Lakes and drop it as heavy "lake effect" snows on the opposite shores. (That's why areas like Buffalo, New York record such high annual snowfall totals). The occasional widespread snowstorm generally begins as a strong low pressure system carried southeastward from the Pacific Northwest by a trough in the jet stream. These storms cross the Rocky Mountains, pick up moisture from the Gulf of Mexico, then are carried back north by the jet stream.

  In the Spring and early Summer, warm, moist Gulf air moves northward, spawning violent thunderstorms and tornadoes as it clashes with the retreating cold Arctic air. Most of the precipitation in this region is produced by thunderstorms. The total annual precipitation is 16 to 32 inches per year in the western half (the area furthest from the Gulf and closest to the rain-blocking western mountains) and 32 to 48 inches in the eastern half.

The north central U.S. is prone to prolonged periods of drought. The severe Summer drought of 1988 resulted when a huge persistent high pressure system anchored itself over central Canada and the northern U.S., steering Pacific storms northward into Canada and preventing moist Gulf air from moving northward.

- The South Central and Gulf States. Warm, moist Gulf air dominates this lower half of the central U.S., which includes the states of Oklahoma, Arkansas, Tennessee, Texas, Louisiana, Mississippi, and Alabama. Although Arctic air occasionally forces its way down into the region, Winter temperatures in the region rarely average below freezing for any month of the year. Summers are slightly warmer than those in the north central U.S., but they last much longer and are sandwiched by much warmer Springs and Falls.

  Precipitation over the area generally averages 40 to 50 inches per year (the exception is western Texas and Oklahoma). Summer is the wettest season, but precipitation falls year around. The regional climate is generally sunny, but the average relative humidity is significantly higher than it is in the southwestern or Rocky Mountain states.

  This region is home to more violent weather—thunderstorms, tornadoes and hurricanes—than any other region of the country.

- The Southern Atlantic States. The climate of this region, which includes Florida, Georgia, South Carolina, North Carolina, Virginia, and West Virginia, features hot, damp Summers and mild Winters. Arctic air can bring below freezing temperatures as far south as northern Florida, but such incursions are rarer than they are in the south central U.S..

  No part of this region averages less than 32 inches of precipitation per year, with several areas averaging over 60 inches. In Summer, much of the precipitation falls from frequent thunderstorms that form in late afternoon. The southern half of the region averages less than 1 inch of snow per year, while the northern half averages under 10 inches.

  The thunderstorms in this area spawn tornadoes, but they tend to be weak. Far more dangerous and destructive are hurricanes, such as Andrew (which devastated South Florida) and Hugo (which slammed into the Charleston, South Carolina area).

- The Middle Atlantic and New England States. The climate in this region, which includes New York, Pennsylvania, Delaware, Maryland, Connecticut, Massachusetts, Rhode Island, Vermont, New Hampshire, and Maine, is wetter and more moderate than that of the north central U.S.. The Summers are hot but less humid than other regions east of the Rockies—thunderstorms are significantly less frequent than they are in the north central, south central, and south Atlantic states. Winters are quite long in the northern half of this region, but the air temperatures and wind chill factors seldom plummet to the bitter depths of those reached in the upper midwest.

  Annual precipitation averages 40 to 60 inches, and is evenly distributed throughout the year. Upstate New York, western Massachusetts, Vermont, New Hampshire and Maine average more than 60 inches of snow per year—more than any other areas of the country except northern Michigan and the highest elevations of the western mountains. Unpleasant mixes of snow, sleet, freezing rain and rain often plague the southern half of this region.

  Devastating thunderstorms and tornadoes are relatively infrequent in this region. But hurricanes moving up the coast have caused significant destruction.

ALASKA—A MINI-CORNUCOPIA OF CLIMATES

Alaska is the largest state, covering an area more than twice the size of Texas. It stretches east-west across three time zones and north-south across an expanse as long as the distance from El Paso, Texas to the Canadian border. Much of this state is covered by mountains that include some of the highest peaks in North America. Large glaciers descend from these mountains almost to sea level.

The interior and northern coast of Alaska have an Arctic or sub-Arctic climate. Barrow, on the shores of the Arctic Ocean, is ice-bound for most of the year. The average daily high in Barrow is below zero four months of the year and is above freezing only four months of the year. Fairbanks, which is representative of inland Alaska, has a long, bitterly cold Winter. However, because of the long hours of daylight, the Fairbanks Summer is surprisingly warm, with the average high temperature reaching 72 degrees in July. Because the air is too cold to hold significant moisture, northern and interior Alaska receives very little precipitation.

Anchorage, which is located in a sheltered bay on the southern coast, has much warmer Winters than inland regions. Average Winter temperatures approximate those of Minneapolis. However, ocean waters keep Summer temperatures quite cool—the average daily high in July is just 65 degrees. Precipitation in Alaska is frequent but light.

The Aleutian Islands experience a very narrow range of temperatures year around. On the average December day on the island of Atka, the high temperature is 37 and the low is 30; on the average July day, the high is 57 and the low is 46. Rainfall is frequent and plentiful year around.

HAWAII—THE WEATHERMAN'S PARADISE

The best way to sum up the weather forecast for a tropical area is: "The weather tomorrow will be the same as the weather today." The television stations in Hawaii, which has a tropical climate, don't have regular weathermen on their news broadcasts. "Sunny, warm, with a chance of an afternoon shower" will do it for almost every day of the year. Hawaii has never had a heating degree day (which means the average temperature for a day has never fallen below 65 degrees) and has never seen a flake of snow.

Precipitation does vary from location to location. Moist air blowing in off the Pacific hits some volcanic mountains, dropping large volumes of precipitation that have reached over 700 inches per year. Other areas, in the rain shadows of these mountains, get as little as 12 to 20 inches per year.

ALL-TIME HIGHEST AND LOWEST TEMPERATURES IN EACH STATE

The all-time high and low temperatures for each state are:

	High	Low		High	Low
Alabama	112°	-24°	Montana	117°	-70°
Alaska	100°	-80°	Nebraska	118°	-47°
Arizona	127°	-40°	Nevada	122°	-50°
Arkansas	XX°	-tk°	New Hampshire	106°	-46°
California	134°	-45°	New Jersey	110°	-34°
Colorado	118°	-60°	New Mexico	116°	-50°
Connecticut	105°	-32°	New York	108°	-52°
Delaware	110°	-17°	North Carolina	109°	-29°
Florida	109°	- 2°	North Dakota	121°	-60°
Georgia	112°	-17°	Ohio	113°	-39°
Hawaii	100°	14°	Oklahoma	120°	-27°
Idaho	118°	-60°	Oregon	119°	-54°
Illinois	117°	-35°	Pennsylvania	111°	-42°
Indiana	116°	-35°	Rhode Island	102°	-23°
Iowa	118°	-47°	South Carolina	111°	-13°
Kansas	121°	-40°	South Dakota	120°	-58°
Kentucky	114°	-34°	Tennessee	113°	-32°
Louisiana	114°	-16°	Texas	120°	-23°
Maine	105°	-48°	Utah	116°	-50°
Maryland	109°	-40°	Vermont	105°	-50°
Massachusetts	106°	-34°	Virginia	110°	-29°
Michigan	112°	-51°	Washington	118°	-48°
Minnesota	114°	-59	West Virginia	112°	-37°
Mississippi	115°	-19°	Wisconsin	113°	-54°
Missouri	118°	-40°	Wyoming	114°	-63°

U.S. WEATHER RECORDS

HIGH TEMPERATURE

U.S. record: 134 degrees Fahrenheit, Death Valley, California, July 10, 1913.

If hanging out by a foundry furnace is your idea of a blast, then plan your next vacation for Death Valley. This 140-mile-long, 6-mile-wide stretch of paradise has the highest summer temperatures in the world—daily highs average 116 in July, and nightime lows are sometimes over 100. However, when the sun heads south in the fall, the temperatures in Death Valley plummet. The highest year-round average temperature is the 78 degrees recorded in Key West, Florida, the southernmost point in the contiguous 48 states.

LOW TEMPERATURES

U.S. record :-80 degrees, Prospect Creek, Alaska, January 23, 1971

U.S. record(Ex. Alaska): -70 degrees, Rogers Pass, Montana,

January 20, 1954

Barrow, Alaska, which will never be confused with Key West, has an average annual mean temperature of just 9 degrees (You bring the suntan oil, I'll bring the volleyball.) In the lower 48 states, Butte, Montana, may be the coldest city—it averages 223 days per year with below-freezing temperatures. Another Montana city, Havre, may have experienced the nation's longest sustained cold snap when it recorded below-zero temperatures for 400 consecutive hours. The most severe cold wave may have occurred in February 1899, when the Mississippi River froze all the way from its source to its mouth, with 2 inch-thick ice clogging the harbor at New Orleans.

If you're looking for a question that is guaranteed to stump any non meteorologist (and maybe even a few of them), try this one: What two states share the distinction of having the lowest all-time high temperature. I bet you'll get one—Alaska—right off the bat. The second is the zinger—it's Hawaii. That's right, the temperature has never climbed above 100 degrees in the Aloha state, while the thermometer has reached as high as 114 degrees in Minnesota and 121 degrees in North Dakota. If you hate hot weather and don't want to leave the continental United States, consider a move to Eureka, California—it's never been warmer than 85 degrees.

Anyone could guess that Hawaii is the one state in which the temperature has never gone below zero—the state record low is 14 degrees, recorded at the top of a 10,000-foot volcanic peak.

RECORD TEMPERATURE SWINGS

In the course of the average year, the range of temperatures recorded in the United States is often as great as 180 degrees—from temperatures as low as -60 degrees to as high as 120 degrees. What's really unusual is when the temperature spans a large chunk of that range in a matter of minutes.

If you think you have trouble deciding how to dress when the weather is changing, imagine yourself in front of the closet in Spearfish, South Dakota, on January 22, 1943. At 7:30 in the morning, the temperature climbed 49 degrees in just two minutes—from -4 degrees to 45 degrees. The temperature eventually reached 54 degrees by 9 a.m.—then a cold front swept back in and the temperature plummeted 58 degrees (back down to -4 degrees) in just 27 minutes.

The cause of this abrupt change was a warm Chinook wind roaring down the eastern slopes of the Rocky Mountains. Another Chinook raised the temperature 83 degrees, from -33 degrees to 50 degrees, over the course of 12 hours on February 21, 1918, in Granville, North Dakota. Yet a third warm wind melted 30 inches of snow as the temperature rose 80 degrees on December 1, 1896, in Kipp, Montana.

Rapid swings from warm to cold temperatures also occur in the winter months, when an Arctic cold front sweeps in to replace unusually warm air. The U.S. records are:

24 hours—100 degrees, from 44 degrees to -56 degrees,
in Browning, Montana, on January 23, 1916

12 hours—84 degrees, from 63 degrees to -21 degrees,
in Fairfield, Montana, on December 24, 1924

2 hours—62 degrees, from 49 degrees to -13 degrees,
in Rapid City, South Dakota, on January 12, 1911

RECORD PRECIPITATION

Mt. Waialeale, on the Hawaiian Island of Kauai is the average annual precipitation champ, at 460 inches per year). The U.S. all-time yearly precipitation record, 739 inches, is held by Kukui, on the Hawaiian island of Maui.

Thunderstorms and tropical storms normally account for the heaviest short-term rainfall totals. Following are some U.S. records:

1minute—1.23 inches Unionville, Maryland.

2 hours, 45 minutes—22.00 inches D'Harris, Texas

12 hours—42.00 inches Alvin, Texas

The longest recorded period without rainfall in the United States was 767 days, from October 3, 1912, to November 8, 1914, in Bagdad, California. Death Valley, with a yearly average 1.63 inches of precipitation, is the driest location in the United States.

High mountain areas of the American West hold all the U.S. snowfall records. They are:

24 hours—76 inches Silver Lake, Colorado

1 storm—189 inches Mt. Shasta Ski Bowl, California

1 season—1,122 inches Paradise Ranger Station, Washington

Outside of the western mountains, the highest average annual snowfalls are recorded in northern Michigan, Upstate New York, and northern New England. Bennetts Bridge, New York, about 30 miles east of Oswego, may have experienced the greatest non-mountain snowfall on January 17, 1959, when 51 inches fell in 16 hours. Among the leaders in annual average are:

Marquette, Michigan 126.0 inches

Sault Ste. Marie, Michigan 116.4 inches

Caribou, Maine 113.3 inches

Syracuse, New York 110.5 inches

The line between the part of the country in which snow is a regular part of winter weather (an average of 12 inches per year or more) and the part of the country in which snow is infrequent runs roughly from Washington, D.C., across the northern borders of North Carolina, Tennessee, Arkansas, Oklahoma, and Texas to the southern border of the Rocky Mountains.

AVERAGE ANNUAL PRECIPITATION BY STATE

The driest states are all in the Southwest, the wettest in the Southeast. Except for the states of Washington and Oregon, precipitation averages rise as you move from west to east. Following is a list of states by average annual rainfall.

Under 10 inches

Nevada

10 to 19 inches

Alaska	Arizona
Colorado	Idaho
Montana	New Mexico
North Dakota	South Dakota
Utah	Wyoming

20 to 29 inches

California	Kansas
Minnesota	Nebraska
Oregon	Texas

30 to 39 inches

Hawaii	Illinois
Indiana	Iowa
Missouri	New York
Ohio	Oklahoma
Vermont	Washington
	Wisconsin

40 to 49 inches

Arkansas	Connecticut
Delaware	Kentucky
Maine	Maryland
Massachusetts	New Hampshire
New Jersey	North Carolina
Pennsylvania	Rhode Island
South Carolina	Virginia
	West Virginia

50 inches or more

Alabama	Florida
Georgia	Louisiana
Mississippi	Tennessee

America's Natural Wonders

I've visited so many places in this great land that I could list pages of national wonders worth visiting. In many ways, picking the ten greatest is sort of like making a list of your ten favorite children—you're bound to offend almost everyone else. However, I also think top-ten lists are kind of fun (Letterman's done okay with his), so here's mine—you can take your potshots when I come to your town to sign my books.

Grand Canyon

The Grand Canyon is about 217 miles long, 4 to 18 miles wide, and more than 1 mile deep. The entire canyon is breathtakingly beautiful, containing towering buttes, mesas, and valleys within its main gorge. A spectacular section of the canyon, together with plateau areas on either side of it, is preserved as the Grand Canyon National Park.

The canyon cuts steeply through an arid plateau region that lies between about 5000 and 9000 ft above sea level. This region, although lacking year-round streams in recent times, is sharply eroded, showing such characteristic forms as buttes; it is interspersed with old lava flows, hills composed of volcanic debris, and intrusions of igneous rock. The plateau area has a general downward slope to the southwest and in its upper reaches is sparsely covered with growths of such evergreens as juniper and pinion. Parts of the northern rim of the canyon are forested. Vegetation in the depths of the valley consists principally of such desert plants as agave and Spanish bayonet. In general the entire canyon area has little soil. The climate of the plateau region above the canyon is severe, with extremes of both heat and cold. The canyon floor also becomes extremely hot in summer, but seldom experiences frost in the wintertime.

The sculpturing of the canyon has been acomplished the downward cutting of the Colorado River, which flows through the lowest portions of the canyon. Other factors have also played a part. The Kaibab Plateau, which forms the northern rim of the canyon, is about 365 m (about 1200 ft) higher than the Coconino Plateau, which forms the southern rim. Water from the northern side has flowed into the canyon, forming tributary valleys, while the streams of the southern plateau flow away in a southerly direction without carving valleys in the canyon walls. The underlying rock beds also have a southwestern slant, with the result that groundwater from the north finds its way into the canyon, but water from the south does not. In the entire canyon region, the rocks have been broken by jointing and faulting, and fractures in the rocks resulting from these processes have contributed to the rapid erosion of the gorge.

The Grand Canyon is of relatively recent origin; apparently the river began its work of erosion little more than a million years ago. Coupled with the downward cutting of the river has been a general rising or upwarping of the Colorado Plateau, which has added its effect to the action of the river. Although the canyon itself is of comparatively recent origin, the rocks exposed in its walls are not. Most of the strata were originally deposited as marine sediment, indicating that for long periods of time the canyon area was the floor of a shallow sea.

My two trips to the Grand Canyon have left me virtually speechless both times. My first trip was with GMA, when we did our broadcast there in November, 1992. I was so impressed that I returned a week later and took a small airplane tour of the canyon. It was beautiful beyond description.

The first Europeans to see the canyon were members of a group headed by the Spanish explorer Francisco Vasquez de Coronado, which set out from New Spain (now Mexico) in February 1540. The sighting was made later that year. Because of the inaccessibility of the canyon, it was not until more than three centuries later that it was fully explored. Beginning about 1850, a series of expeditions commanded by officers of the U.S. Army surveyed the canyon and the surrounding area. The first passage of the canyon was accomplished in 1869 by the American geologist John Wesley Powell and ten companions, who made the difficult journey through the length of the gorge in four rowboats.

Niagara Falls

One of the world's most memorable natural sights, and one of my personal favorites, it consists of two cataracts: the Canadian, or Horseshoe, Falls, 161 ft. high, on the Canadian side of the river, and the American Falls (167 ft high), on the U.S. side. The waterfalls are separated by Goat Island, New York. The crestline of the crescent-shaped Canadian

Falls, which carries about nine times more water than the U.S. cataract, is about one half mile long, and the fairly straight crest of the American Falls measures about 1,001 ft. A small section of the American Falls near Goat Island is also known as Bridal Veil Falls.

Niagara Falls was formed about 12,000 years ago, when glaciers retreated north, allowing water from Lake Erie to flow over the Niagara Escarpment, a ridge that extends from southern Ontario to Rochester, New York. Since that time, erosion has slowly pushed the waterfall about about 7 miles upstream, forming the Niagara Gorge. At present the Canadian Falls is receding at an average yearly rate of about about 5 feet, and the American Falls is being cut away at an annual pace of about about 6 inches. The Canadian Falls erodes at a faster rate mainly because it carries more water. In 1954 a considerable portion of the American Falls broke off, creating a large talus, or rock slope, at the base of the cataract. In order to study ways of preventing further rockfalls and to remove some of the talus, the American Falls was successfully shut off for several months in 1969 by a dam that was constructed between the U.S. mainland and Goat Island.

Samuel de Champlain, a French explorer, probably visited Niagara Falls in 1613. Father Louis Hennepin, a Flemish monk, is known to have visited the waterfall in 1678; he later published an eyewitness description of it. I have seen the falls three times (both sides), and each time I am more awe-stricken than before by the profound power of nature.

The Niagara's large volume of flow, averaging about about 194,940 cubic feet per second, plus its steep drop, give the river great power potential. The waterpower probably was tapped first in 1757, when Daniel Chabert Joncaire built a sawmill on the upper river. In 1853 work started on a hydraulic canal to divert the waters of the upper river to drive machinery in mills and factories situated below Niagara Falls. In 1875 the first flour mill powered by the canal water was opened, and in 1881 the first hydroelectric generator was installed along the waterway. The first large-scale hydroelectric facility, the Edward Dean Adams Power Plant, was opened on the U.S. side in 1896.

Mammoth Cave

Mammoth Cave is a vast limestone cavern that has more than 194 miles of charted passageways extending through five levels and is the longest known cave system in the world. The corridors and chambers are festooned with stalactites, stalagmites, and drapelike formations noted for their fantastic configurations. These formations contain crystals of calcite, gypsum, and other minerals that lend them color. Flowing through the cave's lowest level, about 350 feet below the surface of the earth, is a subterranean stream, the Echo River. This river, three-quarters of a mile long with a maximum width of about 200 ft contains rare forms of fauna, including blindfish and colorless and eyeless crayfish. The caverns remain at a constant temperature of 54 degrees farenheit.

Mammoth Cave was discovered by pioneers in 1798. Archaeological discoveries indicate early Indian occupancy of the cave. The mummified body of a pre-Columbian man was found in the cave in 1835. During the War of 1812, the Rotunda, the first large chamber of the main cave, was mined for saltpeter (niter) from which gunpowder was manufactured. In addition to the cave, the park contains a region of forests and hills traversed by the Green River, which receives the Echo River and other underground streams. The area is 81.9 square miles.

The Everglades

The Everglades is a vast freshwater marsh, covering much of southern Florida. Formed by centuries of constant overflow from Lake Okeechobee after heavy rains, it extends south from the lake for about 110 miles and has a total area of almost 5000 square miles. The marsh lies in a basin of limestone, which appears a few miles north of Cape Sable at the southernmost tip of Florida and extends north to Lake Okeechobee. Tidal bays, lakes, and connecting waterways, varying in depth from 1 to 7 ft, and sometimes considerably less, honeycomb the wilderness of swamp, savanna, and virgin forest that make up the Everglades. Several small streams, such as the Miami, Little, and New rivers on the east and the Shark and Harney on the southwest, drain the region, but none flows into it. Most of the marsh, however, is covered with saw grass that rises 3 to 10 ft above the surface of the water, and is so dense that passage is effectively barred except for the natural water lanes.

Cypress, mangrove thickets, palms, live oaks, bays, and lush vegetation cover the numerous islets (known as hammocks) that are found in the Everglades. The highest land of the Everglades is only about 7 ft above sea level, and in the summer rainy season large areas of land become swamp or are covered by water. The Everglades comprises one of the wildest and most inaccessible areas in the U.S. Wildlife is abundant and largely protected within the Everglades National Park. Inhabitants of the Everglades are almost entirely limited to several hundred Seminole. snake venom and frog legs from this region are marketed commercially throughout the U.S. In the late 19th and early 20th cen-

THE GEOGRAPHY BOOK

turies large drainage programs were undertaken; canals were dug to lower the level of Lake Okeechobee and keep it from overflowing into the Everglades. On drained areas south of the lake, farmers grow fruit, vegetables, and sugarcane. Two major highways cross the Everglades.

Cape Cod

Cape Cod is a peninsula in southeastern Massachusetts, crossed at its base by the Cape Cod Canal. The cape is surrounded by Cape Cod Bay to the north; the Atlantic Ocean to the east; Nantucket Sound to the south; and Buzzards Bay to the southwest. The cape is deeply indented; in shape it resembles a flexed arm. It extends eastward about 35 miles and then northward about 30 miles. The width varies from about 20 miles between the towns of Sandwich and Woods Hole at the neck to a few hundred meters at the tip near the town of Provincetown. The cape is sandy, hilly, and thickly forested in spots and contains many lakes and ponds. Cape Cod is well known as a popular summer resort and has excellent facilities for swimming, fishing, and boating. Fishing is an important industry, especially in Provincetown, which has one of the largest and safest harbors on the Atlantic seaboard. Cranberries are the chief crop. Other communities on Cape Cod are Barnstable, Yarmouth, Orleans, Falmouth, Bourne, and Hyannis. The cape has a number of lighthouses, including Chatham Light at Chatham, and Cape Cod Light at Truro. Cape Cod was found and named in 1602 by the English explorer Bartholomew Gosnold, who was impressed by the abundance of codfish in the surrounding waters. In 1620 the Pilgrims dropped anchor in Provincetown Harbor. Cape Cod National Seashore covers much of the northeastern part of the peninsula.

Death Valley

Death Valley is an arid, depressed desert region, in southeastern California. It was given its name by one of 18 survivors of a party of 30 attempting in 1849 to find a short cut to the California gold fields. Much of the valley is below sea level, and near Badwater at 282 ft below sea level, is the lowest point in the western hemisphere. Death Valley National Monument (established 1933) has an area of 3230.9 sq. miles and incorporates the valley and surrounding mountains.

The valley is from (4 to 16 miles wide and about 140 miles long and is almost entirely enclosed by mountain ranges, volcanic in origin, bare and brilliantly colored. The Panamint Range on the west, which rises to a maximum altitude of 11,045 ft in Telescope Peak, shuts out the moist Pacific winds. On the east are the peaks of the Amargosa Range.

The summer temperatures in Death Valley, one of the hottest regions known, exceed 125 degrees farenheit in the shade and rarely fall below 70. The National Weather Service recorded 134 degrees farenheit in 1913, the highest temperature ever recorded in the U.S. Average rainfall in a normal year is less than 2 inches. Sandstorms and dust whirlwinds of several hours' duration are common.

Several watercourses enter the valley, among them the Amargosa River from the south and Furnace Creek from the east, but it is only after heavy rains, which are rare, that they contain water. The lowest parts of the valley floor are salt flats, devoid of vegetation; higher portions contain a mixture of sand and salt grains, occasionally forming dunes. The western side of the valley floor is bordered by stunted mesquite, and in a marsh in the northern section a growth of tall, coarse grass is found; the east and west slopes have a sparse vegetation of cacti and desert shrubs and grasses. Animal life is confined to a few species of desert reptiles, such as horned toads and lizards, and such mammals as rabbits, rats, and the desert bighorn sheep.

Gold has been found in Death Valley, and silver, copper, and lead have been taken in paying quantities. The famous borax deposits of Death Valley were first mined in the 1880s.

Yellowstone National Park

Yellowstone National Park, which is located in northwestern Wyoming, and extends into southwestern Montana and eastern Idaho, was established in 1872 as the oldest and largest national park in the United States. Located in the central Rocky Mountains, it is known for its spectacular geysers, hot springs, waterfalls, and canyons. In the heart of the park is a broad plateau, which has an average elevation of about 8,000 feet and is surrounded by lofty and rugged mountain ranges. The Yellowstone River traverses the region from south to north, flowing into Yellowstone Lake and then through the famous Grand Canyon of the Yellowstone. The river descends into the canyon in two spectacular falls about 110 and 312 feet in height. The yellow canyon walls rise abruptly from the river to elevations as high as 1100 feet. the park covers an area of 3,500 square miles.

The best-known features of the park are its more than 3,000 geysers and hot springs, the greatest concentration of such phenomena in the world. Old Faithful is the most celebrated of the park's geysers. It erupts for about 4 minute (at intervals of between 37 and 93 minutes) in a column of steam and hot water that rises as high as 170 ft; approximately 10,000 to 12,000 gallons of water are expelled at each eruption. Other geysers include the Giant, the highest, which erupts at irregular intervals, throwing up a jet of hot water more than 200 feet in height; and the Giantess, which erupts for more than 4 hours, about twice a year.

Hot springs in the park are even more numerous than the geysers. Minerals present in the waters of some of the hot springs have been deposited on the surrounding ground, building up cones and terraces. The most striking example is Mammoth Hot Springs, which contains terraces as high as 300 feet. Algae thrive in the pools of warm water that collect on the terraces, giving them a brilliant coloration. Also present in the park are mud volcanoes. These are mounds formed by the issuing from the earth of hot water mixed with fine rock matter. Other points of interest include Tower Falls (132 ft) Golden Gate Canyon, and Obsidian Cliff, a volcanic black glass formation 5,165 feet high.

Yellowstone National Park is one of the greatest wildlife refuges in the world. Among the many animal species found here are grizzly bear, elk, antelope, moose, bison, mountain sheep, lynx, and otter. More than 200 species of birds, including eagle, pelican, and the rare trumpeter swan, have been sighted in the park. Vast forests, largely coniferous, cover most of the terrain and provide an undisturbed wilderness habitat for wildlife. Area: 3,468.5 sq. miles.

Crater Lake

Crater Lake National Park, which is located in southern Oregon, was established 1902. The park embraces an area on the crest of the Cascade Range and includes Crater Lake, the second deepest lake (1932 ft) in the western hemisphere. The lake occupies the crater of a prehistoric volcano, Mount Mazama, which geologists believe erupted violently about 6,600 years ago, blowing off much of its 12,000 ft height. Wizard Island, a small island near the lake's western edge, is the top of a volcano, now extinct, that formed after the destruction of the main volcanic cone.

The lake, which has an area of 20 sq. miles, is known for its intense blue color, a result of its great depth. It is encircled by lava cliffs that rise about 500 to 2,000 ft above the water's surface. A scenic drive has been built along the rim. Crater Lake was formed and is replenished by snow and rainwater; it is neither fed nor drained by streams. Evaporation and seepage account for the annual water loss. The lake has been stocked with trout, and fishing is permitted. Also in the park are Castle Crest Wildflower Garden and the volcanic landscape of Pumice Desert. The lake covers 286.3 sq. miles.

Glacier National Park

Glacier National Park , located in northwestern Montana, was established in 1910. Straddling the Continental Divide, the park is distinguished by its extraordinary beauty. It has nearly 50 glaciers, more than 200 glacier-fed lakes, broad glacier-carved valleys, and precipitous peaks, with a maximum elevation of 10,448 ft atop Mount Cleveland. The eastern and western sides of the park are connected by the Going-to-the-Sun Highway, which crosses the Continental Divide through Logan Pass (6664 ft). On the eastern slopes of the divide are stands of spruce, fir, and lodgepole pine; the moister western slopes are thickly forested with ponderosa pine, larch, fir, hemlock, and cedar. Animal life includes grizzly bear, elk, moose, cougar, mountain caribou, bobcat, mountain goat, bald eagle, and osprey. The park is linked with the adjacent Waterton Lakes National Park in Alberta, to form the Waterton-Glacier International Peace Park (established 1932). The area of the park is 1583.8 sq. miles.

Great Smoky Mountains National Park

Great Smoky Mountains National Park, located western North Carolina and eastern Tennessee, was established in 1930 and has since become the most visited of all national parks.. The Great Smoky Mountains extend the entire length of the park and contain some of the highest peaks of eastern North America, including Clingmans Dome (6642 ft). The park is noted for its luxuriant and diverse vegetation. Nearly 40% of its area is covered with virgin forest, and a total of 130 native species of trees have been identified. Canada hemlock, red spruce, silver bell, yellow buckeye, mountain ash, and other trees grow to great heights. The peaks themselves are covered with forests of spruce and fir. Rhododendron, dogwood, laurel, and flame-colored azalea are among the flowering plants that grow throughout the park. A wildlife sanctuary, the park harbors black bear, bobcat, deer, red and gray fox, wild turkey, and ruffed grouse. The many streams are noted for their rainbow and brook trout and smallmouth bass. The Appalachian Trail runs along the crest of the mountains through the park. The area of the park is 813 square. miles.

America's Bounty

A Look At Our Natural Resources

As I've said before, nature is primarily responsible for the prosperity of our nation. Taken together, our natural resources are unmatched by any other nation. Unfortunately, misuse or overuse of these resources can threaten our well-being in the future. Let's take a look at some of the most important of these resources and see how we stand.

Petroleum

It's pretty hard to imagine modern life without petroleum (I'd be peddling across the George Washington Bridge around 4 a.m. every morning). Fuel distilled from petroleum give us mobility on the ground and in the air, heats most of our homes, generates lots of electricity, and powers the equipment that builds and maintains our cities. Petroleum is also a vital ingredient in the manufacture of medicines and fertilizers, foodstuffs, plastic ware, building materials, paints, and cloth.

When you think about how important petroleum is to the world economy, it's surprising to realize that it's widespread use didn't begin until the early twentieth century. And it's downright frightening to realize it's very likely that the worldwide supply of oil will be gone in the next three or four decades.

How Is Oil Formed and How Does It Get to Our Fuel Tanks?

The gas you pump into your tank every few days is really the remains of tiny marine organisms and plants that died and drifted to the ocean floor hundreds of millions of years ago. These organic remains mix with sand and silt, which is all packed together as more sediment piles up on top. Eventually, when the pressure gets high enough, the temperature in the rocks goes up several hundred degrees, hardening the sediment into rock and cooking the dead organisms into crude oil and natural gas.

When the oil and gas are formed, they start drifting upward again because they're less dense than the salt water that's also trapped in the rocks. Some of the oil reaches the surface—oil bubbling from the ground has been used for torches and as a sealant for thousands of years. Lots of it remains trapped in porous rock like shale. And some of it is trapped under denser rock, where it pools into reservoirs. Colonel Edwin L. Drake found such a reservoir when he drilled into the earth in Oil Creek, Pennsylvania 1859, and petroleum gushed from the world's first oil well.

The reason that oil gushes up when a well is drilled is that pressure builds up when the upward movement is stopped, especially because natural gas is normally mixed with the crude. However, the natural pressure soon dissipates, which means the oil stops rising on its own. Today, even by using such techniques as pumping water into wells to increase the pressure, wells are usually abandoned when just 25% to 35% of the oil has been recovered.

Although the consistency of crude oil varies from liquid that flows easily to gooey sludge, all of it has to be refined, or broken down into heavier and lighter components. Today, out of every 42 gallon barrel of oil, we obtain about 21 gallons of gasoline, 3 gallons of jet fuel, 9 gallons of gas oil and distillates, 4 gallons of lubricants, and 3 gallons of heavier residues.

How Much Oil Do We Use and How Much Do We Have?

In terms of energy, the United States is like the monster that devoured Cleveland—we use two and a half times as much as any other nation. In terms of petroleum, that translates into about 17 million barrels of oil per day, or 6.2

billion barrels per year. We have to import about 7.6 million barrels per day, or about 45% of our needs.

The reason we need to buy so much oil is that our reserves of the precious fluid are about 23 billion barrels. Without imports our wells would soon run dry. Fortunately, known reserves in the world are about 1 trillion barrels. Two-thirds of these reserves, or 660 billion barrels, are in Middle Eastern countries. Saudi Arabia has about 260 billion barrels and Kuwait another 100 billion—figures that explain our quick reaction to the Iraqi invasion of Kuwait.

The first oil wells drilled in the U.S. were in Pennsylvania, where pools of oil bubbled up from beneath the ground for centuries before drilling commenced. These fields have long been exhausted. About half our current reserves are under the North Slope of Alaska, with much of the rest located in the Gulf Coast states or offshore in the Gulf of Mexico.

What Are The Chances of Finding More Oil?

The continental U.S. has been so thoroughly explored that chances of finding huge reserves of crude oil are very slim. There are three possibilities for extending our oil supply:

- Technology has recently been developed that can increase the percentage of oil pumped from certain wells from about 30% to 60% . Doubling the output from all known wells could add decades to our oil supply.
- Although offshore drilling is difficult, improved technology could lead the way to the discovering of additional major fields.
- An immense amount of oil is trapped in shale—the state of Colorado alone has oil shale equal to five times the total world reserves. Right now, it's far too expensive to extract the oil from shale, but new technology could tap this immense supply.

Natural Gas

Natural gas is created by the same process as oil, and it is usually found in the same locations. However, our reserves of natural gas are about 165 trillion cubic feet, about twice as much in relation to total world reserves as our oil reserves—our reserves rank sixth among world nations. The major problem with using natural gas is that a system of pipelines must be built and maintained to bring it to factories and homes.

Coal

Coal, like petroleum, is formed from plants. But instead of microorganisms, coal comes from giant ferns and other vegetation that thrived in swamps that covered much of the world between 350 and 250 million years ago. This green stuff died, fell to the bottom of the water, and gradually lost its oxygen and hydrogen as it decomposed. This spongy, partially decomposed material with a high carbon content is called peat. Dried and compressed, peat is used for fuel in Ireland and other European countries.

However, peat isn't very efficient as fuels go, because it still has excess water and ash. Over millions of years, mud and sand pile up on peat deposits, adding heat and pressure. Eventually, peat turns into a hard substance we call coal.

There are three types of coal: lignite has the least carbon; bituminous (soft) coal, which is by far the most common, has a higher carbon content; and anthracite (hard) coal has the most carbon and highest heating value. Pressure and heat are used to transform coal into graphite, which is nearly pure carbon. Coal can also be liquified to form an oil substitute—the Republic of South Africa meets all its oil needs by liquifying coal.

As an energy source, there's good news and bad news about coal. The good news is that we have lots of it in the U.S.—17% of the world's reserves. Coal is also relatively inexpensive to mine. The bad news is that coal, like other fossil fuels, produces carbon dioxide when it's burned, leading to global warming. It also releases sulfur and nitrogen, which contribute to the formation of acid rain. Finally, some mining techniques badly scar the land.

Today, we get about 50% more of our annual energy needs from U.S. mined coal than we do from petroleum pumped in the U.S.. About 86% of all coal is used to generate electricity. Most of the coal mined in the U.S. comes from Appalachia (Pennsylvania, West Virginia, Kentucky, Tennessee, Ohio, and Alabama), the central Midwest (Illinois and parts of Indiana and Kentucky), and a Corn Belt field (parts of Iowa, Missouri, Kansas, and Oklahoma). Vast reserves in many of the Mountain States have not yet been extensively mined.

The key to increased use of our vast coal reserves is continuing development of technology that allows coal to be burned cleanly, reducing emissions of hydrocarbons, sulfur, and nitrogen.

Farmland

Food, along with water and air, is one of the three essential elements of life. And no nation of the world can match the United States in the ability to not just feed its population, but provide immense surpluses for export. The U.S. has almost 1 billion acres of farmland, including a large portion of the most fertile soil in the world. That represents about 47% of the total land area of this country. Approximately 399 million acres make up cropland resources. Almost 83% of cropland is cultivated, including about 57 million acres used for wheat, about 74 million acres used for corn, and about 62 million acres used for hay. More than 50% of croplands are prime farmland, the best land for producing food and fiber.

The nation has nearly another 1 billion acres of nonfederal rural land currently being used for pastures, range, forest, and other purposes. About 68 million acres of this land are suitable for conversion to cropland if needed.

Farming Regions of the U.S.

The U.S. has ten major farming areas. They vary by soil, slope of land, climate, and distance to market, and in storage and marketing facilities.

The states of the Northeast and the Lake states are the country's principal milk-producing areas. Climate and soil there are suited to raising grains and forage for cattle and for pastures. Broiler farming is important to Maine, Delaware, and Maryland. Fruits and vegetables are also important to the region.

The Appalachian region is the major tobacco-producing area of the nation. Peanuts, cattle, and dairy production also are important.

Beef cattle and broilers are the major livestock products farther south in the states of the Southeast; fruit and vegetables and peanuts are also grown. Florida has vast citrus groves and winter vegetable production areas.

In the Delta states, principal cash crops are soybeans and cotton. Rice and sugarcane are grown in the more humid and wet areas. With improved pastures, livestock production has gained importance in recent years. It also is a major broiler-producing region.

The Corn Belt, extending from Ohio through Iowa, has rich soil, good climate, and sufficient rainfall for excellent farming. Corn, beef cattle, hogs, and dairy products are of primary importance. Other feed grains, soybeans, and wheat also are grown.

The northern and southern Plains, extending north and south from Canada to Mexico and from the Corn Belt into the Rocky Mountains, are restricted by low rainfall in the western portion and by cold winters and short growing seasons in the north. But about 60% of the nation's winter and spring wheat grows in the Plains states. Other small grains, grain sorghums, hay, forage crops, and pastures help make cattle important to the region. Cotton is produced in the southern part.

The Mountain states provide yet a different terrain. Vast areas are suited to cattle and sheep. Wheat is important in the north. Irrigation in the valleys provides water for hay, sugar beets, potatoes, fruits, and vegetables.

The Pacific region includes California, Oregon, and Washington plus Alaska and Hawaii. In the northern mainland, farmers raise wheat, fruit, and potatoes. Dairy, vegetables, and some grain are important to Alaska. Many of the more southerly farmers have large tracts on which they raise vegetables, fruit, and cotton, often under irrigation. Cattle are raised throughout the region. Hawaii grows sugarcane and pineapple as its major crops.

Agricultural Resources

The total land area of the United States is about 2.27 billion acres, of which about 47% is used to produce crops and livestock. The rest is distributed among forest land (29%) and urban, transportation, and other uses (24%).

The history of agriculture in the U.S. since the Great Depression has been one of consolidation and increasing efficiency. From a high of 6.8 million farms in 1935, the total number declined to 2.1 million in 1991 on a little less than the same area, about 982 million acres. Average farm size in 1935 was about 155 acres; in 1991 it was about 467 acres.

About 4.6 million people lived on farms in 1990, based on a new farm definition introduced in 1977 to distinguish between rural residents and people who earned $1000 or more from annual agricultural product sales. The farm population continues to constitute a declining share of the nation's total; about 1 person in every 54, or 1.8%, of the nation's 250 million people were farm residents in 1990.

Total value of land and buildings on U.S. farms in 1990 was $658 billion, substantially less than the value in 1980. Value of products sold was $170 billion per year. Overall net farm income was more than $46 billion in 1989, of which government subsidies accounted for 23%.

Outstanding farm debt in 1989 was $146 billion, of which about 55% was owed on real estate. Interest payments on the mortgage debt were about $7.6 billion per year.

In 1980 a report based on projections by the U.S. government stated that in the next 20 years world food requirements would increase tremendously, with developed countries requiring most of the increase, and food prices would double. Less than five years later, however, the U.S. farmer was enveloped in a major crisis caused by exceptionally heavy farm debts, mounting farm subsidy costs, and rising surpluses. A number of farmers were forced into foreclosure.

Agricultural Exports

The U.S. is the world's principal exporter of agricultural products. In 1989 the value of produce exported was about $39.7 billion, including roughly $1.5 billion in donations and loans to developing nations. We're the number one exporter of both corn and wheat, which along with rice are the world's most important grains.

Top Ten States in Farm Acreage

Texas	130,000,000 acres	Montana	60,000,000 acres
Kansas	48,000,000 acres	Nebraska	47,000,000 acres
New Mexico	44,000,000 acres	South Dakota	43,000,000 acres
North Dakota	40,000,000 acres	Arizona	36,000,000 acres
Oklahoma	34,000,000 acres	Colorado \ Kansas	33,000,000 acres

Livestock Produced Annually in the United States

Cattle	101,000,000	Diary Cows	9,038,000
Hogs	57,938,000	Sheep	7,071,000

SECTION

THREE

Where and How We Live

Several times in this book, I've mentioned that geography is as much about people as it is about places. We've seen that maps are the representations that help us tell the story of the places in America. Now, we're going to take a look at what helps us tell the story of you and the other people in America—statistics.

Now, I suspect that the mention of that word has put half of you to sleep. But like maps, numbers are boring only when they're presentated that way. If they're used properly, they can tell many fascinating stories about who we are and how we live.

For example, you may be interested to know that you are one of approximately 115 billion people who have lived in the last 2.5 million years. You're also one of about 8 billion people (6% of those who have ever lived) whose name will have survived in books, on monuments, or in public records.

Another example: people don't earn as much as you think they do. Less than one-third of all workers earn more than $30,000 per year, and only 8% earn more than $50,000 per year. Half of all full-time, year-around workers earn more than $400 per week. And a lot of us live on the edge—half of all Americans have less than $5,000 in assets. On the other side, the richest 1% of Americans owned 31% of the corporate stock and one-quarter of all U.S. real estate.

And what about your love life? The average American has his or her first puppy love at age 13 and the first "serious" romance four years later. You'll fall in love an average of 6 times in your life. About 25% of you are love-prone, having fallen in love 10 or more times. When romances end, it's the woman who does the breaking up 70% of the time.

Your leisure time(what's that?) has dropped, on the average, 32% in the last decade, from 26 to 17 hours. One major reason is that we're working longer hours to make a living, from 40.6 to 48.8 hours per week. Women have, on the average, three fewer leisure hours per week. Perhaps because we work so hard, we spend a majority of leisure hours watching television (save a portion of that for me).

I hope this gives you a taste of the fascinating insight into our lives that numbers can provide. Before we go on to some other interesting snapshots of American life, you probably want to know where these statistics come from. The overwhelming majority are compiled by the U.S. government. Every ten years, the U.S. Census Bureau contacts every American (or almost everyAmerican) to fill out those long forms. But they also collect lots of information between official census counts through scientific surveys of about 40,000 Americans. The National Center for Health Statistics compiles birth, death, marriage, divorce, and hospital records to provide information about our health. The U.S. Bureau of Labor Statistics finds out who's hired, who's fired, and what kind of work we do. Our old friends at the Internal Revenue Service provide information (confidential) on what people make and how they spend their money. Other agencies provide information on additional subjects that affects how government structures its programs and spends our money.

Are these statistics accurate? The answer is, as accurate as possible. You can trust them to provide almost as accurate a portrait of people as maps do of the towns, cities, and states in which we live. I hope you enjoy the following pages.

The Settling of America

Who We Are And How We Got Here

Picture this: the population of all six New England states packing up and moving out, leaving Boston, Providence, Hartford, Portland, and every other city and village as ghost towns. It sounds incredible, until you realize that the equivalent number of people from other nations left their homes and arrived on the shores of America nearly every decade between 1890 and 1930. Even today, the United States admits almost 1 million new residents per year. No wonder we live in, by far, the most ethnically and racially diverse country on earth.

I believe that this diversity, even more than our natural resources, is the reason for America's prosperity and power. Think about the courage it would take for you to leave your home, your family and friends, and your culture to travel thousands of miles to an un familiar land. The people whom other nations contributed to America tended to be those with the most ambition, initiative, and personal fortitude. In their new land, they had to work harder and be smarter to carve a niche. For more than three hundred years, waves of new energy and new ideas have helped fuel our economic, political, and social progress.

That's why I find the recent increase in racial and ethnic tensions and the growing hostility toward immigrants so disturbing. I hope that when you learn how the story of our land is truly the story of our people, you'll join me in celebration rather than ruing our diversity.

A Little Background on Human Evolution

Evolutionary science tells us that we humans are descended from apes that evolved about 20 million years ago and were the ancestors of such modern creatures as gorillas, chimpanzees, and orangutans. Although there are many differences between us and our distant relatives, there are two that are so important and obvious that they are used to track our evolution—walking on two legs and an increase in brain size.

Climate changes in Africa replaced vast areas of forest with equally vast grasslands. Moving across grasslands is easier and safer when walking on two legs instead of four. Walking on two legs comfortably requires changes in the spinal column, pelvis and legs, and we have found fossils indicating we had bipedal ancestors by about 5 million years ago.

These ancestors survived by gathering food and by hunting, both of which are more easily accomplished by groups who cooperate with each other and use tools. An increase in brain size helped bring about the evolution of both practices by about 2.5 million years ago. At some later time, further increases in brain size resulted in the development of language.

As our ancestors became more intelligent and adaptable, they were able to accomplish something few other living species can—they could migrate to and thrive in different environments. About 1 million years ago, our distant relatives left central Africa for the warmer parts of Europe and Asia, and about 500,000 years ago, they moved on to the temperate climates. The first Homo sapiens appeared somewhere between 300,000 and 200,000 years ago, and modern man (Homo sapiens sapiens) evolved around 90,000 years ago. They spread throughout Africa, Europe, and Asia.

The First Americans

Separated from Africa, Asia, and Europe by broad oceans, North America witnessed none of this evolution. It wasn't until somewhere after 50,000 B.C. that the first humans from Asia made their way across a land bridge that came into existence across the Bering Strait, which now separates Russia from Alaska's Aleutian Islands. These earliest inhabitants were Stone Age people, who lived by hunting and gathering, using implements not unlike those known from Southeast Asia. They were later supplanted by other migrants with more advanced tools who in a relatively short period of time made the incredible trek all the way through North America and into South America. By about 10,000 years ago, some of these people were learning to domesticate plants and to build permanent settlements, like the peoples in other cradles of civilization such as the Tigris-Eurphrates and Nile River valleys. Great prehistoric civilizations rose in Central America and Peru, and a plethora of American Indian tribes developed sophisticated cultures.

The Age of Exploration

We all know now that Columbus wasn't the first European to discover America. The Vikings reached Greenland by about 1100 A.D., and an Icelandic trader named Bjarni Herjolfsson was probably the first to sight the mainland in 986. Leif Ericson is believed to have landed somewhere between Labrador and New England, a belief supported by the discovery of a Viking-type settlement dating back to about 1000 A.D. at L'Anse-aux-Meadows in Newfoundland. Why didn't the Vikings stay? The answer is, bad weather. They had sailed the North Atlantic during an unusually warm period in recent history. However, this warm spell was followed by frigid weather known as the "Little Ice Age" that made long sea journeys too dangerous.

Columbus led the way nearly 500 years later on a more southern route that took advantage of the westerly trade winds that regularly blow north of the equatorial region. Columbus landed on Cuba and Hispaniola, but the first man to reach the mainland was Italian navigator John Cabot, who was sailing for England when he touched Labrador or New England in 1498. The Spanish founded the first permanent settlement at St. Augustine, Florida in 1565, and the first permanent English settlement was built at Jamestown, Virginia in 1607.

Dividing Up The New World

Three countries dominated the exploration and settlement of the New World—England, Spain, and France. English claims in the New World were based on John Cabot's voyage. After ousting the Dutch from the Middle Atlantic region, they controlled the entire area from Maine to Georgia. As rulers of a world-wide Empire based on trade, the British encouraged settlement of the New World, and the growth in the English colonies was the primary reason that the population of what is now the U.S. jumped form 50,000 in 1650 to 250,000 in 1700.

The Spanish had conquered and exploited Central and South America, and from this base they claimed Florida and what is now the American southwest. However, the Spanish colonial philosophy was based on ruthless economic exploitation of the natives combined with a prohibition of settlement or trade with any foreigners. Native American tribes were so spread out in Spanish-controlled parts of the U.S. that there was little incentive for settlement. The European population of these areas was tiny in the seventeenth and eighteenth centuries.

The French explored and claimed parts of the present-day United States by traveling from their base in Canada down the Mississippi River. Their main interest was fur trading, not building large settlements. They were able to fend off claims by other nations largely because of their alliance with Indian tribes. However, France lost all of its territory east of the Mississippi River in the French and Indian War, which ended in 1763.

The U.S. is Born

The 13 colonies that joined together to form the United States of America acquired all the British territory by winning the Revolutionary War. Vermont, Kentucky, Tennessee and Ohio, all carved from British territory, were the next states admitted.

The first giant acquisition by the U.S. came in 1803, when our country bought the vast French lands west of the Mississippi in the Louisiana Purchase. The Southern States (except Florida) and the Midwestern States were carved primarily from this territory. The U.S. bought Florida from Spain in 1819, but most of the remaining territory in the present U.S. was controlled by Mexico, which gained control after it gained independence from Spain in 1821. Texans were the first to challenge Mexico, winning their freedom in 1836. Texas joined the Union in 1845, triggering the Mexican-American War that gave the U.S. control of the Southwest, California, and much of the Mountain States. Alaska was purchased from Russia in 1867, and Hawaii was annexed in 1898.

Some Famous European Explorers of North America

Year	Explorer	Nationality and Employer	Discovery or Exploration
1497	John Cabot	Italian-English	Newfoundland or Nova Scotia
1498	John and Sebastian Cabot	Italian-English	Labrador to Hatteras
1500-02	Gaspar Corte-Real	Portuguese	Labrador
1513	Vasco Nunez de Balboa	Spanish	Pacific Ocean
1513	Juan Ponce de Leon	Spanish	Florida
1519	Alonso de Pineda	Spanish	Mouth of Mississippi River
1519	Hernando Cortes	Spanish	Mexico
1524	Giovanni da Verrazano	Italian-French	Atlantic coast-New York harbor
1534	Jacques Cartier	French	Canada, Gulf of St. Lawrence
1536	A.N. Cabeza de Vaca	Spanish	Texas coast and interior
1539	Francisco de Ulloa	Spanish	California coast
1539-41	Hernando de Soto	Spanish	Mississippi River near Memphis
1539	Marcos de Niza	Italian-Spanish	Southwest (now U.S.)
1540	Francisco V. de Coronado	Spanish	Southwest (now U.S.)
1540	Hernando Alarcon	Spanish	Colorado River
1540	Garcia de L. Cardenas	Spanish	Grand Canyon of the Colorado
1542	Juan Rodriguez Cabrillo	Portuguese-Spanish	San Diego harbor
1565	Pedro Menendez de Aviles	Spanish	St. Augustine
1576	Martin Frobisher	English	Frobisher's Bay, Canada
1577-80	Francis Drake	English	California coast
1582	Antonio de Espejo	Spanish	Southwest
1584	Amadas & Barlow (for Raleigh)	English	Virginia
1585-87	Sir Walter Raleigh's men	English	Roanoke Is., N.C.
1595	Sir Walter Raleigh	English	Orinoco River
1603-09	Samuel de Champlain	French	Canadian interior, Lake Champlain
1607	Capt. John Smith	English	Atlantic coast
1609-10	Henry Hudson	English-Dutch	Hudson River, Hudson Bay
1634	Jean Nicolet	French	Lake Michigan; Wisconsin
1673	Jacques Marquette, Louis Jolliet	French	Mississippi S to Arkansas
1682	Robert Cavelier, Sieur de La Salle	French	Mississippi S to Gulf of Mexico
1789	Alexander Mackenzie	Canadian	Canadian Northwest

The People of America

The arrival of the Europeans was a catastrophe of gigantic proportions for the Native American populations. Disease, mistreatment, and warfare decimated their populations to such an extent that the descendants of the original inhabitants make up less than 1% of the population. That means that the ancestors of more than 99% of us were immigrants to this land.

Before the Revolutionary War, most of the new residents of this country came from two places. One was Great Britain. The English colonies were all the on the Atlantic Coast, making them easy to reach, and the British government encouraged the development of colonies that could provide agricultural commodities and a market for English manufactured goods. The colonies also provided a home for those whose religious views conflicted with those of the Crown.

The second area of origin was Africa, but these immigrants did not come of their own free will. The Spanish had imported slaves to their Caribbean, Central American, and South American colonies as early as 1519. Because large plantations had labor requirements far in excess of the few English colonists willing to do the work, English landowners began following the Spanish example after the formation of the Royal African Company in 1663. Far more than the Spanish, the English attempted to strip their slaves of their native African cultures in the belief that it would make them easier to handle.

The Population of America in Colonial Days

1630	4,600	1720	466,200
1650	50,400	1740	905,600
1670	111,900	1750	1,170,800
1690	210,400	1770	2,148,100
1700	250,900	1780	2,780,400

Source: U.S. Bureau of the Census

A History of U.S. Territorial Acquisitions

Acquisition	Date	Area acquired (sq. mi)
United States	x	3,618,770
Territory in 1790	1790	891,364
Louisiana Purchase	1803	831,321
Purchase of Florida	1819	69,866
Texas	1845	384,958
Oregon	1846	283,439
Mexican Cession	1848	530,706
Gadsden Purchase	1853	29,640
Alaska	1867	591,004
Hawaii	1898	6,471
Other areas:		
Puerto Rico	1898	3,515
Guam	1898	209
American Samoa	1899	77
Virgin Islands of the U.S.	1917	132
Pacific Islands, Trust Territory of the 5	1947	533
No. Mariana Islands	1947	184

Source: Bureau of the Census, U.S. Department. of Commerce

Because I am descended from from some of those millions of people uprooted from their homelands and brought to this country as slaves the subject is both fascinating and painful to me. Because we're talking about geography in this book, I just want to share one observation. Take a look at a world map and calculate how long the voyage is from Africa to the southern United States. Then ponder the brutally inhuman conditions in which the human cargo was transported. I have often been, and still am in awe of the physical and emotional strength and endurance my African ancestors demonstrated in surviving this incredible ordeal.

There were three waves of immigration stimulated by the westward expansion of the United States in the nineteenth century. Between 1810 and 1840 came the first wave of Irish, English, and Germans. Between 1840 and 1870, the new residents were overwhelmingly from the Scandinavian countries. Between 1870 and 1910 came the southern and eastern Europeans, including the Italians, Poles, and Greeks.

By the time World War I broke out, the great westward expansion was over, and the U.S. government began passing laws to curb immigration. In recent decades, the largest number of immigrants have come from Third World countries, especially Latin America, the Far East, and the Philippines.

Official Census of the U.S. Population, 1790 - 1990

1790	3,929,000	1860	31,443,321	1930	123,202,624
1800	5,308,000	1870	38,558,371	1940	132,164,569
1810	7,240,000	1880	50,189,209	1950	151,325,798
1820	9,638,453	1890	62,979,766	1960	179,323,175
1830	12,860,702	1900	76,212,168	1970	203,302,031
1840	17,063,353	1910	92,225,496	1980	226,542,203
1850	23,191,867	1920	106,021,537	1990	248,709,873

Source: Bureau of the Census

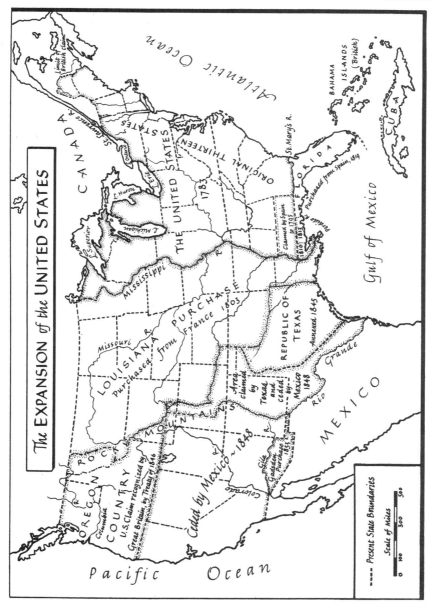

Where We Came From

Immigration by Country of Last Residence: 1820-1990
Source: U.S. Immigration and Naturalization Service (thousands)

Country	Total
All countries	56,994
Europe	37,101
Austria	4,343
Hungary	211
Belgium	146
Czechoslovakia	371
Denmark	37
Finland	787
France	7,083
Germany	5,119
Great Britain	704
Greece	4,723
Ireland	5,373
Italy	375
Netherlands	754
Norway	606
Poland	501
Portugal	285
Spain	1,246
Sweden	359
Switzerland	3,444
USSR	136
Yugoslavia	294
Other Europe	4,547
Asia	6,019
China	897
Hong Kong	302
India	456
Iran	177
Israel	138
Japan	462
Jordan	74
Korea	642
Lebanon	958
Philippines	1,026
Turkey	412
Vietnam	459
Other Asia	879
The Americas	13,068
Argentina	131
Brazil	98
Canada	4,296
Colombia	296
Cuba	748
Dominican Rep.	510
Ecuador	155
El Salvador	296
Guatemala	137
Haiti	235
Honduras	91
Mexico	3,888
Panama	94
Peru	121
West Indies	1,209
Other Americas	798
Africa	334
Australia and New Zealand	147
Other Oceania	57
Unknown or Not Reported	2

From Caves to Cleveland

10

A Quick Look at U.S. Cities

In the Introduction to this book, I explained how I did most of my traveling as a child through books. I didn't get a chance to go on my first real trip until I was thirteen years old, when my parents, my brother, and I drove to East Orange, New Jersey to visit my uncle. For weeks, my father and I sat down to go over maps to plan our route. Once the long-awaited departure day arrived, I was appointed navigator. I can still remember the thrill of pointing out each landmark and each state border. But the highlight of the trip came when we reached a certain point on the New Jersey Turnpike where the New York City skyline appeared on the horizon across the Hudson River. I'll never forget the thrill of identifying the Empire State Building soaring majestically above the other spires.

To this day, the same view still gives me a rush. Despite all the urban problems, there remains something magical about cities that draws us like a magnet. Perhaps it is because us humans evolved as social animals. And the rise of our civilization dates back to the time when we established permanent settlements that had populations far greater than the extended families in which more primitive hunter-gatherer peoples live. Living in large groups allowed the division of labor—while some people provided food, others could produce tools, organize a government, or study nature. Trade sprang up between cities along with systems of writing to keep track of that trade. People created maps to define boundaries, indicate trade routes, or locate natural resources. The history of any part of the world, including the United States, is to a great extent the history of its cities.

What Is a City?

The word city is derived from the Latin word *civitas,* which means a community that administers its own affairs. In ancient Greece such an independent community was called a city-state; it consisted of a chief town and its immediate neighborhood. During the Middle Ages a city was usually identical with a cathedral town; accordingly, when King Henry VIII of England established new bishoprics in boroughs, he made these into cities. In modern Great Britain, city is merely a complimentary title conferred by the monarch on important towns.

In the United States a city is a chartered municipal corporation. Charters are granted by state governments according to requirements prescribed by the legislature of that state; a city must usually attain a certain population before it can be granted a charter.

The History of Cities

Cities began to evolve in prehistoric times when groups of nomadic hunters and foragers developed a settled agricultural life. In order to protect themselves and their food supplies from raids by predatory nomads, they built their dwellings within a walled area or in a naturally fortified place, such as the acropolis of ancient Greek cities. Because the availability of water was also an important consideration, these settlements were usually located along a river. Such settlements led to specialization and the division of labor. Markets developed in which artisans could exchange their specialties for other types of goods. A growing priesthood contributed to intellectual life. Thus, cities were responsible not only for the rise of commerce and industry but also of art and learning, and they played an essential role in the emergence of all great civilizations. Among the most notable cities of the ancient world were, in the order of their development, Thebes, Memphis, Babylon, Nineveh,

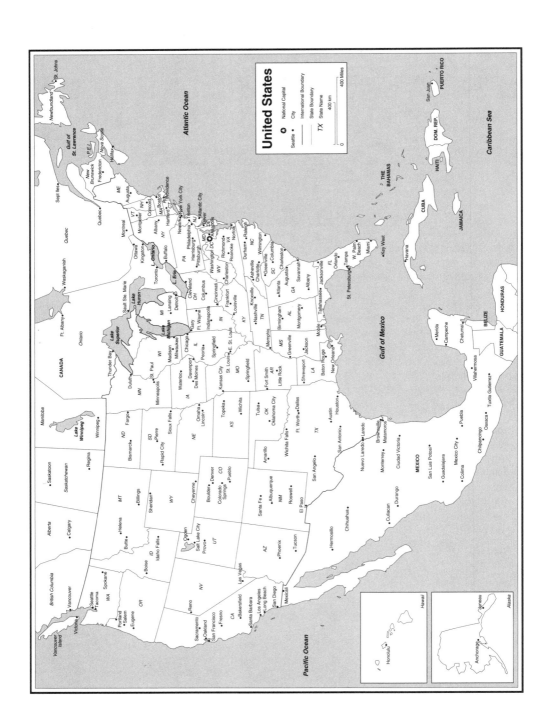

United States

- ✪ National Capital
- • City
- International Boundary
- State Boundary
- TX State Name

0   400 km
0   400 Miles

Susa, Tyre, Carthage, and Jerusalem. Alexandria is said to have contained more than 500,000 inhabitants, and Rome was still larger. As capital of the Eastern Roman Empire, Constantinople succeeded Rome as the principal city in Europe. In the Islamic East, during part of the Middle Ages, Baghdad, Dimashq, and Cairo led in population; Cordoba was the greatest city of the Islamic West and, for a time, of all Europe.

The development of cities in Europe was a feature of the breakup of feudalism. At the beginning of the 16th century Europe had 6 or 7 cities of 100,000 or more inhabitants; at the end of the century it had 13 or 14 such cities. During the 17th century, although the population of Europe remained stationary, that of the cities increased.

U.S. Cities

Cities were crucial to the development of this new nation. In our descriptions of the U.S. states, we profile more than 150 of the most important American cities. In this century, these cities, and the suburban areas that have grown up around them, have attracted an increasing percentage of our population at the expense of urban areas. Today, more than 75% of the population of our country lives in urban areas.

In 1850, the United States had six cities with populations of 100,000 or more. New York, with 696,000 people, was nearly four times the size of second place Baltimore. The other cities on the list were Philadelphia, Boston, New Orleans, and Cincinnati (the gateway to the west). By 1900, three cities had over one million residents. It's no surprise that New York was once again first with 3,437,202 people. In second place was newcomer Chicago, which had jumped from 29,000 people in 1850 to 1,698,000 in 1900, and many other cities in the newly opened Midwest experienced similar growth patterns.. Philadelphia had 1,000% more people, at 1,293,000. In total, the U.S. had 31 cities with populations of 100,000 to 250,000, and 19 with 250,000 or more.

In contrast, by 1990, cities with populations of 100,000 to 250,000 numbered 143 and cities with 250,000 or more numbered 64. While New York was still the largest, Los Angeles, which had just 102,000 people in 1900, had jumped to second place with 3,485,000 people. Most of the fastest growing cities in recent decades have been in the Southwest and West.

Metropolitan areas consist of major cities and the smaller cities and suburban areas around them. These areas are defined by the U.S. Census Bureau, and their boundaries expand as commuters settle farther and farther from the central city. The 1990 census figures show that the nation has 39 metropolitan areas of at least 1 million population,, and in these areas reside 124.8 million people, or 50.2% of the U.S. population. The 1950 census showed only 14 metropolitan areas of 1 million people, and their combined population of about 45 million amounted to less than 30% of the national total. The census shows that the U.S. population living in all metropolitan areas totals 192,725,741, an increase of just over 20 million (11.6%) since 1980. Ninety percent of the nation's growth in the 1980s took place in metropolitan areas.

A county is an area of government that normally includes a city and some of its suburbs. However, some counties have an area identical to that of the city. And New York City contains five counties, four of which (Queens, Kings [Brooklyn], New York [Manhattan], and the Bronx) have populations of more than 1,000,000. Los Angeles County, the nation's largest, also had the largest numeric increase, 1.4 million, followed by San Diego and Maricopa (Phoenix) each with over 600,000, and San Bernardino and Riverside each with more than 500,000.

Counties with 1990 Population over 1 Million

Source: U.S. Bureau of the Census,

County	1990 population	County	1990 population
Los Angeles, CA	8,863,164	New York (Manhattan)	1,487,536
Cook, IL	5,105,067	San Bernardino, CA	1,418,380
Harris, TX	2,818,199	Cuyahoga, OH	1,412,140
San Diego, CA	2,498,016	Middlesex, MA	1,398,468
Orange, CA	2,410,556	Allegheny, PA	1,336,449
Kings (Brooklyn), NY	2,300,664	Suffolk, NY	1,321,864
Maricopa, AZ	2,122,101	Nassau, NY	1,287,348
Wayne, MI	2,111,687	Alameda, CA	1,279,182
Queens, NY	1,951,598	Broward, FL	1,255,488
Dade, FL	1,937,094	Bronx, NY	1,203,789
Dallas, TX	1,852,810	Bexar, TX	1,185,394
Philadelphia, PA	1,585,577	Riverside, CA	1,170,413
King, WA	1,507,319	Tarrant, TX	1,170,103
Santa Clara, CA	1,497,577	Oakland, MI	1,083,592
		Sacramento, CA	1,041,219

Population of U.S. Cities

Source: U.S. Bureau of the Census .(100 most populated cities ranked by April, 1990 .census)

Rank	City	1990	Rank	City	1990
1	New York, NY	7,322,564	51	Wichita, Ks	304,011
2	Los Angeles, Ca	3,485,398	52	Santa Ana, Ca	293,742
3	Chicago, Ill	2,783,726	53	Mesa, Az	288,091
4	Houston, Tx	1,630,553	54	Colorado Springs, Co	281,140
5	Philadelphia, Pa	1,585,577	55	Tampa, Fl	280,015
6	San Diego, Ca	1,110,549	56	Newark, NJ	275,221
7	Detroit, Mi	1,027,974	57	St. Paul, Mn	272,235
8	Dallas, Tx	1,006,877	58	Louisville, Ky	269,063
9	Phoenix, Az	983,403	59	Anaheim, Ca	266,406
10	San Antonio, Tx	935,933	60	Birmingham, Al	265,968
11	San Jose, Ca	782,248	61	Arlington, Tx	261,721
12	Indianapolis, In	741,952	62	Norfolk, Va	261,229
13	Baltimore, Md	736,014	63	Las Vegas, Nv	258,295
14	San Francisco, Ca	723,959	64	Corpus Christi, Tx	257,453
15	Jacksonville, Fl	672,971	65	St. Petersburg, Fl	238,629
16	Columbus, Oh	632,910	66	Rochester, NY	231,636
17	Milwaukee, Wi	628,088	67	Jersey City, NJ	228,537
18	Memphis, Tn	610,337	68	Riverside, Ca	226,505
19	Washington, DC	606,900	69	Anchorage, Ak	226,338
20	Boston, Ma	574,283	70	Lexington-Fayette, Ky	225,366
21	Seattle, Wa	516,259	71	Akron, Oh.	223,019
22	El Paso, Tx	515,342	72	Aurora, Co	222,103
23	Nashville-Davidson, Tn	510,784	73	Baton Rouge, La	219,531
24	Cleveland, Oh	505,616	74	Stockton, Ca	210,943
25	New Orleans, La	496,938	75	Raleigh, NC	207,951
26	Denver, Co	467,610	76	Richmond, Va	203,056
27	Austin, Tx	465,622	77	Shreveport, La	198,525
28	Fort Worth, Tx	447,619	78	Jackson, Ms.	196,637
29	Oklahoma City, Ok	444,719	79	Mobile, Al	196,278
30	Portland, Or	437,319	80	Des Moines, Ia	193,187
31	Kansas City, Mo	435,146	81	Lincoln, Ne	191,972
32	Long Beach, Ca	429,433	82	Madison, Wi	191,262
33	Tucson, Az	405,390	83	Grand Rapids, Mi	189,126
34	St. Louis, Mo	396,685	84	Yonkers, NY	188,082
35	Charlotte, NC	395,934	85	Hialeah, Fl	188,004
36	Atlanta, Ga	394,017	86	Montgomery, Al	187,106
37	Virginia Beach, Va	393,069	87	Lubbock, Tx	186,206
38	Albuquerque, NM	384,736	88	Greensboro, NC	183,521
39	Oakland, Ca	372,242	89	Dayton, Oh	182,044
40	Pittsburgh, Pa	369,879	90	Huntington Beach, Ca	181,519
41	Sacramento, Ca	369,365	91	Garland, Tx	180,650
42	Minneapolis, Mn	368,383	92	Glendale, Ca.	180,038
43	Tulsa, Ok	367,302	93	Columbus, Ga	179,278
44	Honolulu, Hi.	365,272	94	Spokane, Wa	177,196
45	Cincinnati, Oh.	364,040	95	Tacoma, Wa	176,664
46	Miami, Fla.	358,548	96	Little Rock, Ar	175,795
47	Fresno, Ca	354,202	97	Bakersfield, Cal	174,820
48	Omaha, Ne	335,795	98	Fremont, Ca	173,339
49	Toledo, Oh	332,943	99	Fort Wayne, In	173,072
50	Buffalo, NY	328,123	100	Newport News, Va	170,045

The Population of Major Metropolitan Areas

New York	18,087,251	Greensboro, NC	942,091
Los Angeles	14,531,529	Birmingham, AL	907,810
Chicago	8,065,633	Jacksonville, FL	906,727
San Francisco	6,253,311	Albany, NY	874,304
Philadelphia	5,899,345	Richmond, VA	865,640
Detroit	4,665,236	West Palm Beach, FL	863,518
Boston	4,171,643	Honolulu, HI	836,231
Washington, D.C.	3,923,574	Austin, TX	781,572
Dallas-Fort Worth	3,885,415	Las Vegas, NV	741,459
Houston	3,711,043	Raleigh-Durham, NC	735,480
Miami	3,192,582	Scranton- PA	734,175
Atlanta, GA	2,833,511	Tulsa, OK	708,954
Cleveland	2,759,823	Grand Rapids, MI	688,399
Seattle-Tacoma, WA	2,559,164	Allentown-Bethlehem-Easton, PA	686,688
San Diego, CA	2,498,016	Fresno, CA	667,490
Minneapolis-St. Paul, MN	2,464,124	Tucson, AZ	666,880
St. Louis, MO	2,444,099	Syracuse, NY	659,864
Baltimore, MD	2,382,172	Greenville-Spartanburg, SC	640,861
Pittsburgh	2,242,798	Omaha, NE	618,262
Phoenix, AZ	2,122,101	Toledo, OH	614,128
Tampa-St. Petersburg	2,067,959	Knoxville, TN	604,816
Denver	1,848,319	El Paso, TX	591,610
Cincinnati	1,744,124	Harrisburg, PA	587,986
Milwaukee	1,607,183	Bakersfield, CA	543,477
Kansas City, MO	1,566,280	New Haven, CT	530,180
Sacramento, CA	1,481,102	Springfield, MA	529,519
Portland	1,477,895	Baton Rouge, LA	528,264
Norfolk-Virginia Beach	1,396,107	Little Rock, AK	513,117
Columbus, OH	1,377,419	Charleston, SC	506,875
San Antonio, TX	1,302,099	Youngstown, OH	492,619
Indianapolis, IN	1,249,822	Wichita, KS	485,270
New Orleans, LA	1,238,816	Stockton, CA	480,628
Buffalo	1,189,288	Albuquerque, NM	480,577
Charlotte NC	1,162,093	Mobile, AL	476,923
Providence RI	1,141,510	Columbia, SC	453,331
Hartford CT	1,085,837	Worcester, MA	436,905
Orlando, FL	1,072,748	Johnson City, TN	436,047
Salt Lake City, UT	1,072,227	Chattanooga, TN	433,210
Rochester, NY	1,002,410	Lansing-, MI	432,674
Nashville, TN	985,026	Flint, MI	430,459
Memphis, TN	981,747	Lancaster, PA	422,822
Oklahoma City, OK	958,839	York, PA	417,848
Louisville, KY	952,662	Lakeland, FL	405,382
Dayton, OH	951,270		

Source: Bureau of the Census, U.S. Dept. of Commerce

The Population of Major Metropolitan Areas
Fastest Growing Cities, 1980-1990

	City	Suburb of	1990	1980	Change
1.	Moreno Valley, Calif.	Riverside	118,779	28,309	319.6%
2.	Mesa, Ariz.	Phoenix	288,091	152,404	89.0
3.	Rancho Cucamonga, Calif.	Los Angeles	101,409	55,250	83.5
4.	Plano, Tex.	Dallas	128,713	72,331	77.9
5.	Irvine, Calif.	Los Angeles	110,330	62,134	77.6
6.	Escondido, Calif.	San Diego	108,635	64,355	68.8
7.	Oceanside, Calif.	Los Angeles	128,398	76,698	67.4
8.	Santa Clarita, Calif.	Los Angeles	110,642	66,730	65.8
9.	Bakersfield, Calif.		174,820	105,611	65.5
10.	Arlington, Tex.	Dallas	261,721	160,113	63.5
11.	Fresno, Calif.		354,202	217,491	62.9
12.	Chula Vista, Calif.	San Diego	135,163	83,927	61.0
13.	Las Vegas, Nev.		258,295	164,674	56.9
14.	Modesto, Calif.		164,730	106,963	54.0
15.	Tallahassee, Fla.		124,773	81,548	53.0
16.	Glendale, Ariz.	Phoenix	148,134	97,172	52.4
17.	Mesquite, Tex.	Dallas	101,484	67,053	51.3
18.	Ontario, Calif.	Los Angeles	133,179	88,820	49.9
19.	Virginia Beach, Va.	Norfolk	393,069	262,199	49.9
20.	Scottsdale, Ariz.	Phoenix	130,069	88,622	46.5
21.	Santa Ana, Calif.	Los Angeles	293,742	204,023	44.0
22.	Stockton, Calif.		210,943	148,283	42.3
23.	Pomona, Calif.	Los Angeles	131,723	92,742	42.0
24.	Irving, Tex.	Dallas	155,037	109,943	41.0
25.	Aurora, Colo.	Denver	222,103	158,588	40.1
26.	Raleigh, N.C.		207,951	150,255	38.4
27.	San Bernardino, Calif.		164,164	118,794	38.2
28.	Santa Rosa, Calif.	San Francisco	113,313	82,658	37.1
29.	Overland Park, Kan.	Kansas City	111,790	81,784	36.7
30.	Vallejo, Calif.	San Francisco	109,199	80,303	36.0
31.	Thousand Oaks, Calif.	Los Angeles	104,352	77,072	35.4
32.	Salinas, Calif.		108,777	80,479	35.2
33.	Durham, N.C.		136,611	101,149	35.1
34.	Austin, Tex.		465,622	345,890	34.6
35.	Laredo, Tex.		122,899	91,449	34.4
36.	Sacramento, Calif.		369,365	275,741	34.0
37.	El Monte, Calif.	Los Angeles	106,209	79,494	33.6
38.	Reno, Nev.		133,850	100,756	32.8
39.	Riverside, Calif.		226,505	170,591	32.8
40.	Chesapeake, Va.	Norfolk	151,967	114,486	32.7
41.	Tempe, Ariz.	Phoenix	141,865	106,919	32.7
42.	Oxnard, Calif.	Los Angeles	142,216	108,195	31.4
43.	Fremont, Calif.	San Jose/Oakland	173,339	131,945	31.4
44.	Colorado Springs, Colo.		281,140	215,105	30.7
45.	Garland, Tex.	Dallas	180,650	138,857	30.1

Source: Bureau of the Census

U.S. Super Statistics

What's the highest, lowest, biggest, smallest, easternmost and westernmost in America.

AREA FOR 50 STATES AND D. OF C.
Total   3,618,770 sq. mi.
Land   3,539,289 sq. mi.   •   Water   79,481 sq. mi.

LARGEST STATE
Alaska   591,004 sq. mi.

SMALLEST STATE
Rhode Island   1,212 sq. mi.

LARGEST COUNTY (EXCLUDES ALASKA)
San Bernardino County, California   20,064 sq. mi.

SMALLEST COUNTY
Kalawo, Hawaii   14 sq. mi.

NORTHERNMOST CITY
Barrow, Alaska   71°17′N.

NORTHERNMOST POINT
Point Barrow, Alaska   71°23′N.

SOUTHERNMOST CITY
Hilo, Hawaii   19°43′N.

SOUTHERNMOST SETTLEMENT
Naalehu, Hawaii   19°03′N.

SOUTHERNMOST POINT
Ka Lae (South Cape), Island of Hawaii   18°55′N.   (155°41′W.)

EASTERNMOST CITY

Eastport, Maine   66° 59′ 02″ W.

EASTERNMOST SETTLEMENT

Lubec, Maine   66° 58′ 49″ W.

EASTERNMOST POINT

West Quoddy Head, Maine   66° 57′ W.

WESTERNMOST CITY

West Unalaska, Alaska   166° 32′ W.

WESTERNMOST SETTLEMENT

Adak, Alaska   176° 39′ W.

WESTERNMOST POINT

Cape Wrangell, Alaska 172° 27′ E.

HIGHEST SETTLEMENT

Climax, Colorado   11,560 ft.

LOWEST SETTLEMENT

Calipatria, California   -185 below sea level

HIGHEST POINT ON ATLANTIC COAST

Cadillac Mountain, Mount Desert Is., Maine   1,530 ft.

OLDEST NATIONAL PARK

Yellowstone National Park (1872), Wyoming, Montana, Idaho   3,468 sq. mi.

LARGEST NATIONAL PARK

Wrangell-St. Elias, Alaska   13,018 sq. mi.

LARGEST NATIONAL MONUMENT

Death Valley, California, Nevada   3,231 sq. mi.

HIGHEST WATERFALL

Yosemite FallsTotal in three sections   2,425 ft.
Upper Yosemite Fall   1,430 ft.
Cascades in middle section   675 ft.
Lower Yosemite Fall   320 ft.

LONGEST RIVER

Mississippi-Missouri   3,710 mi.

HIGHEST MOUNTAIN

Mount McKinley, Alaska   20,320 ft.

LOWEST POINT

Death Valley,   282 ft below sea level

(Source: U.S. Geological Survey; U.S. Bureau of the Census)

Who We Are and Where We Live

12

In 1990, there were 248,709,000 Americans. This included a shade under 200 million whites and a shade under 30 million blacks, who made up 12.1% of the population. Almost 10 million Americans were Asian, Native American, or other races.

Our average age was 32.9 years, the highest in American history and a figure about 10 years higher than the average age of the population of the world. Americans age 65 and older made up 12.6% of the population, compared with just 8% of the population in 1950. That percentage is expected to double to one-quarter of the population by the year 2030 (when I'll be 83!). If I do live that long, I'll be part of a minority—there are only 57 men for every 100 women among people ages 75-84.

We live in nearly 92 million households, a little over half of which are headed by a married couple. One out of nine households is headed by a single female. The number of Americans living alone has doubled in the last twenty years, largely because of the increased number of older Americans, who are more likely to live alone.

Here's a look at who we are:

Total Population	
Total Population	248,708,873
Male	121,239,418
Female	127,470,455
Under 5 years	18,354,443
5 to 17	45,249,989
18 to 20	11,726,868
21 to 24	15,010,898
25 to 44	80,754,835
45 to 54	25,223,086
55 to 59	10,531,756
60 to 64	10,616,167
65 to 74	18,106,558
75 to 84	10,055,108
85 years and older	3,080,165
Under 18	63,604,432
Percentage of total	25.6%
65 and older	31,241,831
Percentage of total	12.6%
Median Age	32.9 year

```
Total Households . . . . . . . . . . . . . . . . . . . . . . . . . .91,947,410
Family households  . . . . . . . . . . . . . . . . . . . . . . .64,517,947
Married couple families  . . . . . . . . . . . . . . . . . . .50,708,322
Percentage of total  . . . . . . . . . . . . . . . . . . . . . . . .55.5%
Other family, male householder  . . . . . . . . . . . . . .3,143,582
Other family, female householder . . . . . . . . . . . . .10,666,043
Non—family households . . . . . . . . . . . . . . . . . . .27,429,463
Householder living alone  . . . . . . . . . . . . . . . . . .22,580,420
Percentage living alone  . . . . . . . . . . . . . . . . . . . . .24.6%
Persons living in households  . . . . . . . . . . . . . . .242,012,129
Average household size  . . . . . . . . . . . . . . . . . . . . . .2.63
Persons living in group quarters
(institutions, military, etc.) . . . . . . . . . . . . . . . . . . .6,697,744
```

Where We Live

One out of every nine Americans lives in California, and one out of every five lives in either New York or the Golden State. Half of you live in one of the eight largest states.

The U.S. population grew about 10% between the 1980 and 1990 Census, but some states topped that by a lot—Nevada grew 50%. Nearly one out of every three Americans is a member of a minority group, but the distribution varies widely from state to state.

Population Trends 1980-1990

State	1990	% Change 1980-90	Minority 1990	% change Minority 1980-90	State	1990	% Change 1980-90	Minority 1990	% change Minority 1980-90
U.S.	248,709,873	9.8%	60,581,577	30.9%	Colo.	3,294,394	14.0%	635,449	27.2%
Cal.	29,760,021	25.7%	12,730,895	61.1%	Conn.	3,287,116	5.8%	532,932	43.2%
N.Y.	17,990,455	2.5%	5,530,266	25.9%	Okla.	3,145,585	4.0%	597,997	31.6%
Tex.	16,986,510	19.4%	6,694,830	37.2%	Ore.	2,842,321	7.9%	262,589	48.3%
Fla.	12,937,926	32.7%	3,462,600	52.3%	Ia.	2,776,755	-4.7%	112,915	24.8%
Pa.	11,881,643	0.1%	1,459,585	13.3%	Miss.	2,573,216	2.1%	949,018	3.5%
Ill.	11,430,602	0.0%	2,880,394	14.5%	Kan.	2,477,574	4.8%	287,050	27.5%
Oh.	10,847,115	0.5%	1,402,493	10.4%	Ark.	2,350,725	2.8%	417,643	2.7%
Mich.	9,295,297	0.4%	1,645,346	11.4%	W.Va.	1,793,477	-8.0%	74,581	-13.3%
N.J.	7,730,188	5.0%	2,011,222	30.7%	Utah	1,722,850	17.9%	151,596	37.1%
N.C.	6,628,637	12.7%	1,657,510	14.1%	Neb.	1,578,385	0.5%	118,290	25.2%
Ga.	6,478,216	18.6%	1,934,791	24.9%	N.M.	1,515,069	16.3%	750,905	21.7%
Va.	6,187,358	15.7%	1,485,708	27.3%	Me.	1,227,928	9.2%	24,571	30.7%
Mass.	6,016,425	4.9%	736,133	66.2%	Nev.	1,201,833	50.1%	255,476	90.5%
Ind.	5,544,159	1.0%	578,917	7.9%	N.H.	1,109,252	20.5%	29,768	97.1%
Mo.	5,117,073	4.1%	668,608	10.5%	Ha.	1,108,229	14.9%	760,585	14.4%
Wis.	4,891,769	4.0%	427,092	42.3%	Id.	1,006,749	6.7%	78,088	35.2%
Tenn.	4,877,185	6.2%	849,554	9.2%	R.I.	1,003,464	5.9%	107,355	71.8%
Wash.	4,866,692	17.8%	645,070	58.8%	Mont.	799,065	1.6%	65,187	24.7%
Md.	4,781,468	13.4%	1,455,359	32.2%	S.D.	696,004	0.8%	61,216	14.9%
Minn.	4,375,099	7.3%	273,833	71.7%	Del.	666,168	12.1%	138,076	24.2%
La.	4,219,973	0.3%	1,443,951	5.8%	N.D.	638,800	-2.1%	37,208	26.1%
Ala.	4,040,587	3.8%	1,080,420	4.1%	D. of C.	606,900	-4.9%	440,769	-7.0%
Ky.	3,685,296	0.7%	307,274	1.7%	Vt.	562,758	10.0%	10,574	39.4%
Ariz.	3,665,228	34.8%	1,039,043	50.2%	Alas.	550,043	36.9%	143,321	47.4%
S.C.	3,486,703	11.7%	1,096,647	10.8%	Wyo.	453,588	-3.4%	40,877	8.7%

Source: U.S. Bureau of the Census, U.S. Dept. of Commerce, 1990 Census   *Note: 'Minority' includes Black, Asian, Hispanics, and other races.

Where We Work and How Much We Earn

13

About 118 million of us had jobs in 1993, while about 8.7 million people who wanted a job were unemployed. About 55 million people age 25 and older were not in the workforce. About 41 million of the working women held down full time jobs. I was surprised to learn that the unemployment rate for men and women was identical. However, you were more likely to be unemployed if you were a teenager, were Afro-American or Hispanic, were a woman heading a family, didn't have a high school education, or were a construction worker. On the opposite end of the spectrum, managers and professionals had the lowest unemployment. Married men and women who lived with a spouse also had low unemployment rates.

HERE ARE OUR OCCUPATIONS BY AGE AND SEX:

Occupation (in thousands)	Total 16 years and over	Men 16 years and over	Women 16 years and over
Total	117,914	64,435	53,479
Managerial and professional specialty	30,657	16,619	14,038
Executive, administrative and managerial	14,839	8,897	5,843
Technical, sales and administrative support	36,675	12,933	23,742
Technicians and related support	3,842	1,954	1,888
Sales occupations	14,191	7,208	6,983
Administrative support, including clerical	18,041	3,771	14,870
Service occupations	16,759	6,288	9,470
Private household	782	29	753
Protective service	1,988	1,897	291
Service, except private household and protective	12,989	4,582	8,427
Precision production, craft, and repair	13,641	12,482	1,159
Mechanics and repairers	4,448	4,289	159
Construction trades	5,147	5,051	98
Other precision production, craft, and repair	4,046	3,142	904
Operators, fabricators, and laborers	17,775	13,249	4,528
Machine operators, assemblers, and inspectors	8,071	4,842	3,228
Transportation and material moving occupations	4,849	4,413	436
Handlers, equipment cleaners, helpers, and laborers	4,855	3,994	881

Source: Bureau of Labor Statistics, U.S. Dept. of Labor

What We Earn

The average annual salary was $23,227 per year. The average annual salary for men was $28,448, while women's average salary was $17,145. Part of the difference is that a much higher percentage of women worked part time, but women working full time still made only 75% of what men earned.

Education level was one key to higher earnings. High school graduates averaged $18,737 per year, $6,000 more than high school dropouts. A bachelor's degree hiked average earnings all the way to $32,629, while those with advanced degrees averaged $48,653.

On an hourly bases, about 4 million workers made $4.25 or under, 40 million made $10 per hour or less, and 22 million made more than $10 per hour.

The average weekly earnings were $445 per week. Here are the average weekly earnings by sex and occupation:

Occupation	Median weekly earnings	
	1991 Men	1991 Women
Managerial and professional specialty	741	519
Executive, administrative, and managerial	737	495
Professional specialty	744	548
Technical, sales, and administrative support	498	351
Technicians and related support	564	442
Sales occupations	499	311
Administrative support, including clerical	456	349
Service occupations	320	243
Private household	(1)	164
Protective service	494	436
Service, except private household and protective	279	244
Precision production, craft, and repair	488	354
Mechanics and repairers	472	541
Construction trades	478	(1)
Other precision production, craft, and repair	517	320
Operators, fabricators, and laborers	391	275
Machine operators, assemblers, and inspectors	403	272
Transportation and material moving occupations	421	320
Handlers, equipment cleaners, helpers, and laborers	316	269
Farming, forestry, and fishing	269	220

(1) Data not shown where base is less than 100,000.

Source: Bureau of Labor Statistics, U.S. Dept. of Labor

Who's Rich and Who's Poor

Who's Rich?

In 1992, 1,965,000 Americans had gross assets of more than $1 million—that's about 8 out of every 1,000 people. When debts were subtracted, about 65% had net worth of $1 million or more.

What Americans Own

The assets of all Americans totaled $15 trillion dollars in 1993. Below is a breakdown of those assets:

Asset	% of net worth
Physical assets	32.8%
Real estate	21.7%
Household good	8.3%
Automobiles	2.35
Financial assets	67.2%
Cash, bank accounts	15.2%
Stocks, bonds	23.1%
Equity in business	15.9%
Life insurance, pension funds	13.0%

Subtracted from the net assets are $2.7 trillion dollars in debt, including $1.45 trillion in home mortgages.

Where Does Our Money Come From?

With all the arguments about welfare and entitlement programs, it's enlightening to take a look at where our money comes from. About three-quarters of our income, on the average, comes from our salaries. But nearly half of all households (47%) contain at least one person receiving a government benefit. Over 33,000,000 people collect social security benefits, eight times as many people as receive welfare benefits, and social security benefits account for 7% of all the income earned by Americans.

However, about 80% of people earning under $7,200 per year receive money from the government, and those checks account, on the average, for 78% of their income. At the other end of the spectrum, only 22% of people earning $60,000 or more receive government money, and that accounts for just 3% of their income.

How Many Women Earn More Than Their Husbands?

Both husband and wife work full time in about 26,000,000 families. Of all working wives, about 18% earn more than their husbands, and another 8% earn at least 80% of their husbands' wages. The women who earned more tended to be under age 45, have a college degree, and have no children.

Here's the number of Americans who receive income from various sources:

Wages and salaries	116,983,000
Interest	100,240,000
Social security	33,006,000
Dividends	18,073,000
Self-employment	12,626,000
Rents	8,773,000
Private pensions	8,116,000
Unemployment compensation	7,046,000
Welfare	4,751,000
Supplemental security income	3,561,000
Child support and alimony	3,446,000
Mortgage interest	3,414,000
Royalties	2,900,000
Veterans pensions	2,807,000
State employee pensions	2,377,000
Federal pensions	1,433,000
Relatives or friends	1,106,000
U.S. military pensions	936,000
Annuities	729,000
Workman's compensation	666,000

What the Rich Own

The wealthiest 10% of Americans own:

51% of money deposited in banks
72% of all stock
70% of all bonds
86% of all tax-free bonds
50% of all property
78% of all Business

Who's Poor

The answer to the question above is likely to be children living with their mother only. Although 12.8% of the population has an income below the poverty level, that figure soars to 35.9% of families headed by a female. That figure jumps to 46% for families headed by black women.

	Number of poor in thousands	Percentage of population
Total poor	31,528	12.8
In families	24,066	11.5
Head	6,784	10.3
Related children	12,001	19.0
Other relatives	5,281	6.6
Unrelated individuals	6,760	19.2
In families with a female householder, no husband present	11,668	35.9
Head	3,504	32.2
Related children	6,808	51.1
Other relatives	1,356	16.3
Unrelated female individuals	4,221	22.2
All other	12,398	7.0
Head	3,280	5.9
Related children	5,193	10.4
Other relatives	3,925	5.5
Unrelated male individuals	2,539	15.7
Total white poor	20,785	10.0
In families	15,179	8.6
Head	4,409	7.8
Female	1,858	25.4
Related children	7,164	14.1
Other relatives	3,606	5.3
Unrelated individuals	5,063	16.9
Total black poor	9,302	30.7
In families	7,704	29.7
Head	2,077	27.8
Female	1,524	46.5
Related children	4,257	43.2
Other relatives	1,370	15.9
Unrelated individuals	1,471	35.2

Source: U.S. Bureau of the Census, Current Population Reports

PERSONAL INCOME PER CAPITA FOR SELECTED COUNTIES, 1992

	Highest per captia Income			**Lowest per capita Income**	
Rank/County	Per capita income 1992	Percent of national average	Rank/County	Per capita income 1992	Percent of national average
1. New York,N.Y	$49,197	244.7%	1.Starr Tex.	$ 6,015	29.9%
2. Sherman, Tex.	36,822	183.1	2. Sharnon,S.Dak.	6,826	34.0
3. Pitkin,Colo.	36,356	180.8	3. Maverick,Tex.	7,687	38.2
4. Marin,Calif.	36,076	179.4	4. Todd, S.Dak.	7,954	39.6
5. Fairfield,Conn.	35,423	176.2	5. Zavala,Tex.	8,094	40.3
6. Westchester,N.Y	34,843	173.3	6. Zapata,Tex.	8,743	43.5
7. Somerset,N.J.	34,580	172.0	7. Dimmit, Tex.	9,213	45.8
8. Sully S.Dak	33,851	168.4	8. Willacy Tex.	9,299	46.3
9. Bergen, N.J.	33,815	168.2	9. Jefferson Miss.	9,435	46.9
10. Morris, N.J.	33,616	167.2	10. McKinley N.Mex.	9,500	47.3
11 Montgomery, MD.	33,614	167.2	11. San Juan Utah	9,609	47.8
12. Arlington, Va.	32,872	163.5	12. Apache,Ariz.	9,623	47.9
13. Nantucket, Mass.	32,828	163.3	13. Presidio,Tex.	9,648	48.0
14. Alexandria City, Va.	32,761	162.9	14. McCreary,Ky.	9,655	48.0
15. Nassau, N.Y.	32,270	160.5	15. Cibola,N.Mex.	9,762	48.6
16. Teton, Wyo.	32,245	160.4	16. Hidalgo,Tex.	9,802	48.8
17. Montgomery, Pa.	31,747	157.9	17. Holmes Miss.	9,805	48.8
18. Fairfax, Fairfax City			18. WestFeliciana,La.	9,863	49.1
and Fall Church, Va.	31,204	155.2	19. Elliott,Ky	9,918	49.3
19. San Francisco, Calif.	30,942	153.9	20. Union Fla.	9,922	49.4
20. Palm Beach, Fla.	30,901	153.7	21. Morgan Ky	9,943	49.5
21. Hartley, Tex.	30,630	152.4	22. Claiborre,Miss.	9,946	49.5
22. Hunterdon, N.J.	30,139	149.9	23. Johnson,Tenn.	9,966	49.6
23. Martin, Fla.	30,005	149.2	24. Wade Hampton		
24. Hamilton, Kans.	29,969	149.1	Census Area, Alaska	9,993	49.7
25. San Mateo, Calif.	29,918	148.8	25. Conejos,Colo.	10,043	50.0

Source: U.S. Bureau of Economic Analysis

Where We're Born and How We Die

The birth rate has been going up in the last decade as we experience a second baby boom. Death rates, on the other hand, have been going down. Here's some facts about where we're born, where we die, and how we die.

Births and Deaths by States and Regions

Area	Live births Number	Rate	Deaths Number	Rate
New England	204,145	15.5	202,027	15.5
Maine	16,211	13.1	16,842	13.8
New Hampshire	16,927	15.0	17,946	16.2
Vermont	8,045	14.0	7,920	14.0
Massachusetts	95,066	16.0	96,457	16.3
Rhode Island	15,666	15.6	15,302	15.3
Connecticut	52,230	16.1	47,560	14.7
Middle Atlantic	594,883	15.7	578,025	15.3
New York	302,084	16.8	291,145	16.2
New Jersey	120,654	15.5	116,619	15.1
Pennsylvania	172,145	14.2	170,261	14.1
East North Central	673,457	15.8	649,459	15.4
Ohio	165,546	15.1	162,793	14.9
Indiana	85,202	15.1	82,764	14.8
Illinois	192,545	16.4	189,129	16.2
Michigan	157,674	16.9	142,673	15.4
Wisconsin	72,490	14.8	72,100	14.8
West North Central	275,609	15.3	267,762	15.0
Minnesota	68,353	15.5	66,593	15.3
Iowa	39,595	13.9	39,241	13.8
Missouri	83,085	16.0	80,126	15.5
North Dakota	10,483	16.0	10,862	16.5
South Dakota	10,912	15.2	10,991	15.4
Nebraska	24,317	15.0	24,317	15.1
Kansas	38,864	15.4	35,632	14.2
South Atlantic	705,114	16.1	678,784	15.7
Delaware	11,728	17.1	11,492	17.1
Maryland	75,557	15.9	67,550	14.4
West Virginia	23,202	12.6	23,079	12.4
District of Columbia	21,912	36.8	22,461	37.2

Virginia	96,665	15.6	93,453	15.3
North Carolina	105,230	15.8	102,817	15.6
South Carolina	56,521	15.9	55,214	15.7
Georgia	114,818	17.6	109,905	17.1
Florida	199,481	15.3	192,813	15.2
East South Central	244,572	15.8	231,994	15.1
Kentucky	56,753	15.2	52,591	14.1
Tennessee	77,821	15.6	76,780	15.5
Alabama	66,935	16.2	60,360	14.7
Mississippi	43,063	16.4	42,263	16.1
West South Central	483,507	17.8	451,625	16.7
Arkansas	35,499	14.7	34,997	14.5
Louisiana	71,913	16.5	68,813	15.7
Oklahoma	46,119	14.3	46,455	14.4
Texas	329,976	19.2	301,360	17.7
Mountain	242,892	17.7	235,645	17.4
Montana	11,482	14.2	11,394	14.1
Idaho	16,418	16.0	15,459	15.2
Wyoming	6,517	13.9	6,491	13.7
Colorado	53,238	16.0	52,863	15.9
New Mexico	28,252	18.3	27,324	17.9
Arizona	68,701	18.9	67,609	19.0
Utah	37,175	21.6	36,208	21.2
Nevada	21,109	18.1	18,297	16.5
Pacific	772,564	19.7	705,189	18.4
Washington	77,034	15.8	73,261	15.4
Oregon	45,851	15.9	43,835	15.5
California	617,704	20.7	557,003	19.2
Alaska	11,506	21.8	11,545	21.9
Hawaii	20,469	18.1	19,545	17.6

Source: National Center for Health Statistics.

Who's Having Babies?

Despite all the publicity about older women having babies, the average age of women who gave birth in 1992 was 25.9 years, only 5 months older than the average age in 1960. the percentage of women having babies who were in their 30s actually dropped from 26.6% in 1960 to 26.2% in 1992. However, college-educated women are waiting later—44% have their first babies in their 30s.

Of all babies born:

42% were first babies
33% were second babies
15% were third babies
6% were fourth babies
4% were fifth or more babies

In 1960, just 26% of all babies born were first babies. You may be surprised to learn that just 8% of women age 40 to 44 had not given birth to a child.

The 15 Leading Causes of Death

	Cause	Number	Death rate	% of total deaths
	All causes	2,162,000	861.9	100.0
1.	Diseases of heart	725,010	289.0	33.5
2.	Malignant neoplasms,			
	including. neoplasms of lymphatic & hematopoietic tissues	506,000	201.7	23.4
3.	Cerebrovascular diseases	145,340	57.9	6.7
4.	Accidents & adverse effects	93,550	37.3	4.3
	Motor vehicle accidents	47,880	19.1	2.2
	Other accidents & adverse effects	45,680	18.2	2.1
5.	Chronic obstructive pulmonary			
	diseases & allied conditions	88,980	35.5	4.1
6.	Pneumonia & influenza	78,640	31.3	3.6
7.	Diabetes mellitus	48,840	19.5	2.3
8.	Suicide	30,780	12.3	1.4
9.	Homicide & legal intervention	25,700	10.2	1.2
10.	Chronic liver disease & cirrhosis	25,600	10.2	1.2
11.	Human immunodeficiency virus infection (AIDS)	24,120	9.6	1.1
12.	Nephritis, nephrotic syndrome, & nephrosis	20,860	8.3	1.0
13.	Septicemia	19,750	7.9	0.9
14.	Certain conditions originating in perinatal period	17,520	7.0	0.8
15.	Atherosclerosis	16,490	6.6	0.8
	All other causes	295,100	117.6	13.6

Source: National Center for Health Statistics

The Median Age of the U.S. Population 1820-2030

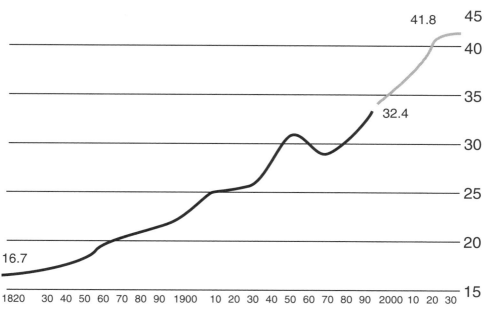

Source: The U.S. Bureau of the Census

Here is the death rate per 100,000 Americans for various causes in 1990:

Cause of death	1990
All causes	861.9
Viral hepatitis	0.7
Tuberculosis, all forms	0.7
Septicemia	7.9
Syphilis	0.0
All other infectious and parasitic diseases	12.8
Malignant neoplasms, including neoplasms of lymphatic and hematopoietic tissues	201.7
Diabetes mellitus	19.5
Meningitis	0.5
Major cardiovascular diseases	366.9
Diseases of heart	289.0
Hypertension	3.7
Cerebrovascular diseases	57.9
Atherosclerosis	6.6
Other diseases of arteries, arterioles, and capillaries	9.6
Acute bronchitis and bronchiolitis	0.2
Influenza and pneumonia	31.3
Influenza	0.8
Pneumonia	30.6
Chronic obstructive pulmonary diseases	35.5
Chronic and unspecified bronchitis	1.3
Emphysema	6.6
Asthma	1.8
Ulcer of stomach and duodenum	2.5
Hernia and intestinal obstruction	2.2
Cirrhosis and chronic liver disease	10.2
Cholelithiasis, cholecystitis, and cholangitis	1.2
Nephritis, nephrosis and nephrotic syn.	8.3
Infections of kidney	0.4
Hyperplasia of prostate	0.1
Congenital anomalies	5.3
Certain conditions in perinatal period	7.5
Symptoms, signs, ill-defined conditions	10.5
All other diseases	69.4
Accidents	37.3
Motor vehicle accidents	19.1
Suicide	12.3
Homicide	10.2
All other external causes	0.9

Source: Natl. Center for Health Statistics, U.S. Depart. of Health and Human Services

Where Our Children Live

16

Only 4.2% of children lived with a never-married parent in 1960; in 1970, the percentage rose to 6.8; by 1980, to 14.6, and as of 1990, it had climbed to 30.6, according to the Bureau of the Census. Babies born to single women in 1988 the latest year for which such statistics were available represented 26% of all American newborns, the highest proportion ever; and most of the mothers were at least 20 years old, according to the Federal Centers for Disease Control. The sharpest increases in birthrates from 1980 to 1988 were among older women, with the birthrate for single women age 15 to 17 rising 29% and the birthrate for women age 30 to 34 rising 52%. The rate among single black women was 89 births per 1,000, compared with 27 for single white women. However, the rate among white women grew 51% from 1980 to 1988, while the rate among black women increased only by 7%. Children living with one divorced parent increased from 23.0% in 1960 to 30.2% in 1970, and 42.4% in 1980; between 1980 and 1990 there was a decline to 38.6%. Children living with 2 parents in 1960 represented 87.7%; in 1970, 85.2%; in 1980, 76.7%; and in 1990, 72.5%.

Baby Boom And Bust:1940-90
(Births per 1,000 U.S. woman aged 15-44)

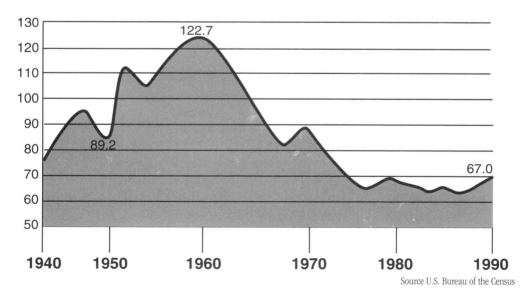

Source U.S. Bureau of the Census

Living Arrangements of Children, 1960-1990

Numbers in thousands. Excludes persons under age 18 who were maintaining households or family groups.

Living arrangements all races	Number of children
Children under 18	64,137
Living with	
Two parents	46,503
One parent	15,867
Mother only	13,874
Father only	1,993
Other relatives	1,422
Nonrelatives only	346
African-American	
Children under 18	10,018
Living with	
Two parents	3,781
One parent	5,485
Mother only	5,132
Father only	353
Other relatives	654
Nonrelatives only	98
Hispanic	
Children under 18	7,174
Living with	
Two parents	4,789
One parent	2,154
Mother only	1,943
Father only	211
Other relatives	177
Nonrelatives only	54

Source: Bureau of the Census

One Person added every 14 Seconds

According to the Census Bureau,
the U. S. population is growing by
approximately 6,300 persons daily.
Here's how: each day there are
10,600 births and 6,200 deaths,
resulting in a natural increase of
4,400; the other 1,900 persons
are added through immigration.
In 1990 the U.S. had a net gain of one
person every 14 seconds.This figure is based
on one birth every eight seconds, one death every
14 seconds, on immigrant every 35 seconds, on
one person leaving the country every three minutes.

Marriage, Divorce and Life Alone

Are more people getting married these days? Are more getting divorced? Who never marries? Who lives alone? For the answers, read on.

Marriages, Divorces, and Rates.

Year	Marriages No.	Rate	Divorces No.	Rate
1895	620,000	8.9	40,387	0.6
1900	709,000	9.3	55,751	0.7
1905	842,000	10.0	67,976	0.8
1910	948,166	10.3	83,045	0.9
1915	1,007,595	10.0	104,298	1.0
1920	1,274,476	12.0	170,505	1.6
1925	1,188,334	10.3	175,449	1.5
1930	1,126,856	9.2	195,961	1.6
1935	1,327,000	10.4	218,000	1.7
1940	1,595,879	12.1	264,000	2.0
1945	1,612,992	12.2	485,000	3.53
1950	1,667,231	11.1	385,144	2.6
1955	1,531,000	9.3	377,000	2.3
1960	1,523,000	8.5	393,000	2.2
1965	1,800,000	9.3	479,000	2.5
1970	2,158,802	10.6	708,000	3.5
1975	2,152,662	10.0	1,036,000	4.8
1980	2,413,000	10.6	1,182,000	5.2
1985	2,425,000	10.2	1,187,000	5.0
1990	2,448,000	9.8	1,175,000	4.7

Source: National Center for Health Statistics

When You Marry For the First Time

In 1890, the estimated median age at first marriage was 26.1 years for men and 22.0 years for women. At that time, a decline in the median age at first marriage began that did not end until 1956, when the median reached a low of 20.1 years for women and 22.5 years for men. The 66-year decline was reversed between 1956 and 1990, as the median returned to the 1890 level of 26.1 years for men and an even higher 23.6 median for women.

Year	Men	Women
1890	26.1	22.0
1900	25.9	21.9
1910	25.1	21.6
1920	24.6	21.2
1930	24.3	21.3
1940	24.3	21.5
1950	22.8	20.3
1955	22.6	20.2
1960	22.8	20.3
1965	22.8	20.6
1970	23.2	20.8
1975	23.5	21.1
1980	24.7	22.0
1985	25.5	23.3
1989	26.2	23.8
1990	26.1	23.9

Source: Bureau of the Census

Who Lives Alone

Between 1970 and 1990, the number of women living alone increased by 91%, and the number of men living along increased by 156%. As of 1990, more than 1 in every 9 adults age 15 and over lived alone, 61% of them women and the number of men fast increasing.

(in thousands) Age	1970	1980	1990
Both sexes	10,851	18,296	22,999
Men	3,532	6,966	9,049
15 to 24 years	274	947	674
25 to 34 years	535	1,975	2,395
35 to 44 years	398	945	1,836
45 to 54 years	513	804	1,167
55 to 64 years	639	809	1,036
65 to 74 years	611	775	1,042
75 years and over	563	711	901
Median age	55.7	40.9	42.5
Women	7,319	11,330	13,950
15 to 24 years	282	779	536
25 to 34 years	358	1,284	1,578
35 to 44 years	313	525	1,303
45 to 54 years	790	901	1,256
55 to 64 years	1,680	2,000	2,044
65 to 74 years	2,204	3,076	3,309
75 years and over	1,693	2,766	3,924

Source: Bureau of the Census, U.S. Dept. of Commerce

Who Never Marries

Between 1970 and 1990 the proportion of 20- to 24-year-olds who had never married increased from 36 to 63% for women and from 55 to 79% for men. The proportion never married for 25- to 29-year-olds tripled for women (from 11 to 31%) and more than doubled for men (from 19 to 45%). Although the proportion never married declines as age increases, significant increases have occurred for persons in their thirties. Between 1970 and 1990, the proportion never married for persons 30- to 34- years old nearly tripled; among those 35- to 39- years old, the proportion never married doubled.

	Women 1960-1990				Men 1960-1990			
	1960	1970	1980	1990	1960	1970	1980	1990
Total, 15 years and over	17.3	22.1	22.5	22.8	23.2	28.1	29.6	29.9
Under 40 years	28.1	38.5	38.8	40.5	39.6	47.7	48.8	50.6
40 years and over	7.5	6.2	5.3	5.3	7.6	7.4	5.7	6.4
15-17 years	93.2	97.3	97.0	98.5	98.8	99.4	99.4	99.8
18 years	75.6	82.0	88.0	92.0	94.6	95.1	97.4	98.5
19 years	59.7	68.8	77.6	88.7	88.7	89.9	90.9	95.2
20-24 years	28.4	35.8	50.2	62.8	53.1	54.7	68.8	79.3
20 years	46.0	56.9	66.5	76.6	75.8	78.3	86.0	90.8
21 years	34.6	43.9	59.7	71.1	63.4	66.2	77.2	87.6
22 years	25.6	33.5	48.3	63.0	51.6	52.3	69.9	80.1
23 years	19.4	22.4	41.7	55.5	40.5	42.1	59.1	72.2
24 years	15.7	17.9	33.5	48.7	33.4	33.2	50.0	66.7
25-29 years	10.5	10.5	20.9	31.1	20.8	19.1	33.1	45.2
25 years	13.1	14.0	28.6	41.9	27.9	26.6	44.3	57.7
26 years	11.4	12.2	22.7	33.8	23.5	20.9	36.5	52.4
27 years	10.2	9.1	22.2	29.6	19.8	16.5	31.5	43.7
28 years	9.2	8.9	16.0	27.3	17.5	17.0	26.8	36.7
29 years	8.7	8.0	14.6	23.5	16.0	13.8	24.0	36.2
30-34 years	6.9	6.2	9.5	16.4	11.9	9.4	15.9	27.0
35-39 years	6.1	5.4	6.2	10.4	8.8	7.2	7.8	14.7
40-44 years	6.1	4.9	4.8	8.0	7.3	6.3	7.1	10.5
45-54 years	7.0	4.9	4.7	5.0	7.4	7.5	6.1	6.3
55-64 years	8.0	6.8	4.5	3.9	8.0	7.8	5.3	5.8
65 years and over	8.5	7.7	5.9	4.9	7.7	7.5	4.9	4.2
Median age	66.1	65.6	65.8					

Source: Bureau of the Census

What You Own 18

Americans have always been concerned with keeping up with the Jones. And who are Mr. and Mrs. Jones? Statistically, Mrs. Jones is 35 years old, stands 5'4" tall, weighs 143 pounds, wears a size 12 dress, has a bra size of 34-36 B, and slips into 7-1/2 shoes. Mr. Jones is almost two years younger, stands 5'9" tall, weighs 173 pounds, wears a 40 regular suits, a shirt with a 15-1/2" neck and a 33" sleeve, and puts on 9-1/2 shoes.

I've compiled some information about what, according to the numbers, you own.

Your Automobiles

At last count, there were about 185,000,000 registered motor vehicles in this country, or slightly more than one for each of our 175,000,000 licensed drivers. We put an average of 11,766 miles on each vehicle, consuming 698 gallons of gasoline. The average age of these vehicles was about 8 years.

The 5.6 million new cars purchased in 1993 were of the following types:

Small 32.8%
Midsize 43.5%
Large 11.1%
Luxury 12.1%

The top ten bestselling cars in 1993 were:

Ford Taurus 360,448
Honda Accord 330,030
Toyota Camry 299,737
Chevrolet Cavalier 273,617
Ford Escort 269,034
Honda Civic 255,579
Saturn 229,356
Chevrolet Lumina 219,683
Ford Tempo 217,644
Pontiac Grand Am 214,761

Your Home

About 64 %of you own your own home, and of those 60 million dwellings, 35 million still carry a mortgage. The average mortgage payment is $722 per month, and the median value of all homes is $89,000. If you rent, you'll be interested to know that the average monthly payment is $469.

We Americans are such mobile people that only one in five of us has lived in the same residence for more than 15 years. The average person moves 11 times in a lifetime, and 20% move during the course of any given year. A little over 60% of Americans live in the state in which they were born. The most settled population is New Yorkers, 96% of whom have lived in the Empire State for the last five years. In contracts, only 41% of Alaskans have lived there for five years or more.

Your Appliances and Electronic Equipment

You might be surprised that only 84% of American homes have refrigerators, but 88% have microwave ovens! Three-quarters of you can wash your clothes at home, but only 53% of you own a clothes dryer.

All but 2% of homes have television sets, with an average of 2.2 per home. But the most ubiquitous piece of electronic equipment is the radio, with an average of 5.5 sets per household.

Here's the percentage of you that own appliances and electronic equipment:

Appliances & electronic equipment and percentage of ownership	
Color television	96%
Black and white television	31%
Clothes washer	76%
Range	58%
Oven (regular)	88%
Microwave oven	79%
Refrigerator	84%
Separate freezer	35%
Dishwasher	45%
Dehumidifier	12%
Air conditioner	
Central	39%
Window	39%
Window or ceiling fan	51%
Whole house fan	10%
Personal computer	16%
Gas appliances	
Range	42%
Oven	41%
Clothes dryer	16%
Outdoor gas grill	26%

Your Pets

There are furry or finny or feathery creatures in 57% of American households. About 37% of homes have a dog, and 26% own a cat, but there are more cats than dogs because the average cat owner has two felines. Americans also own 27 million birds and one-quarter billion fish.

Here is a list of the ten most popular dog breeds:

1. Laborador Retrievers
2. Rottweilers
3. German Shepherds
4. Cocker Spaniels
5. Golden Retrievers
6. Poodles
7. Beagles
8. Dachshunds
9. Dalmatians
10. Shetland Sheepdogs

Religion In Our Life

Your Religion

The 1991 Yearbook of American and Canadian Churches reported a total of 147,607,394 members of religious groups in the U.S. 59.3% of the population; membership rose 1.5% from the previous year. While it's easy to see which groups have the most members, exact comparisons are difficult, because some churchs count membership from birth, while others don't consider children full members until they reach age 13.

Group	Members
Adventist churches:	
Advent Christian Ch. (346)	25,400
Primitive Advent Christian Ch. (9)	350
Seventh-day Adventists (4,193)	701,781
American Rescue Workers (20)	2,700
Anglican Orthodox Church (40)	6,000
Baha'i Faith (1,700)	110,000
Baptist churches:	
Amer. Baptist Assn. (1,705)	250,000
Amer. Baptist Chs. in U.S.A. (5,833)	1,549,573
Baptist General Conference (792)	133,742
Baptist Missionary Assn. of America (1,339)	229,315
Conservative Baptist Assn. of America (1,126)	210,000
Duck River (and Kindred) Assn. of Baptists (85)	8,632
Free Will Baptists (2,517)	204,489
Gen. Assn. of Regular Baptist Chs. (1,582)	216,468
Natl. Baptist Convention of America (11,398)	2,668,799
Natl. Baptist Convention, U.S.A. (26,000)	5,500,000
Natl. Primitive Baptist Convention (616)	250,000
No. Amer. Baptist Conference (276)	42,629
Progressive National Baptist Convention (655)	521,692
Seventh Day Baptist General Conference (86)	5,200
Southern Baptist Convention (37,739)	14,907,826
Brethren (German Baptists):	
Brethren Ch. (Ashland, Ohio) (126)	13,155
Church of the Brethren (1,079)	151,169
Fellowship of Grace Brethren (319)	39,481
Brethren, River:	

Brethren in Christ Ch. (189)	17,240
Buddhist Churches of America (67)	19,441
Christadelphians (850)	15,800
The Christian and Missionary Alliance (1,829)	265,863
Christian Catholic Church (6)	2,500
Christian Church (Disciples of Christ) (4,113)	1,052,271
Christian Churches and Churches of Christ (5,579)	1,070,616
Christian Congregation (1,456)	108,881
Christian Methodist Episcopal Church (2,340)	718,922
Christian Nation Church U.S.A. (5)	200
Christian Union (114)	6,000
Churches of Christ (13,375)	1,626,000
Churches of Christ in Christian Union (250)	9,674
Churches of God:	
Chs. of God, General Conference (350)	33,909
Ch. of God (Anderson, Ind.) (2,338)	199,786
Ch. of God (Seventh Day), Denver, Col. (140)	5,000
Church of God in Christ (9,982)	3,709,661
Church of Christ, Scientist (3,000)	
Church of God by Faith (106)	4,500
Church of the Nazarene (5,158)	561,253
Conservative Congregational Christian Conference (180)	28,413
Eastern Orthodox churches:	
Albanian Orth. Diocese of America (2)	714
American Carpatho-Russian Orth. Greek Catholic Ch. (71)	20,000
Antiochian Orth. Christian Archdiocese of No. Amer. (160)	350,000
Diocese of the Armenian Ch. of America (66)	450,000
Bulgarian Eastern Orth. Ch. (13)	86,000
Coptic Orthodox Ch. (42)	165,000
Greek Orth. Archdiocese of N. and S. America (535)	1,950,000
Orthodox Ch. in America (440)	1,000,000
Patriarchal Parishes of the Russian Orth. Ch. in the U.S.A. (38)	9,780
Romanian Orth. Episcopate of America (37)	65,000
Serbian Eastern Orth. Ch. (68)	67,000
Syrian Orth. Ch. of Antioch (Archdiocese of the U.S.A. and Canada) (28)	30,000
Ukrainian Orth. Ch. of America (Ecumenical Patriarchate) (27)	5,000
Ukrainian Orthodox Church in the U.S.A. (107)	87,745
The Episcopal Church in the U.S.A. (7,372)	2,433,413
Reformed Episcopal Church in America (78)	6,274
American Ethical Union (Ethical Culture Movement) (21)	3,212
Evangelical Church (185)	16,113
Evangelical Congregational Church (155)	34,779
The Evangelical Covenant Church (592)	89,014
Evangelical Free Church of America (1,040)	165,000
Evangelical associations:	
Apostolic Christian Chs. of America (80)	11,450
Apostolic Christian Ch. (Nazarean) (48)	2,799
Friends:	
Evangelical Friends Alliance (217)	24,095
Friends General Conference (505)	31,690

Friends United Meeting (543)	54,155
Grace Gospel Fellowship (50)	4,500
Independent Fundamental Churches of America (705)	73,809
Jehovah's Witnesses (9,141)	825,570
Jewish organizations:	
Union of Amer. Hebrew Congregations (Re- form) (839)	1,300,000
Union of Orthodox Jewish Congregations of America (1,000)	1,000,000
United Synagogue of America (Conservative) (850)	2,000,000
Latter-day Saints:	
Ch. of Jesus Christ (Bickertonites) (63)	2,707
Ch. of Jesus Christ of Latter-day Saints (Mormon) (9,049)	4,175,400
Reorganized Ch. of Jesus Christ of Latter Day Saints (1,048)	190,183
Lutheran churches:	
Ch. of the Lutheran Brethren of America (114)	12,625
Ch. of the Lutheran Confession (69)	8,738
Evangelical Luthern Church in America (11,067)	5,238,798
Evangelical Lutheran Synod (125)	21,544
Assn. of Free Lutheran Congregations (193)	26,870
Latvian Evangelical Lutheran Church of America (56)	12,865
Lutheran Ch.-Missouri Synod (5,990)	2,609,025
Protestant Conference (Lutheran) (9)	1,065
Wisconsin Evangelical Lutheran Synod (1,198)	419,312
Mennonite churches:	
Beachy Amish Mennonite Chs. (99)	8,872
Evangelical Mennonite Ch. (26)	4,026
General Conference of Mennonite Brethren Chs. (128)	17,065
General Conference Mennonite Church (1,023)	92,682
The General Conference Mennonite Ch. (220)	33,982
Hutterian Brethren (77)	3,988
Mennonite Ch. (1,034)	92,517
Old Order Amish Ch. (785)	70,650
Old Order (Wisler) Mennonite Ch. (36)	9,731
Methodist churches:	
African Methodist Episcopal Ch. (6,200)	2,210,000
African Methodist Episcopal Zion Ch. (6,060)	1,220,260
Evangelical Methodist Ch. (130)	8,282
Free Methodist Ch. of North America (1,066)	75,869
Fundamental Methodist Ch. (13)	733
Primitive Methodist Ch., U.S.A. (85)	8,244
Reformed Methodist Union Episcopal Ch. (18)	3,800
Southern Methodist Ch. (137)	7,572
United Methodist Ch. (37,514)	8,979,139
Missionary Church (290)	26,332
Moravian churches:	
Moravian Ch. Northern Province (100)	31,248
Moravian Ch. in America Southern Province (55)	21,341
Unity of the Brethren (26)	4,336
Moslems	6,000,000+
New Apostolic Church of North America (497)	37,201
North American Old Roman Catholic Church (133)	62,611

Old Catholic churches:
Christ Catholic Ch. (12) 1,394
Pentecostal churches:
Apostolic Faith (Portland, Ore.) (50) 4,100
Assemblies of God (11,192) 2,137,890
Bible Church of Christ (6) 6,500
Bible Way Church of Our Lord Jesus Christ World Wide (350) 30,000
Church of God (Cleveland, Tenn.) (5,763) 582,203
Church of God of Prophecy (2,119) 73,430
Congregational Holiness Ch. (174) 8,347
Gen. Council, Christian Ch. of No. Amer. (104) 13,500
Intl. Ch. of the Foursquare Gospel (1,404) 203,060
National Gay Pentecostal Alliance (2) *
Open Bible Standard Chs. (330) 46,000
Pentecostal Assemblies of the World (550) 4,500
Pentecostal Church of God (1,165) 90,870
Pentecostal Free-Will Baptist Ch. (141) 11,757
United Pentecostal Ch. Intl. (3,592) 500,000
Polish Natl. Catholic Church of America (162) 282,411
Presbyterian churches:
Associate Reformed Presbyterian Ch. (Gen. Synod) (182) 38,274
Cumberland Presbyterian Ch. (743) 90,906
Evangelical Presbyterian Ch. (160) 54,781
Orthodox Presbyterian Ch. (188) 19,094
Presbyterian Ch. in America (1,600) 217,374
Presbyterian Ch. (U.S.A.) (11,489) 2,886,482
Reformed Presbyterian Ch. in No. Amer. (68) 5,174
Reformed churches:
Christian Reformed Ch. in N. America (712) 225,699
Hungarian Reformed Ch. in America (27) 9,780
Protestant Reformed Chs. in America (21) 4,544
Reformed Ch. in America (928) 330,650
Reformed Ch. in the U.S. (34) 3,778
The Roman Catholic Church (23,500) 57,019,948
The Salvation Army (1,022) 445,566
The Schwenkfelder Church (5) 2,461
Social Brethren (40) 1,784
Natl. Spiritualist Assn. of Churches (120) 3,406
Gen. Convention, The Swedenborgian Church (50) 2,423
Unitarian Universalist Assn. (1,010) 182,211
United Brethren:
Ch. of the United Brethren in Christ (252) 25,462
United Christian Ch. (12) 420
United Church of Christ (6,388) 1,625,969
Universal Fellowship of Metropolitan Community Chs. (195) 22,296
Vedanta Society (13) 2,500
Volunteers of America (607) 36,634
The Wesleyan Church (1,650) 110,027

SECTION FOUR

America's States and Territories

Recently I had dinner at the house of some friends who have a nine year old son. The boy's fourth grade class was scheduled to take a geography test the next day, but his parents were having a lot of trouble getting him away from the table and upstairs to study the state capitals. Finally, with a big sigh, he stood up and grumbled, "I wish I lived in England—they only have one capital."

I probably felt the same way when I was in fourth grade; and I'm certain that I muttered similar sentiments when I first became a weatherman. You see, that huge map of the United States which appears behind me (and behind every other weatherperson) has no state names on it. And there were many times during my first weeks as a weatherman when I hesitated for fear that I would point to Kansas and call it Nebraska, or point to Wyoming and identify it as Colorado. In order to avoid terminal embarrassment — and unwanted phone calls from viewers all too eager to point out my mistakes — I gave myself a crash course in geography.

And after I refreshed my memory about where all the states are, I realized again how lucky we are to have them. The very brave men, women, and children who first settled this country left their native lands to venture into the unknown largely because of a central government (usually run by a king) that didn't tolerate their religious or political beliefs. Fortunately, the New World was large enough for many different groups to form communities—Puritans in Massachusetts, Baptists in Rhode Island, Quakers in Pennsylvania, etc. Unfortunately, as the colonies grew, the British crown demanded more and more control. With no where left to go, colonists of every ilk united and rebelled. As we all know, the British were defeated and our independence was won.

Having accomplished the task, it's not surprising that the colonists were unwilling to place their fate in the hand of a strong, new central government. It's not surprising that it took five years after the peace treaty was signed to draft a Constitution that contained enough protections to satisfy the majority. The key : the Constitution vested great powers in the individual states. And it required that no provisions could be added to the Constitution without the states' approval. The degree of sovereignty that rests in the states is one major difference between our government and those of most European countries.

Eventually, the states developed their individual identities. I can attest to that, because I have had the pleasure of visiting 49 of the 50 states (and by the time you read this, I probably will have visited all 50). To my delight, I have found that each state has a distinctive geography, history, culture, and economy. And the more I've learned about the states, the more those lines on my weather map have come to life. For example, that curvy, snaky line which serves as a boundary for some 10 states no longer looks to me like the work of a jittery cartographer — I know that it was carved into the land by a mighty river called the Mississippi. I've learned that some borders were drawn in straight lines to settle boundary disputes, while others reflect cultural and political differences existing at the time of their drawing. Now, when I look at my map, I see places and events and people that tell a story.

In the following pages, I've done my best to encapsulate the important facts about each state, as well as the major U.S. territories. I hope this spurs you to further travel and study.

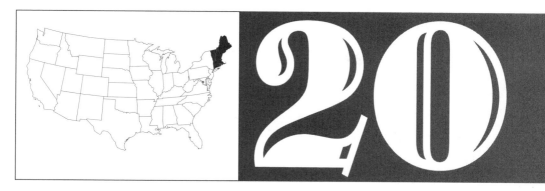

NEW ENGLAND

Comparing the New Engalnd States

STATE	POPULATION	SQUARE MILES	PER CAPITA INCOME
National rankings in parenthesis	(1993 U.S. Bureau of the Census)	(U.S. Geological Survey)	(U.S. Bureau of Labor Statistics)
Massachusetts	6,012,268 (13)	10,555 (44)	$17,224 (6)
Connecticut	3,277,216 (27)	5,544 (48)	$20,289 (1)
Maine	1,239,448 (39)	39,387 (39)	$12,957 (31)
New Hampshire	1,125,310 (41)	9,351 (46)	$15,959 (9)
Rhode Island	1,000,012 (43)	1,545 (50)	$14,989 (14)
Vermont	575,691 (49)	9,615 (45)	$13,987 (23)

Leaves crunching under foot and blowing in a balmy breeze as I stroll across a village green towards a white-steepled church—that's what comes to mind when I hear the words "New England." I also think about the waves breaking on the rocky coasts of Maine, skiers plummeting down the slopes of Vermont, or wise old codgers sitting around a wood stove in a small town general store.

New England calls forth different images because it is an area of considerable geographic diversity. But what binds it together is a common culture based on "old fashioned" values such as hard work, thrift, and Yankee ingenuity. Settlers in New England found harsh weather and relatively infertile soil. To survive, they made themselves into merchants and manufacturers, and succeeded spectacularly well—today, New England is still the wealthiest area of the country. And that same independence that allowed them to survive made them staunch opponents of British rule. New England was the birthplace of democracy in the New World and the cradle of the American Revolution.

EARLY EXPLORATION OF NEW ENGLAND

Both Henry Hudson and Giovanni da Verrazano sailed by New England. The first French and British explorers came in the early 1600s, and in 1614, Captain John Smith sailed by an area so inviting that he named it "New England."

Shortly afterward, the first settlers came, driven from England by religious persecution. And it was religious persecution practiced by those first settlers that drove others to colonize Connecticut, Rhode Island and other parts of the area.

My on-the-road assignments for *Good Morning America* have taken me to virtually every corner of New England, and I am continually amazed by the harmonious blend of the old and the new — the cities are as commercially vibrant as those of any other region, while the small towns and rural areas have largely retained their old world character and charm.

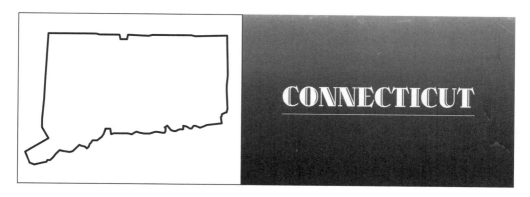

CONNECTICUT

Connecticut is the southernmost of the New England states. It is bordered on the north by Massachusetts, the east by Rhode Island, the south by Long Island Sound, and the west by New York. The third smallest state, Connecticut runs 90 miles east-west and 75 miles north-south. The coastline is dotted with many small offshore islands.

Hartford, the state capital, is in the center of the state about 110 miles northeast of New York City on I-684/84 and about 110 miles southwest of Boston on I-90/84. Bridgeport, the largest city, is on the shoreline 54 miles northeast of New York City and 125 miles southwest of Providence on I-95.

THE LAY OF THE LAND

The name Connecticut comes from an Indian word meaning "place beside the long river." That river is the Connecticut, which begins in Canada and flows southward to Long Island Sound, bisecting the state and forming the Connecticut Valley. This valley, 25 to 35 miles wide, contains very fertile soil.

In the northwest corner of Connecticut is a section of the Taconic Mountains known as the Northwest Hills or the Litchfield Hills. To the south is an expanse of rolling hills that is part of the Western New England Uplands. The hills begin to level off to the Coastal Lowlands, flat, sometimes marshy terrain that extends the entire length of the state's shoreline on Long Island sound. Beyond the Connecticut Valley to the east is an area of Eastern New England Uplands, which are covered with hills somewhat lower than the Western Uplands.

Numerous lakes and boulders seemingly dropped from the sky at random are evidence that this land was carved by glaciers during the recent ice ages. The state has two major rivers in addition to the Connecticut: the Housatonic, which empties into Long Island Sound at Bridgeport, and the Thames, which empties into the sound at New London.

HOW CONNECTICUT BECAME A STATE

The Connecticut River Valley was first visited by the Dutch explorer Adriaen Bloch in 1614, and occasional fur traders worked the territory over the next two decades. Massachusetts claimed control over the area, into which its residents began to move. In 1635, Edmund Winslow established a community named Windsor on the Connecticut River, and shortly afterwards Thomas Hooker founded Newtown near present day Hartford.

In 1636, Massachusetts gave Connecticut self-government. In 1662, the residents of the state obtained a royal charter that gave the state all the land west to the Pacific Ocean! That charter was obviously too ambitious, but Connecticut did continue to control a settlement in Pennsylvania's Wyoming Valley until 1754.

With their long tradition of independence, Connecticut residents didn't want to be included in James I's Dominion of New England, in which New England, New York, and New Jersey were combined in 1685. Ordered to surrender the state's charter, angry rebels hid the document in the trunk of an oak tree that became famous as the Charter Oak.

The Dominion ended when James I lost his crown in the Glorious Revolution of 1688. But the state's tradition of independence continued. Anti-British sentiment ran strong in the state long before the Revolutionary War. Connecticut militiamen fought several battles against the British and Connecticut sailors hampered British shipping off shore. At the post war Constitutional Convention, it was the Connecticut Compromise (which put forth a two part legislature with equal state representation in the Senate and proportional representation in the House) that led to the passage of the Constitution. Connecticut entered the Union as the fifth state on January 9, 1788.

CONNECTICUT'S CITIES

HARTFORD (139,739) was born when a minister named Thomas Hooker left the Massachusetts Bay Colony with his congregation to form what he named a "New Towne" on the Connecticut River. Samuel Stone was Hooker's assistant pastor, and the name of the settlement was changed in honor of Hertford, England, Stone's birthplace. In 1639, Hartford became part of Connecticut, and in 1662, it became the capital. The city's location made it a prominent trading center. Hartford was co-capital of the state with New Haven from 1701-1873. Today, the city is known as major center of the insurance industry. Other major industries include metal products, firearms, and aerospace technology.

BRIDGEPORT (141,686) was founded in 1639 as a fishing village. It was named for a draw bridge in its harbor. The city was the birthplace and residence of the famous American showman P.T. Barnum, and Bridgeport recently completed a new museum built in his honor. Manufacturing is the heart of the economy now, with machine parts, fabricated metals, helicopters, and electronic products the major industrial products.

NEW HAVEN (130,474) was founded by English Puritans at the mouth of the Quinnipiac River. It was named either because it was a "new harbor" or after Newhaven, England. New Haven was governed as an independent country until 1665, when it joined the rest of the state. In 1716, a college that had begun as the Collegiate School in another community in 1701 moved to New Haven and took the name of its benefactor, English merchant Elihu Yale. The city thrived in Colonial America as a center of whaling and the West Indian trade. Although its best known today as the home of Yale, it has a diverse manufacturing economy that includes printing and publishing, firearms, rubber boots, paper, and chemicals.

CONNECTICUT'S PEOPLE

Connecticut's 3,277,316 population ranks 27th, which makes it one of the more densely populated states. Almost four out of five Connecticut residents live in urban areas. Whites constitute 87% of the population, while blacks make up 8.3% The people are proud of being politically progressive and independent; the state elected an independent governor running against a Republican and a Democrat for the 1991-1995 term.

CONNECTICUT'S ECONOMY

Connecticut residents enjoy the highest per capita incomes of any state. Partly because so many executives in New York-based companies lived in Connecticut and faced long commutes to work, a significant number of corporations relocated to the state over the last three decades.

Agriculture plays a much smaller role in today's economy than it did in the state's early years. Connecticut was an early tobacco-growing state, and the leaf is still cultivated. However, the major farm products are nursery plants, Christmas trees, vegetables, corn, and hay.

Defense spending has been a cornerstone of the state's manufacturing industries over the last few decades. Connecticut makes helicopters, nuclear submarines, aircraft parts and engines, and it's also the home of the U.S. Coast Guard Academy. Other major industries include firearms, ball bearings, hand tools, optical equipment, brass, clocks, chemicals, pharmaceuticals, and processed foods. Hartford is a major insurance center and Bridgeport has an important banking community.

It's relatively uncrowded beaches, quaint towns, and tourist attractions like the restored 19th century shipping village Mystic Seaport brought nearly $3 billion to the state's coffers in 1993.

Places to Visit

Connecticut offers a wide range of places to see and visit. One of it's newest attractions is Mystic Seaport (203-572-0711), a recreated New England fishing village that is America's largest maritime museum. Mystic Seaport has also served as the location for one of my many GMA "remotes". Right down the road is the Mystic Marinelife Aquarium (203-536-3323). And if your nautical appetite is still not satisfied, you can tour the U.S. Nautilus, the first nuclear sub, at the Submarine Force Museum (203-449-3174) in New London. Among other major attractions are Yale's Peabody Museum of Natural History in New Haven (203-432-5050), Hartford's Wadsworth Atheneum (203-247-9111), Norwalk's Maritime Center (800-243-2280), and the Foxwoods Resort and Casino (203-885-3000) in Ledyard. Tourist Information: Connecticut Department of Economic Development, 865 Brook St., Rocky Hill, CT 06067 (800-282-6863/203-258-4200).

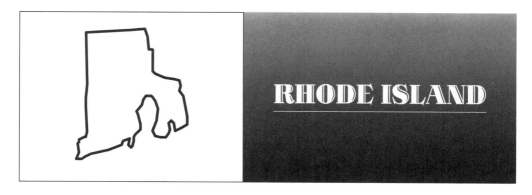

RHODE ISLAND

"Little old" Rhode Island, the nation's smallest state, covers an area just twice the size of the city of Jacksonville, Florida. The state, rectangular in shape, is bordered by Massachusetts on the north and east, on the south by Rhode Island Sound, and on the east by Connecticut. The state's dimensions are 47 miles north-south and 40 miles east-west. It's 40 miles of coastline are dotted by 35 islands, the largest of which is Block Island.

Providence, the capital and largest city, is 209 miles northeast of New York City and 50 miles southwest of Boston on I-95.

THE LAY OF THE LAND

Rhode Island has just two types of landforms. The area near Rhode Island Sound is part of the Seaboard Lowlands. Beaches and salt marshes give way to low hills and sandy plains. The soil is not very fertile. Farther inland is a section of the Eastern New England Uplands, an area of rolling hills dotted with small lakes.

If you look at Rhode Island on a map, its most distinguishing physical feature is Narragansatt Bay, which cuts into the state like a wedge. The bay contains several islands, and its numerous inlets give the state 384 miles of shoreline. Rhode Island also has numerous small lakes, ponds, and fast-flowing rivers.

HOW RHODE ISLAND BECAME A STATE

Giovanni da Verrazano sailed into Narragansatt Bay in 1524, but the first known settlement was founded at Valley Falls by William Blackstone in 1635. Roger Williams, who was expelled by the Massachusetts Bay Colony for criticizing its religious inflexibility, founded Providence in 1636 on land he purchased from the Narragansatt Indians. In 1639, Williams and his followers started America's first Baptist church. Portsmouth, Newport, and Warwick were also founded between 1638 and 1643.

Massachusetts claimed sovereignty over the area, but Williams traveled to England and in 1643 was given a charter for these communities, which called themselves the Providence Plantation. Williams established a government based on freedom of religion, and in the 1650s came a significant number of Quakers and Jewish immigrants from Barbados. In 1663, King Charles II granted a charter for the entire area. The name Rhode Island came from either the Dutch word for "red" (referring to the area's red soil) or from the Greek island of Rhodes.

Conflict with the Indians was finally resolved when Rhode Island troops helped the colonists win King Philip's War in 1676. Rhode Island became a thriving commercial center for whaling, naval stores, molasses, preserved meats and cider. British trade regulations in the 1760s led to widespread smuggling, creating an atmosphere of great tension. In 1772, a group of Rhode Islanders lured the British customs ship Gaspée to shore and burned it. On May 5, 1775, Providence burned a huge mound of captured tea. Almost a year later, on May 4, 1776, representatives gathered to declare Rhode Island the first free republic in the New World.

Much of the state was occupied during the Revolutionary War. Afterwards, the very independent state was reluctant to agree to the restrictions imposed on self rule by the proposed constitution. On May 29, 1790, Rhode Island was the last of the original 13 colonies to join the Union. The smallest state has the longest official name—"The State of Rhode Island and Providence Plantations."

RHODE ISLAND'S CITIES

PROVIDENCE (160,728) was founded in 1636 on land purchased by Roger Williams. He named it in gratitude for God's "providence" in providing his followers with a new home. The city became not only a major port and trading center, but the home of some of America's first industries. It became the sole capital of the state in 1900. The city became a major textile center, and in 1824 experienced what may have been this country's first strike when women weavers walked off the job. Today, Providence is the leading center of the costume jewelry business and is also well known for its silverware.

NEWPORT (28,227), which is located on an island in Narragansatt Bay, was founded in 1639 as a "new port" by Antinomians expelled from Massachusetts. Significant numbers of Quakers and Sephardic Jews (Newport is the site of the nation's first synagogue) arrived in the 1650s. The city became a major trading center and was co-capital of the state until 1900. In the nineteenth century, the city became America's most fashionable resort for the ultra-rich, and magnificent homes such as The Breakers, The Elms, Chateâu-sur-Mer, and Marble House, are open to the public today. Newport is the home of the America's Cup sailing race, the U.S. Naval War College, and the International Tennis Hall of Fame and Museum.

WARWICK (85,423), which is on Narragansatt Bay just south of Providence, was founded by Samuel Gorton in 1642. Six years later it was renamed for Robert Rich, 2nd earl of Warwick. It is a manufacturing center for machinery, jewelry, primary and fabricated metals, printed materials, and the insurance and health industries.

RHODE ISLAND'S PEOPLE

Rhode Island's 1,000,012 represent a rich ethnic mix. The state has a tradition for welcoming immigrants, and even today, nearly 10% of the population is foreign-born. Because of its small size, Rhode Island is densely populated and is one of America's most urbanized states.

RHODE ISLAND'S ECONOMY

Rhode Island was one of the birthplaces of America's industrial revolution in the eighteenth century, and it remains today one of the most industrialized states. Textiles and jewelry were traditionally the leading industries. Today, they are joined by fabricated metals, precision instruments, printed materials, rubber and plastic goods, industrial machinery, primary metals, electronic goods, transportation equipment, chemicals, and processed foods.

The size of the state precludes large farms. However, the state does grow vegetables, potatoes, sod, hay, and poultry.

Rhode Island, once a whaling center, still has a significant fishing fleet. Tourists drawn to the state's beautiful beaches account for nearly $2 billion in annual revenues.

Places to Visit

In Newport (800-326-6030), one of America's oldest seaports, you can tour fabulous mansions, visit the International Tennis Hall of Fame, bask on beautiful beaches, head out to sea from the greatest sailing city in the world or enjoy one of the world's greatest and most famous Jazz festivals each summer. Block Island (401-466-2982), 13 miles off the coast, features freshwater ponds, beaches, and narrow roads perfect for strolling. Tourist information: Rhode Island Tourism Division, 7 Jackson Walkway, Providence, RI 02903 (800-556-2484/401-277-2601)

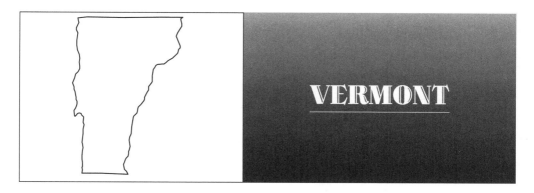

VERMONT

Vermont is the only New England state without a seacoast, but it more than makes up for it with spectacular scenery that makes it obvious why it was named for the French phrase "vert mont," or "green mountain." Vermont is bordered on the north by the Canadian province of Québec, on the east by New Hampshire, on the south by Massachusetts, and on the west by New York. The Connecticut River forms the entire eastern boundary and the western boundary passes through Lake Champlain. The state is rectangular in shape, running 160 miles north-south and 85 miles east-west.

Burlington, the largest city, is 102 miles south of Montreal, Canada on I-89 and about 210 miles north of Springfield, Massachusetts on I-91/89. Montpelier, the capital, is 39 miles southeast of Burlington on I-89.

THE LAY OF THE LAND

The Green Mountains are the backbone of the state, running north to south through the center of the state in a band 25 to 40 miles wide. These mountains average 1,000 feet in elevation, have been eroded to nearly flat summits, and are heavily forested. In the northeast corner of the state are a portion of the White Mountains, which are higher and more rugged. In the southeast is a slice of the ancient Taconic Mountains. The narrow Valley of Vermont separates the Green Mountains and the Taconic Mountains.

The rest of the state consists primarily of Eastern New England Uplands, hilly country cut by fast-flowing streams. The land is relatively flat bordering the Connecticut River and Lake Champlain.

Lake Champlain is the fourth largest lake entirely within the boundaries of the United States, having an area of 431 square miles and a length of 121 miles. Vermont has numerous small streams and rivers and many small lakes.

HOW VERMONT BECAME A STATE

Samuel de Champlain explored the lake that bears his name in 1609. In 1666, the French, led by Capt. de la Motte, built a fort on La Motte Island in Lake Champlain. To counter the French, the British built a fort at Chimney Point. In 1724, the British built the first permanent settlement, Fort Drummer, on the site of the present-day Brattleboro in the southeast corner of the state.

For most of the eighteenth century, Vermont was the focal point of a power struggle between New Hampshire and New York. In 1741, King George II appeared to issue a charter making Vermont part of New Hampshire, and settlers with land grants issued by that state poured in. In 1764, King George III set New Hampshire's western border at the Connecticut River, giving the territory to New York. New York promptly filed suits to eject holders of New Hampshire grants from their land.

The dispute escalated to armed conflict when a settler from Connecticut named Ethan Allan organized a group he called the Green Mountain Boys to prevent New York from implementing its evictions. But when war broke out with the British, Ethan Allan joined his Green Mountain Boys with the forces of General Benedict Arnold to capture Fort Ticonderoga on May 10, 1775, giving the Americans one of their first victories.

However, Vermont also took the opportunity to assert its right to self-government. Vermont governed itself as an independent nation beginning in 1777, although it wasn't officially recognized by the Continental Congress or New York State. Vermont even entered negotiations to become part of Canada, although that was partly a ploy to gain leverage in its negotiations to become a separate U.S. state. New York finally gave up its claims and Vermont entered the Union as the fourteenth state on March 4, 1791.

VERMONT'S CITIES

BURLINGTON (39,127), on the shores of Lake Champlain, was settled in 1773 and named for the Burling family, prominent local settlers. The city was the home of Ethan Allen, who died here in 1789. Burlington was an American military base and the site of several skirmishes in the War of 1812. Today, it's leading products are electrical equipment, steel and wood items, maple syrup, business machines, and textiles.

MONTPELIER (8,247), on the Winooski River, has the lowest population of any state capital. The city was founded in the 1780s and named for Montpelier, France. It became the state capitol in 1805, and the Greek Revival style State Capitol was built in 1859. Montpelier is the center of an agricultural and granite-quarrying region.

RUTLAND (18,230) is on Otter Creek in the center of a year-around resort area. The site was settled in 1770 and probably named for Rutland, Massachusetts. It was the site of the state legislature from 1784-1804. Rutland has been the center of a marble-quarrying area since the 1820s.

VERMONT'S PEOPLE

With just 575,691 people, Vermont is the most sparsely populated Eastern state. It is also one of the most rural, with only 32% of its population living in urban areas. Vermont has the second smallest percentage of black residents (.35%). The state's residents have a reputation for being independent and outspoken on the subject of human liberty. Its passionate opposition to slavery caused one Georgia congressman to suggest that the state be cut off from the rest of the country and allowed to drift off to sea. Vermont was so outraged by Nazi atrocities that it declared war on Germany before the United States Government did.

VERMONT'S ECONOMY

Vermont has traditionally had an agricultural economy. Because it lacks broad expanses of fertile soil, 82% of its farm income comes from livestock and livestock products, especially milk, beef cattle, eggs, and poultry. The leading crops are hay, apples, and maple syrup.

Marble and granite have been quarried in the state for at least a century and a half, and the state is also a producer of asbestos and talc.

In the manufacturing section, Vermont is famous for its wood burning stoves. Other leading industries are electronic equipment, industrial machinery, printed materials, fabricated metals, and stone and wood goods.

Vermont's ski slopes are an important reason that the state earns nearly $2 billion in tourist income annually.

Places to Visit

For many people, Vermont means great skiing at such famous areas as Mt. Snow, Stratten, Bromley, Stowe, Smuggler's Notch, and Killington (for ski conditions, call 802-229-0531) But the small towns that dot the Green Mountains are also perfect summer and fall destinations. Lake Champlain (802-863-3489) features beaches, water sports, and a history dating back to 1609. And even if you don't participate in water sports, you can treat yourself to a visually spectacular experience by simply taking a ferry ride across Lake Champlain, which I have done many times. Tourist information: Vermont Department of Travel and Tourism, 134 State St., Montpelier, VT 05603 (802-828-3236)

NEW HAMPSHIRE

New Hampshire is appropriately known as "The Granite State," not only because of its many stone mountains but because of the hardiness of its people. This New England state is bordered on the north by the Canadian province of Québec, on the east by Maine and the Atlantic Ocean, on the South by Massachusetts, and on the west by Vermont. The Connecticut River forms the entire western boundary, while the Piscataqua River is part of the southeastern border. The state has 13 miles of coastline on the Atlantic. The state runs 180 miles north-south and 90 miles east-west.

Manchester, the largest city, is 58 miles northwest of Boston on I-93 and 102 miles southwest of Portland, Maine on Route 101 and I-95. Concord, the state capital, is 17 miles north of Manchester and 148 miles southeast of Burlington, Vermont on I-89.

THE LAY OF THE LAND

When I think of New Hampshire, I think of mountains—the White Mountains, to be specific. The reason is that in the Presidential Range of these mountains is Mount Washington, the windiest location in the United States and possibly in the entire world. In fact, it is not uncommon in winter to experience wind gusts of 120-160 miles per hour. I know, I've been there. It's also high and cold enough to have the only Alpine climate in the eastern United States. The rest of this rugged terrain is periodically cut by gaps in the mountains known as notches. In Franconia Notch, wind and water has carved a giant stone face known as the Old Man of the Mountain.

The southwestern portion of the state is part of the New England Uplands. Plugs of granite more resistant to erosion than the surrounding rocks has produced isolated mountains such as Mt. Monandnock—which in turn provides the name for these kinds of peaks (monandnocks). In southeast New Hampshire is a flat stretch of the Seaboard Lowlands.

The glaciers left the state with many lakes, the most famous of which is Lake Winnepesakee. Over 83% of New Hampshire is covered by forests.

HOW NEW HAMPSHIRE BECAME A STATE

Martin Pring in 1603 and Samuel de Champlain in 1605 both sailed up the Piscataqua River. John Smith, founder of Jamestown, also visited the area in 1614. In 1620, the area was granted to the Council of New England, formerly the Plymouth Company. In turn, they granted the area, which included present day New Hampshire and Maine, to Sir Ferninando Gorges and John Mason. Settlements were founded at present day Portsmouth and Dover, both on the Piscataqua River, in 1623.

In 1629, the Province of Maine was divided, with the New Hampshire part being given to John Mason. But in 1641, all the settlements in the area were taken under the administrative wing of Massachusetts. It wasn't until 1679 that John Mason's son was able to persuade the British government that Massachusetts had usurped the land, and it was made a separate royal province. It was named for Hampshire, England. However, from 1699 to 1741, the province was governed by the royal governor of Massachusetts.

During this time, New Hampshire waged several bitter boarder disputes: one with Massachusetts, the second with New York over control of Vermont, and the third with Québec. The boundary dispute with Massachusetts was resolved in 1741 when New Hampshire became separately governed. Vermont became an independent territory during the Revolutionary War. But the dispute with Canada wasn't settled until 1840, when famous New Hampshire native Daniel

Webster negotiated the Webster-Ashburton Treaty.

New Hampshire's state motto, "Live free or die," eloquently summed up its attitude towards the British colonial government. On January 5, 1776, New Hampshire became the first state to set up a provisional independent government. Although no battles were fought in New Hampshire, the state sent large numbers of troops to assist in the revolution. New Hampshire officially "created" the United States on June 21, 1788, when it became the ninth state to ratify the Constitution—nine states had to join the Union before it was officially a nation.

NEW HAMPSHIRE'S CITIES

MANCHESTER (99,567) was settled in 1722 on the Merrimack River at the Amoskeag Falls. It was incorporated in 1751 as Derryfield, and it became a fishing and lumbering town. In 1810, the first cotton mill was opened and the city was renamed for Manchester, England. Textiles were the foundation of the economy until 1835, when the largest company in town went bankrupt. However, the city successfully diversified.

CONCORD (36,006), also on the Merrimack River, was established as a trading post in 1659. In 1725, it became the Penacook Plantation, then it was Rumford in 1733. In 1741, both New Hampshire and Massachusetts claimed the area. In 1762, it was awarded to New Hampshire and named Concord for the unanimity with which the residents sought to escape Massachusetts control. It became the capital of New Hampshire in 1808. It's economy was based on the printing industry and the manufacture of stagecoaches—the Concord coach helped conquer the West.

NASHUA (79,662) was founded on the Merrimack and Nashua Rivers in 1650. In the mid-nineteenth century, textile mills were built. The city experienced a boom when it was connected to Boston via the Merrimack River and the Middlesex Canal. Today, it is attracting an increasing amount of high-tech industry moving from the Boston area.

NEW HAMPSHIRE'S PEOPLE

New Hampshire (1,125,310 pop.) is experiencing the fastest growth rate of the New England states as people seek a more tranquil environment than Boston and other major metropolitan areas. However, the state still has a decidedly rural flavor (just 51% of the population lives in urban areas). Traditionally, New Hampshire residents have demanded a voice in their government, and the state legislature has 400 members, the most of any state—proportionally, the U.S. Congress would have 100,000 members!

NEW HAMPSHIRE'S ECONOMY

With only 17% non-forested land, New Hampshire has little room for crops. However, the state does produce dairy products, eggs, beef cattle, apples, and maple syrup. Christmas trees are a major source of income, and the state has a significant logging income.

New Hampshire has been traditionally a major textile state. It has recently diversified into such industries as industrial machinery, precision instruments, electronic equipment, rubber and plastic goods, fabricated metal products, paper goods, and printed materials.

The White Mountains are the chief lure for tourists who annually spend in excess of $3 billion in the state.

Places to Visit

Among the highlights of a visit to the spectacular White Mountains are climbing to the summit of Mount Washington by car or cog railroad (800-367-3364) and the Old Man in the Mountain, a human profile carved by erosion into a cliff in Franconia Notch. New Hampshire's 18 mile long seacoast (800-221-5623) has historic towns and hard white sand beaches. Western New Hampshire is home to Dartmouth College and Mt. Monadnock (603-532-8035), one of the world's most climbed mountains. Tourist information; New Hampshire Office of Travel and Tourist Development, Box 856, Concord, NH 03302 (800-944-1117/603-271-2343)

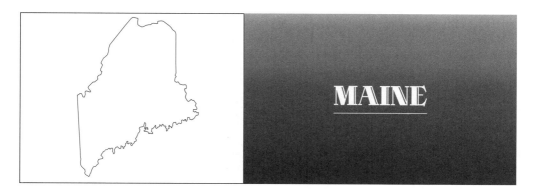

MAINE

Maine is a state of vast forests covering an area larger than the other five New England states combined. The state is bordered by the Canadian province of New Brunswick on the north and east, the Atlantic Ocean on the south, New Hampshire on the west, and the Canadian province of Québec on the northwest. The Saint John and Saint Francis Rivers make up part of the northern border, the Saint Croix River is part of the southeastern border , and the Salmon Falls River is part of the southwestern boarder. Maine's West Quoddy Head is the easternmost point of land in the United States. Maine has 225 miles of coastline dotted with thousands of islands, and that same coastline experiences the world's highest tides.

Portland, the largest city, is 115 miles north of Boston on I-95 and 102 miles northeast of Manchester, New Hampshire on I-95 and Route 101. Augusta, the state capital, is 50 miles north of Portland on I-95.

THE LAY OF THE LAND

For all its size, Maine has just two basic types of terrain. The Seaboard Lowland along the coast is cut with so many inlets that it's known as "the land of a hundred harbors." These harbors gave birth to numerous fishing villages. The coast features many sharp, jagged rocky headlands that were created by glaciers. Most of the offshore islands are the tops of submerged hills. The inland part of the Lowland consists of rolling hills.

The rest of the state is part of the New England Uplands. This vast forest is a rough, hilly landscape that generally has poor soil. One prominent feature is eskars, ridges of glacial gravel deposited by streams that ran underneath the glaciers.

Thanks again to those glaciers, Maine has more than 5,100 rivers and streams and 2,200 lakes. The largest is Moosehead Lake in the northwestern part of the state.

HOW MAINE BECAME A STATE

Pierre du Guast landed on Dochest Island and founded a community called Saint Croix in 1604, and a second settlement was built on Mount Desert Island. Both soon failed. In 1607, 120 settlers landed on Sagadahoc Peninsula at the mouth of the Kennebec River, and during the short life of their colony they built the Virginia, the first ship launched in the New World. In 1614, Captain John Smith visited the area. The land that included present day New Hampshire and Maine was granted to Sir Ferdinando Gorges and John Mason in 1620. It was called the Province of Maine, probably because it was the "mainland." The first permanent settlement, Monhagen, was built in 1622. Although trappers and loggers continued to work the area, Gorges and Mason did little to develop it.

In 1658, Massachusetts assumed jurisdiction over the Province of Maine, and it officially became a part of Massachusetts in 1691. The area was very sparsely settled, partly due to hostility from the Indians that lasted until the 1760s.

Logging was such an important industry to the residents that there was great resentment when King George III ordered that all tall trees belonged to the Crown for use as ship masts. In 1775, just 5 days before the Battle of Bunker Hill, the residents of Machiasport captured a British ship sent to collect the masts and killed the captain. This incident was the first naval battle of the Revolutionary War. Later British efforts to obtain masts also failed, hampering British naval operations.

As part of Massachusetts, Maine joined the Union in 1788. A movement to become a separate state began in 1785 but was really kicked into high gear in 1816 with the Brunswick Convention, which demonstrated how widespread the desire for independence was. The Massachusetts Supreme Court approved the Act of Separation in 1819, but Maine's petition to join the Union was held up in Congress by southern senators reluctant to admit a non-slave state. Finally, in 1820, northern and southern senators reached the Missouri Compromise, which admitted Maine as a free state and Missouri as a slave state. Maine became the 23rd state on May 15, 1820. However, the northern border with Canada was not settled until the signing of the Webster-Ashburton Treaty of 1942.

MAINE'S CITIES

PORTLAND (64,358) is a deep-water harbor on Casco Bay. The first settlement on the site was founded in 1632. Settlements here had several names and were destroyed by the Indians in 1676 and the French and Indians in 1690. In 1775, this area was part of Falmouth and was damaged by the British. Finally, in 1786, it was separated from Falmouth and named for Portland, England. The city became a fishing, shipping, and commercial center, and was the capital of the new state from its admission to the Union in 1820 to 1832. Portland is famous as the birthplace of the poet Henry Wadsworth Longfellow, whose home is now a museum.

AUGUSTA (21,235) was established as a trading post by the Plymouth Colony as early as 1628. In 1754, a fort was built on the site as a supply depot for British troops fighting the French and Indian War. Over the years, it was named Cushnoc, Kouissinoc, Hallowell, and Harrington. Finally, in 1797, it was named for Pamela Augusta Dearborn, daughter of Henry Dearborn, a Revolutionary War soldier and U.S. Representative. It became the state capital in 1832. Augusta produces computer products, paper products, and fabricated steel.

BANGOR (31,643) was founded on the Penobscot River in 1769, and it became the center of a great lumbering area. It was named by a minister for the "Bangor Hymn." Today, it remains the commercial and medical center for the timber-producing and resort areas of northern Maine. It is also home to the paper industry.

MAINE'S PEOPLE

Maine's 1,239,448 people contain the third smallest percentage of minorities of any U.S. state. Just 45% of the state's population lives in urban areas. The northern part of the state is very sparsely populated.Maine residents are also fiercely proud of their seafood. On a recent "live remote" at the annual ouster festival, I mentioned that I did not like shellfish. Well, I was told that I would not be allowed to leave the state without tasting the oysters. I tried them—I liked them!

MAINE'S ECONOMY

Fishing, lumbering, and trapping were the primary industries for most of Maine's history. The soil in the state is generally poor. Except for its famous white potatoes, Maine has a relatively small agricultural income. Dairy products, eggs, hay, and apples are the most significant cash products after potatoes.

Forestry remains a major industry in Maine, producing both pulp for paper and lumber. Fishing has also remained an economic staple, with lobsters the most valuable catch.

Manufacturing has become an increasingly important part of the economy of the state. Maine produced the first nuclear submarine, and ship building is still important. Other major industries are paper and paper products, footwear and leather goods, lumber, electrical equipment, processed foods, printed materials, and industrial machinery.

Places to Visit

Acadia National Park (207-288-3338), which encompasses Mount Desert Island and the historic town of Bar Harbor,. offers some of the most spectacular scenery in the entire U.S.—I loved watching the sunrise over the ocean from the summit of Cadillac Mountain. Maine's southern coast offers sandy beaches and historic towns like Kennebunk/Kennebunkport (203-967-0857) along with more than 200 factory outlet stores such as the famous headquarters of L.L. Bean (800-341-4341). Tourist Information: Maine Publicity Bureau, 325B Water St., P.O. Box 2300, Hallowell, ME (800-533-9595/207-623—363)

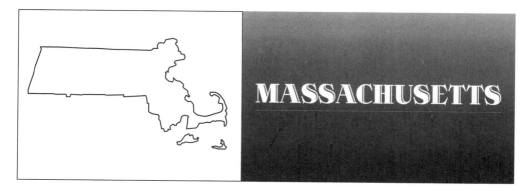

MASSACHUSETTS

Massachusetts, the cultural and financial center of New England, has a population almost as large as the other five New England states combined. It is bordered on the north by Vermont and New Hampshire, on the east and southeast by the Atlantic Ocean, on the south by Rhode Island and Connecticut, and on the west by New York. Massachusetts, which is shaped like a rectangle with a fish hook, has an average east-west length of 190 miles and an average north-south width of 50 miles. The state has 192 miles of coastline, and offshore are two famous resort islands, Nantucket and Martha's Vineyard.

Boston, the capital and largest city, is 222 miles northeast of New York and 115 miles south of Portland, Maine on I-95. It is 166 miles east of Albany on I-90. Worcester, the second largest city, is 42 miles southwest of Boston on I-90 and 63 miles northeast of Hartford on I-84.

THE LAY OF THE LAND

Massachusetts has six regions. Cape Cod, Nantucket, and Martha's Vineyard, are flat, sandy areas of the Atlantic Coastal Plain. The Seaboard Lowlands provide a transition from the mainland coast to the interior. The Massachusetts Lowlands have a large number of drumlins, elongated hills created by glaciers that include the famous Bunker Hill. Depressions in the lowlands are cultivated as cranberry bogs.

Most of the rest of the state part of the New England Uplands. This large area is bisected by the broad Connecticut Valley Lowlands, which has the states most fertile soil. Both sections of Uplands have undulating hills, but the eastern part tends to be less rugged than the western part.

Western Massachusetts has the lower end of the Green Mountains in the north and the northern end of the Taconic Mountains in the south. The two ranges are separated by the Berkshire Valley.

HOW MASSACHUSETTS BECAME A STATE

Bartholomew Gosnold became the first European to set foot on Massachusetts when he landed at Provincetown on Cape Cod in 1602. Capt. John Smith sailed by in 1614. On December 21, 1620, a ship named the Mayflower landed a group of Pilgrims at Plymouth Rock, and they founded the Plymouth Colony, the oldest settlement in the state. Despite losing half their population in the first winter, the Pilgrims established a self-governing unit based on the famous Mayflower Compact. In the late 1820s, Puritans began to arrive, and in 1830, John Winthrop led 1,000 more who established communities at Boston, Charleston, Lynn, Medford, Roxbury, and Watertown. This large group was the first wave of an influx called "The Great Migration." Although the Puritans and Pilgrims had fled religious intolerance in England, they practiced religious repression in the New World. In 1660, they hanged several Quakers and they drove many others into exile. They named the area Massachusetts, which comes from an Indian term for "great mountain place."

Efforts to expand inland were contested by the Indians. But armed conflicts in 1637 and 1676 removed most of the threat. The Massachusetts Bay Colony won control over the Province of Maine and attempted to exercise control over New Hampshire. In 1684, Massachusetts' charter was revoked by the Crown, but it was restored eight years later.

Boston was the center of resistance to such onerous British regulations as the Stamp Act and the Sugar Act in the 1760s. On March 5, 1770, British troops opened fire on colonial hecklers, killing five men. This Boston Massacre inflamed anti-British tensions. Three years later, Boston residents dumped tea into the harbor during

Massacre inflamed anti-British tensions. Three years later, Boston residents dumped tea into the harbor during the equally famous Boston Tea Party. In April, 1775, a local militia confronted the British on Lexington Green as they attempted to raid an arsenal. Later, in Concord, the "shot heard 'round the world" was fired that launched the Revolutionary War.

The Americans won their first big victory in June, 1775, in the Battle of Bunker Hill. In March, 1776, the British evacuated Boston and no further major battles were fought in the state. Massachusetts became the sixth state on February 6, 1788.

MASSACHUSETTS' CITIES

BOSTON (574,283) is a magnificent harbor at the mouth of the Charles River on Massachusetts Bay. The city was founded by Puritan John Winthrop in 1630 and named for Boston, England. It became the springboard for settlement of the New England interior. By 1750, it was the most important seaport and trading center in New England, and it later became the cradle of the American Revolution. Today, Boston is the commercial and cultural center of New England, as well as the higher education capital of the United States. Long known for its textile industry, Boston has become a mecca for research and high technology.

WORCESTER (169,759), on the Blackstone River, was founded as the Quinsegamond Plantation in 1623. In 1684, it was renamed for Worcester, England. It became an armory for British fighting against the Indians in 1707. Worcester's growth as a commercial and transportation center was accelerated in 1828 when it gained a water link with Providence, Rhode Island with the completion of the Blackstone Canal. Today, Worcester is a center of bio-tech research, and it has a diverse manufacturing base.

SPRINGFIELD (156,983) was founded on the Connecticut River in 1636 by William Pynchon. In 1640, it was named for Springfield, England. An arms depot was established in 1777. In 1786, a group of farmers suffering because of a post-Revolutionary War depression formed an armed uprising known as Shays Rebellion, and they unsuccessfully attempted to raid the arms depot. Springfield was best known for its Springfield rifles.

MASSACHUSETTS' PEOPLE

With 6,012,268 people, Massachusetts is the 13th most populous U.S. state. The state served as a gateway for immigrants in the eighteenth century, with the Irish having a particularly large impact on the area. Today, the state has a diverse ethnic mix. Massachusetts has always been predominately an urban state, and 84% of the population lives in urban areas today.

MASSACHUSETTS' ECONOMY

From its earliest days, the foundations of the Massachusetts economy were commerce, fishing, and shipbuilding. Agriculture was never especially important. Today, about three-quarters of the state's farm income comes from crops such as nursery plants, cranberries, hay, apples, tobacco, and potatoes. The state also has a number of dairy farms.

Today, Massachusetts is still the nation's second leading fishing state (after Alaska). Boston, Gloucester, and New Bedford are the major ports.

Massachusetts was a pioneer in the textile, shoe-making, and iron and steel industries. Today, these older industries have been joined by advanced research centers and high-tech electronic companies that have given the Route 128 corridor outside of Boston the nickname "the Silicon Valley of the East." Boston is also a major finance, medical and insurance center.

Places to Visit

You're walking in the footsteps of American heroes when you follow Boston's Freedom Trail, a 1.5 mile trek that includes the Old North Church and the Boston Common. Other attractions in Boston (617-536-4100) include the New England Aquarium, The USS Constitution, and the Museum of Science. Cape Cod (518-362-3225) has 70 miles of gorgeous beaches and resort communities, while Cape Ann (508-745-2268)features historic Salem and Gloucester, America's oldest seaport. The Tanglewood Music Festival (413-637-1940) highlights the summer season in western Massachusett's scenic Berkshire Mountains. And no trip to the Boston area would be complete without at least a distant view of the campus of historic Harvard University. Tourist information: Massachusetts Office of Travel and Tourism, 100 Cambridge St., Boston, MA 02202 (800-447-6277/617-727-3201)

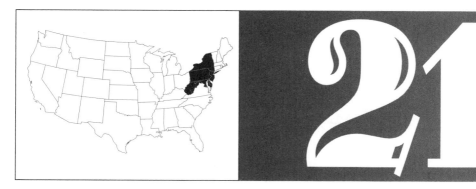

THE MID-ATLANTIC STATES

Comparing the Mid-Atlantic States

STATE	POPULATION	SQUARE MILES	PER CAPITA INCOME
National rankings in parenthesis	(1993 U.S. Bureau of the Census)	(U.S. Geological Survey)	(U.S. Bureau of Labor Statistics)
New York	18,197,154 (2)	54,471 (27)	$16,501 (7)
Pennsylvania	12,048,271 (5)	45,308 (33)	$14,068 (21)
New Jersey	7,879,164 (9)	8,722 (47)	$18,714 (3)
Maryland	4,964,898 (18)	12,407 (42)	$17,730 (4)
West Virginia	1,820,137 (35)	24,231 (41)	$10,520 (49)
Delaware	700,269 (46)	2,489 (49)	$15,854 (10)

The primary characteristic of the states we group as the Middle Atlantic States is that they are located on the east coast between New England and the South. Unlike those other two areas, the Middle Atlantic States don't share distinctive cultural characteristics—in fact, the words "middle Atlantic" tend to bring to my mind an image of a ship riding endless waves rather than a specific geographic area.

That doesn't mean, however, that the Middle Atlantic states aren't a vital and important region of our country. These six states have rich histories and, for the most part, vibrant economies. The Middle Atlantic area is home to three of the nation's ten most populous states, four of its ten most prosperous, two of its top five cities, and one of the top three ports. These states may also be the most geographically diverse in the United States, with some states having as many as seven distinct regions. These states border on great rivers and the Great Lakes, as well as the Atlantic Ocean and the St. Lawrence Seaway. They are home to ethnic groups as diverse as Hasidic Jews and the Pennsylvania Dutch.

Now, you may think I'm getting a little carried away, because I work in New York City and live in New Jersey. But after you read about these states, I think you'll be as fascinated by them as I am.

THE HISTORY OF THE MIDDLE ATLANTIC STATES

The first European to visit this area was Giovanni da Verrazano, a French navigator who was commissioned by King Francis I to sail to the New World to discourage Spanish settlement. In 1524, he arrived at a point off the coast of North Carolina and sailed north all the way to Nova Scotia. In the process, he became the first to enter New York Bay.

A rather mysterious figure named Henry Hudson did him one better. Nothing is known about Hudson until 1607, when an English company hired him to find a northern passage to Asia. He made two voyages for the English, penetrating far into the Arctic but failing to achieve the goal set for him. When the English fired him, he turned to the Dutch East India Company, which gave him a ship called the "Half Moon" with a half Dutch, half English crew. On this voyage, he sailed up the river now called the Hudson as far as present-day Albany, and entered the Chesapeake and Delaware Bays as well. After the success of this voyage, the English government insisted Hudson sail only for his homeland. However, his crew mutinied on his fourth voyage and he was set adrift in a lifeboat, never to return.

Hudson's famous third voyage set the stage for Dutch settlement of the Middle Atlantic area from New York south to Maryland. However, the Dutch were evicted by the British in 1664. From that point, the history of the Middle Atlantic States focuses on the British and their relationship with the colonial settlers.

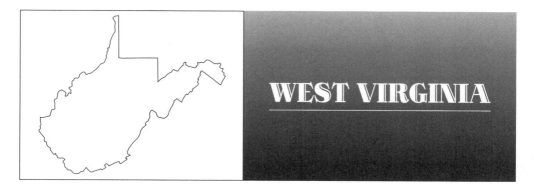

WEST VIRGINIA

West Virginia, which has been admitted to the Union twice, has a history as distinctive as its spectacular mountainous terrain. The state is bordered by Ohio on the northwest, Pennsylvania on the north, Maryland on the northeast, Virginia on the east and south, and Kentucky on the southwest. The Potomac River makes up part of the northeast border, the Tug Fork and Big Sandy rivers are part of the southwest border, and the Ohio River defines the northwest border with Ohio. West Virginia is roughly oval-shaped, with a larger panhandle on the northeast and a small northern panhandle so narrow that the city of Weirton runs from the Ohio to the Pennsylvania border. The extreme dimensions of the state are 265 miles east-to-west and 235 miles north-to-south.

Charleston, the capital, is 250 miles south of Cleveland on I-77, 374 miles southwest of Washington, D.C. on I-79, and 177 miles northeast of Lexington on I-64. Wheeling is 50 miles southwest of Pittsburgh and 137 miles northeast of Columbus, OH on I-70.

THE LAY OF THE LAND

It's not hard to figure out why West Virginia is nicknamed the Mountain State when you learn that it has the highest mean elevation of any state east of the Mississippi. A Valley and Ridge region in the east gives way to the rugged Allegheny Plateau, which covers more than two-thirds of the state. West Virginia has very few areas of fertile soil, but has rich mineral resources.

West Virginia has numerous rivers and streams but no large lakes.

HOW WEST VIRGINIA BECAME A STATE

West Virginia has many ancient burial mounds and earthworks built by the Adena culture, which flourished between 1000 b.c. and 100 b.c., then disappeared. The first European party to explore the area was led by Walter Austin 1641, and Thomas Batts and Robert Fallum followed in 1671. The first permanent settlement was established at Shepherdstown in 1732, and George Washington and Daniel Boone were among the others to probe farther into the mountains of an area that was part of the Virginia Colony.

The settlers who poured into the area in the 1750s and 1760s were very different from the leading citizens of the rest of Virginia. West Virginians were subsistence farmers, hunters, and craftsmen who were overwhelming anti-slavery and who resented a political system dominated by a handful of wealthy plantation owners. Before and during the Revolutionary War, West Virginians were occupied fighting the Indians and the British, and they were proud when the entire area of Virginia joined the Union.

However, tensions began to rise as slavery became the predominant political issue. In the late 1850s, the famous militant abolitionist John Brown established a camp for runaway slaves in the West Virginia mountains and gathered a small armed force that he saw as the nucleus of an anti-slavery army. In 1859, Brown led his men on an attack on the U.S. Army Arsenal and Harpers Ferry. The wounded leader was tracked down by troops led by Col. Robert E. Lee, tried for treason, and hanged at Charleston. This famous episode inflamed passions on both sides of the slavery issue.

West Virginians were outraged when Virginia seceded from the Union in 1861. They met at the Wheeling Convention to repudiate the secession and declare themselves the separate state of Kanawha (for the Kanawha River, which runs through Charleston). The name didn't last, but the area was admitted to the Union on June 20, 1863 as the 35th state.

WEST VIRGINIA'S CITIES

CHARLESTON (57,287) is in an area of southeastern West Virginia that Daniel Boone once represented in the Virginia Assembly. The settlement was built in 1788 by Col. George Clendenin and named Charles Town in honor of his father. In 1818, the name was shortened to Charleston. The city won the honor of being the state capital in 1788—as the story goes, a popular traveling circus performing in the city influenced the outcome. Charleston is a thriving commercial center today. A prime tourist attraction is the Sunrise Museums, built in two old mansions.

WHEELING (34,882), in the northern panhandle on the Ohio River, was the site of Fort Henry, built in 1769. As the western terminus of the Cumberland Road, it was the gateway to the west for thousands of settlers. The city was incorporated in 1836 and took its name from a Delaware Indian term for "place of the head," which reportedly refers to the severed head of a murdered settler. As the site of the Wheeling Convention that declared the area's independence from Virginia, it was the first state capital.

HUNTINGTON (54,844) is located at the junction of the Ohio River near the junction with the Big Sandy River. The site was built in 1871 as the western end of the Chesapeake and Ohio Railroad, and it was named for Collis P. Huntington, the railroad's president. Today, it's main industries are nickel alloys, railroad and mining equipment, glassware, and motor vehicle parts.

WEST VIRGINIA'S PEOPLE

West Virginia's population (1,820,137) dropped more than 8% between 1980 and 1990 (the largest decline of any state). One major reason for the decline is that many young people have been leaving the state because of limited employment opportunities in rural areas where two-thirds of the population lives. Losing residents of child-bearing age has resulted in West Virginia having the nation's lowest birth rate and lowest percentage of the population age 5 and under. About 96% of West Virginia residents are white.

WEST VIRGINIA'S ECONOMY

Agriculture has never been an important element of West Virginia's economy, as it has in almost all other eastern states. The state's relatively few farms produce primarily cattle, hogs, chickens, and dairy products.

Coal underlies two-thirds of the state, and mining has traditionally been West Virginia's leading industry. The state produces 15% of the nation's bituminous coal as well as significant amounts of natural gas and petroleum.

Small factories were producing glass, iron, and salt in West Virginia as early as the 1780s. Today, the state's leading industries are primarily those that make use of its natural resources, such as iron and steel plants, aluminum, chemicals, glass, pottery, wood products, and industrial machinery.

West Virginia has large trade and service industries, and the state has been the beneficiary of considerable federal spending and installations. Tourism also brings annual revenues of $2.5 billion.

Places to Visit

Harpers Ferry State Park (304-535-6223) is the center of a historic area in eastern West Virginia at the junction of the Shenandoah and Potomac Rivers. The waters of White Sulphur Springs, home of the world famous Greenbrier resort (800-624-6070), are renowned for their health value. The state has many scenic drives, such as the Highland Scenic Highway, the Highland Trace, and the Midland Trail, while many state parks have beautiful lodges and attractions such as ski slopes or golf courses. Tourist information: West Virginia Division of Tourism and Parks, 2101 Washington St. E., Charleston, WV 25305 (800-225-5982/304-348-2286)

DELAWARE

Delaware may be the second smallest U.S. state, but it has a rich history and a thriving economy that is the envy of many larger states. The state is located on the Delmarva Peninsula and is bordered on the north by Pennsylvania, the east by the Delaware River, Delaware Bay, and the Atlantic Ocean, and on the south and west by Maryland. Delaware has 28 miles of coastline. Its extreme dimensions are 96 miles north-to-south and just 36 miles east-to-west.

Wilmington, the largest city, is about 140 miles south of New York, 30 miles south of Philadelphia, and 70 miles north of Baltimore on I-95.

THE LAY OF THE LAND

Most of Delaware is on the Delmarva Peninsula, which separates Chesapeake Bay on the west from Delaware Bay and the Atlantic Ocean on the east. Part of the Atlantic Coastal Plain, Delaware is the nation's lowest lying state, with a mean elevation of just 60 feet above sea level. The state's highest point is an unnamed hill 442 feet above sea level. The northern tip of the state, which is separated from the rest by the Chesapeake & Delaware Canal, is slightly higher land on the Piedmont Plateau.

The state has no large lakes, but has large swamps in the southern part. Delaware has many smaller rivers and enjoys 381 miles of shoreline.

HOW DELAWARE BECAME A STATE

Henry Hudson sailed past the area in 1609, and a year later Samuel Angall entered the bay and named it for Thomas West, the 3rd baron de la Warr, the first Colonial governor of Virginia. The Dutch established a 1631 settlement that was destroyed by Indians. In 1638, Sweden hired Dutchman Peter Minuit to transport two boatloads of settlers to the area. They landed near present-day Wilmington at a place called "The Rock," which is referred to as the "Plymouth Rock of Delaware." This joint Swedish-Dutch settlement was named Fort Christina and the colony was called New Sweden.

In 1655, Peter Stuyvesant, governor of New Amsterdam, sent a Dutch force that ended Swedish control and renamed the colony New Amstel. In 1664, the Duke of York sent a large military force that took control of New Amsterdam (changing its name to New York). The English gained control of New Amstel as well. The area became a colony of New York with the capital at New Castle.

In 1682, the King of England included Delaware with Pennsylvania in a charter granted to William Penn. The new Pennsylvania government organized the area as the Low Counties, and retained control until the British government took the area as crown colony in 1704. Delaware now had its own legislature and was a thriving seaport and trading center.

Because of its prosperity, Delaware was more reluctant than many of the other colonies to challenge British rule. But Caesar Rodney, a Delaware delegate to the Continental Congress, got up from his deathbed to ride from Wilmington to Philadelphia to cast the deciding vote in favor of independence. Delaware troops participated in several important Revolutionary War battles. On December 7, 1787, Delaware became the First State by being the first state to ratify the new Constitution.

DELAWARE'S CITIES

WILMINGTON (71,529) was founded by Swedish and Dutch settlers as Christinahamn (later Christina) in 1638.

The area was seized first by the Dutch and then by the English, who placed it in the control of William Penn. A group of Quakers settled in the area and when it was officially laid out as a settlement in 1731, it was known as Willingham. In 1639, the name was changed to Wilmington in honor of Spencer Compton, the earl of Wilmington. In 1802, a man named du Pont founded a gunpowder mill outside of town, the beginning of a chemical empire. Today, Wilmington is a thriving industrial and financial services center located directly on the New York-Washington corridor.

DOVER (27,630), located on the Saint Johns River, was the site of a 1631 Dutch settlement that was wiped out by the Indians. William Penn laid out the town in 1717 and named it for Dover, England. Dover became the state capitol in 1777. Today, , the state government and nearby Dover Air Force base are the focus of the economy.

DELAWARE'S PEOPLE

Delaware's 700,269 people are a cosmopolitan mix that includes a significant Asian and Hispanic population. Even though Delaware is such a small state, 27% of its people live in rural areas. Delaware has one of the nation's highest percentages of people in the workforce.

DELAWARE'S ECONOMY

As headquarters of E.I. du Pont de Nemeurs & Co., Delaware is one of the nation's leading producers of chemicals and chemical products. Other major industries include synthetic textiles, motor vehicle pats, processed foods, precision instruments, rubber and plastic good, and industrial equipment.

Very favorable state laws have made Delaware the legal home of many of the major U.S. corporations. These headquarters have spawned very large financial, legal, and business services industries.

Delaware's early economy was based on agriculture, and today the state's farms are the third most productive per acre in the U.S. Livestock, especially chickens and poultry products, are the major sources of income.

Finally, Delaware's attractive ocean beaches are the cornerstone of a lucrative tourism industry.

Places to Visit

You can't visit Wilmington (800-422-1181) without admiring the achievements of the DuPont Family. The Hagley Museum takes you back to the DuPont company's beginnings as a gunpowder factory in 1802, while the 196 room Winterthur Museum and the Nemours Mansion and Gardens display some what the DuPont wealth purchased. Rehobeth Beach (800-441-1329) is the focus of the Atlantic coastline that draws vacationers from all over the east coast. And for a trip back in time to colonial days, visit historic Georgetown Tourist information: Delaware Tourism Office, 99 Kings Hwy., Box 1401, Dover, DE (800-441-8846/302-739-4271)

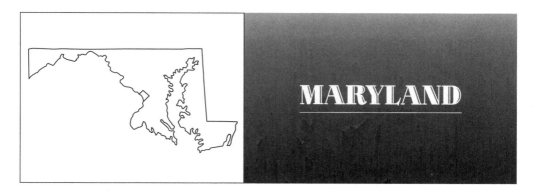

MARYLAND

Maryland is one of America's most oddly drawn states. Pennsylvania is the northern neighbor. The state's Eastern Shore region is an L-shaped section of the Delmarva Peninsula—Delaware and a 31 mile stretch of Atlantic Ocean coastline make up its eastern border, while the southern tip of the peninsula belongs to Virginia. Chesapeake Bay, which looks like a huge wedge of blue on the map, separates the Eastern and Western Shores. The Potomac River makes up the rest of the southern and western borders with Virginia and West Virginia. The District of Columbia is a 68 square mile enclave carved out of the southern part of the Western Shore. West of Hagerstown, a section of the state shaped like a bird's wing thrusts into West Virginia—at one point this section is 1 mile wide, the narrowest of any state. This "bird wing" gives Maryland an extreme 200 mile east-west length, while the north-south dimension is 125 miles.

Baltimore, the largest city, is 203 miles southwest of New York, 102 miles southwest of Philadelphia, and 45 miles northeast of Washington, D.C. on I-95. Cleveland is 380 miles west on I-70/76. Annapolis, the capital, is 28 miles south of Baltimore on I-97.

THE LAY OF THE LAND

About half the state is a part of the Atlantic Coastal Plain. The Eastern Shore is flat and low-lying, while the part of the Plain on the Western Shore is more rolling and tree-covered. Neither part contains especially fertile soil.

Another one-quarter of the state is part of the Piedmont Plateau. From east to west, rolling hills give way to a flatter highland. A strip of the Blue Ridge Mountains runs between Frederick and Hagerstown. On the western "bird wing," a Valley and Ridge region with narrow, sharply crested ridges gives way to the Allegheny Mountains, which feature more rounded summits and broader valleys.

While the state has only 31 miles of coastline, the many inlets of the Chesapeake Bay give it 3,190 miles of shoreline, the 9th most of any state. Maryland's two major rivers are the Potomac and the Susquehanna.

HOW MARYLAND BECAME A STATE

Several explorers, including John Smith, visited the area, but the first settlement was established on Kent Island by William Claiborne in 1622. In 1632, King Charles I gave the area to George Calvert, the first Baron Baltimore, and Lord Baltimore named it in honor Henrietta Maria, wife of Charles I. George Calvert died shortly afterwards, and his son Cecilius Calvert, the second Baron Baltimore, decided to settle the area as a model of religious tolerance. In 1634, a group of settlers established St. Mary's near the mouth of Chesapeake Bay, and in 1635 the state's first assembly was elected. Puritans who were expelled from other colonies began to settle in the area after 1643.

For the next 75 years, the Baltimores, who were Roman Catholics, struggled for control of the area with Protestants who were initially led by William Claiborne. Claiborne expelled Lord Baltimore in 1652, but the Calvert's control was recognized again in 1657 by Oliver Cromwell. In 1675, Charles Calvert, the third Baron Baltimore, took control and lost a struggle with William Penn over the area that is now Delaware. After the Glorious Revolution of 1688 brought Protestants to the crown of England, Maryland became a royal colony. Finally, in 1715, control was returned to Charles Calvert, the fifth Baron Baltimore.

The most lively political issue of the next half century was a border dispute with Pennsylvania. Finally, the British surveyors Charles Mason and Jeremiah Dixon settled the dispute in 1767 by drawing the border at latitude N 39 43'—the famous Mason-Dixon line.

Maryland was the sight of much dissatisfaction with British taxes and regulations—in 1772, residents burned a tea ship in protest. Maryland joined the Continental Congress and participated in the Revolutionary War. Maryland entered the Union was the seventh state on April 28, 1778.

MARYLAND'S CITIES

BALTIMORE (736,014), located at the mouth of the Patapsco River where it empties into Chesapeake Bay, is one of the nation's leading ports. The area was populated first by Swedish settlers, but the first actual settlement was built in 1729 and named for the Barons Baltimore. The city soon became a major tobacco port as well as a flour-milling and ship-building center. In 1814, during the War of 1812, the British fleet bombarded Fort McHenry in Baltimore Harbor, and an observer named Francis Scott Key wrote the "Star Spangled Banner" to commemorate what he saw. Today, Baltimore is a thriving city with a revitalized downtown area symbolized by Camden Yards, the new home of the Baltimore Orioles, and by a beautiful and popular shopping, entertainment, and aquarium complex known as the Inner Harbor. I spent two of the most interesting and enjoyable years of my television career living and working in Baltimore during the mid-1970s.

ANNAPOLIS (33,195) was founded on the south bank of the Severn River in 1643 by Puritans expelled from Virginia. It was named Providence. The city became a busy port and trading center. The State House, built between 1772-1780, is the oldest continually used capitol building. The Continental Congress met there and George Washington resigned from the Army there in 1783. Today, Annapolis is home to the U.S. Naval Academy.

FREDERICK (40,148) was settled in 1745 and named for Frederick Calvert, the sixth Baron Baltimore. The city became a transportation and shipping center for a dairy-farming and corn and wheat growing area. Today, the city's leading industries are electrical and electronic equipment, biomedical devices, hardware, pumps, and control devices.

MARYLAND'S PEOPLE

Maryland's 4,964,898 people make it the 19th most populous state. Almost 93% of the state's residents live in metropolitan areas, making it one of the most urban states. The state has a diverse population that is 71% white, 25% black, and includes a large Asian minority.

MARYLAND'S ECONOMY

Eighteenth and nineteenth century Maryland had an economy based on tobacco, fishing, lumber, and shipbuilding. Today, it's agricultural income is relatively small. About two-thirds of the farm income is derived from broiler chickens, eggs, dairy products, and other livestock, while corn, soybeans and tobacco are the leading crops. Chesapeake Bay was once home to a huge crab, clam, and oyster fishing industry, but pollution and overfishing have drastically reduced the annual catch.

Manufacturing became the leading element of the economy in the late nineteenth century. The leading industries today are precision instruments, printed material, processed foods, transportation equipment, primary metals, and chemicals. Baltimore is one of the nation's top twenty ports.

Maryland's proximity to the nation's capital makes government and service industries the state's largest employers. Research and development is the fastest growing area of the economy—in Baltimore alone there are 61 federal research centers.

Places to Visit

Baltimore's renaissance is symbolized by Camden Yards, the new home of Orioles, which is attracting record crowds. But this vibrant city also features such attractions as the National Aquarium, Ft. McHenry, the Baltimore Museum of Art, and Babe Ruth's birthplace. The Chesapeake Bay area features Annapolis (410-269-6125), the capital and and home of the U.S. Naval Academy and Ocean City (800-626-2326), a popular beach community. Civil War buffs will love the Antietam National Battlefield (301-432-5124) in western Maryland. Tourist information: (800-282-6632)Maryland Office of Tourist Development, 217 E. Redwood St., Baltimore, MD 21202 (800-543-1036/410-333-6611)

NEW JERSEY

Because its the ninth largest state in population, most people don't realize that New Jersey (now my home state) is the nation's fourth smallest in area. The state, which is shaped sort of like a stocking hanging from the mantle, is bordered on the northeast by New York, the east by the Atlantic Ocean, the south by Delaware, and the west and northwest by Pennsylvania. The Hudson River forms the border with New York, while the Delaware River determines the border with Delaware and Pennsylvania. The state runs 165 miles north-south, but only 60 miles east-west. New Jersey has 130 miles of coastline on the Atlantic Ocean.

Newark, the largest city, is just across the Hudson River from New York City, and is 100 miles north of Philadelphia on I-95. Trenton, the state capital, is 76 miles southwest of New York City, 30 miles north of Philadelphia, and 68 miles north of Wilmington, Delaware on I-95.

THE LAY OF THE LAND

About three-fifths of the state is part of the Atlantic Coastal Plain. The area near the coat is marshy and is cut by several bays, including Grant Bay, Barnegat Bay, and Sandy Hook Bay. A sharp rise in elevation marks the beginning of the Piedmont Plateau region, which includes the Wachtung Mountains. North of the Plateau are the New Jersey Highlands, which feature mountains separated by broad, flat valleys. In the northwest is the Appalachian Valley and Ridge region that includes the famous Delaware Water Gap.

New Jersey has many lakes, of which Hopatcong, Mohawk, and Greenwood are the largest.

HOW NEW JERSEY BECAME A STATE

The Dutch and the English both claimed the area. In 1624, the Dutch built Fort Nassau on the Delaware River, and in 1630 added Fort Pavonia. The English forced the Dutch out of the area in 1641, and King Charles II gave all the of the lands between the Delaware and Connecticut Rivers to his brother James, the Duke of York. The Duke in turn assigned the southern portion of this grant to George Carteret and James, Lord Berkeley. It was named after the Isle of Jersey in the English Channel, Carteret's birthplace. The two proprietors issued a constitution known as the Concessions and Agreement of 1665, which granted freeholders the right to elect an assembly and exercise freedom of religion.

In 1674, Lord Berkeley sold his half to a group of Quakers, and the area was divided into East Jersey and West Jersey. After Carteret died in 1681, his holdings were purchased at auction by a group headed by William Penn, proprietor of Pennsylvania. Finally, in 1702, East and West Jerseys were united as a royal colony. For a while, New York and New Jersey were governed together, but in 1738, Lewis Morris became New Jersey's first governor.

There were strong feelings between settlers who supported the British and those who wanted independence. However, under the leadership of Governor William Franklin, Benjamin Franklin's son, the state joined the Continental Congress in declaring the new nation. Several important Revolutionary War battles were fought in New Jersey, including the Battle of Trenton (1776), Princeton (1777), Monmouth (1778) and Princeton was the capital of the U.S. from June to November, 1787. New Jersey became the third state to join the Union on December 18, 1787.

NEW JERSEY'S CITIES

NEWARK (275,221), located at the junction of the Passaic River and Newark Bay, is the state's largest city and is part of the New York-New Jersey Port Authority that is the nation's third largest. The city was settled in 1666 by Puritans from Connecticut and it was named for Newark-on-Trent, England. The leather industry began in the area in the late 18th century, and the area became a leading industrial center after the arrival of the railroad in 1830. Today, Newark is a major trade, insurance, financial and manufacturing center. I have been a guest of the Mayor of Newark, and I'm impressed with the vigor with which the city's highly publicized urban problems are being attacked.

TRENTON (88,675), the state capital, is located on the Delaware River north of Philadelphia. In 1679, Mahler Stacy, an English Quaker, built a grist mill on the site. In 1714, his son sold the mill to William Trent, who built a city in 1721 and gave it his name. On Christmas Day, 1776, George Washington crossed a near impassable Delaware River to surprise and defeat British troops headquartered in Trenton. Today, the city has a diverse manufacturing economy that includes rubber, plastic, and metal goods, processed foods, printed materials, and ceramics. The city is proud of the State House, built in 1792.

ATLANTIC CITY (37,986), located on the Atlantic Ocean, was in an area first settled in 1803. After it became the eastern terminus of the Camden and Allentown Railroad in 1854, it became a beach resort famous for its 6 mile long boardwalk. The city was revitalized as a tourist attraction when the New Jersey legislature authorized casino gambling in 1978, and Atlantic City has since become the "Las Vegas" of the east.

NEW JERSEY'S PEOPLE

When you fit 7,879,164 people, the ninth largest population, into the 4th smallest state, you end up with a population density of 1,049.9 people per square mile, the highest in the U.S. New Jersey is a melting pot, with substantial black, Asian, and Hispanic populations. About 89% of the state's population lives in urban areas, making it among the nation's least rural. Despite its heavily urban character, New Jersey still boasts of some of the richest and most beautiful farmland in the U.S., thus the nickname "The Garden State."

NEW JERSEY'S ECONOMY

Given its proximity to such major metropolitan areas such as New York and Philadelphia, it's not surprising that New Jersey has a great many truck farms that supply tomatoes, vegetables, eggs, and dairy products. The state's farms also produce corn, hay, and beef cattle.

Manufacturing has been the foundation of the state's economy for more than a century. The leading industries are chemicals, processed foods, industrial machinery, precision instruments, electrical and electronic equipment, and primary metals such as iron and steel. New Jersey has major ports, and is a center of the insurance and financial services industries.

Tourism also contributes an impressive $17 billion to the state's economy. The major tourist attraction is the Jersey Shore area, which includes Atlantic City.

Places to Visit

Atlantic City (800-262-7395) is Las Vegas with beautiful sandy beaches and a boardwalk. The rest of the Jersey Shore includes 127 miles of public beaches that stretch from Sandy Hook to Cape May, and is dotted with communities that cater to family vacations. Vernon Valley/Great Gorge (201-827-3900) features great skiing in northwestern New Jersey, while historic Princeton (609-921-7676) was once the capital of the U.S. If you're visiting New York City, hop across the Hudson River to visit the Liberty Science Center in Jersey City. Tourist information: New Jersey Division of Travel and Tourism, 20 W. State St., CN-826, Trenton, NJ 08625)800)- 537-7397/609-292-2470)

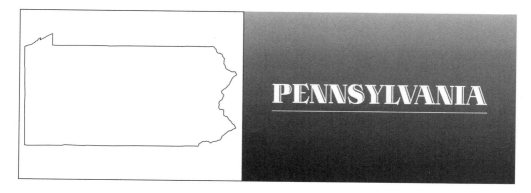

PENNSYLVANIA

The Commonwealth of Pennsylvania is a roughly rectangular state located at a crossroads of several different geographic regions of this country. Pennsylvania is bordered by New York on the north and northeast, New Jersey on the east, Delaware, Maryland and West Virginia on the south, and West Virginia and Ohio to the west. The Delaware River makes up the state's eastern boundary. The state's extreme dimensions are 170 miles north-south and 308 miles east-west. Pennsylvania has no coastline on the Atlantic, but has 40 miles of shoreline on Lake Erie.

Philadelphia, the largest city, is 101 miles southwest of New York City and 147 miles northeast of Washington D.C. on I-95. It's 319 miles west to Pittsburgh on I-70/76. Pittsburgh, the state's other major metropolitan area, is 157 miles southeast of Cleveland on I-76 and 187 miles northeast of Columbus, Ohio on I-70.

THE LAY OF THE LAND

Pennsylvania students studying the geography of their state have to work twice as hard as students in most other places—the state has seven major geographic regions that cross its borders. In the northwest corner, a narrow, flat region of the Eastern Great Lakes Lowlands gives way to the Appalachian Plateau (sometimes called the Allegheny Plateau), which covers about half the state's total area. This region is characterized by narrow valleys giving way to flat-topped peaks. The Pocono Mountains are in the northeastern part of this region, with the Allegheny Mountains in the south. The end of this region drops off sharply from 2,000 to 1,200 feet in an area called the Allegheny Front, one of the wildest areas east of the Mississippi.

The Allegheny Front gives way to the Appalachian Ridge and Valley Region. The best known part of this region is the Great Valley, which includes the Lebanon Valley and the Cumberland Valley. This region ends at a narrow band of mountains. The southern part is an extension of the Blue Ridge Mountains and the northern part, called the Blue Mountains, is the southern extension the New England Uplands.

Moving toward the southeast, we come to the Piedmont Plateau, a broad stretch, which includes Pennsylvania Dutch country, that contains the state's most fertile land. Finally, the southeastern corner of the state is part of the flat Atlantic Coastal Plain.

Pennsylvania has numerous lakes, including more than 300 in the Appalachian Plateau. The state has several important rivers, including the Delaware, the Lehigh, and the Schuylkill. The Ohio River is formed in Pittsburgh by the junction of the Allegheny and Monongahela Rivers.

HOW PENNSYLVANIA BECAME A STATE

Both the Dutch West Indies Company and the New Sweden Company sent settlers into southeastern Pennsylvania in the early 1600s. England evicted the Dutch in 1664, and Quakers from New Jersey also moved westward into the area in the 1670s. In 1681, King Charles II granted a royal patent for the area to William Penn, a member of the Society of Friends, commonly called Quakers. The King is said to have coined the term Pennsylvania from Penn's name and "sylvan," the Latin word for "woods."

Penn immediately drew up a document called the Frame of Government, a written contract with the settlers that guaranteed freedom of religion and trial by jury. In 1682, Penn founded Philadelphia and signed a contract of friendship with the Indians. Penn's charter was revoked in 1692 but was restored in 1699. Penn replaced the Frame of Government with the Charter of Privileges.

Settlers from Scotland, Ireland, and England moved into the territory, along with a large number of Germans, especially members of the Amish and Mennonite religious sects. The latter came to be known as the Pennsylvania

Dutch, a corruption of "Deutsch", meaning "German." Philadelphia was the chief city of British North America.

English control of western Pennsylvania was challenged by the French during the 1700s until their defeat in the French and Indian War. A bitter border dispute with Maryland was settled in 1767 when the famous Mason-Dixon line was drawn by two English surveyors.

The state also became politically divided between an anti-British faction headed by Benjamin Franklin and a pro-British faction headed by Joseph Galloway. Galloway called the First Continental Congress to foster goodwill between the two parties; instead, the Congress declared a commercial boycott on British goods. The Declaration of Independence was signed in Philadelphia on July 4, 1776, and the Constitutional Convention was held there in 1787. Pennsylvania became the second state to join the Union on December 12, 1787.

PENNSYLVANIA'S CITIES

PHILADELPHIA (1,585,597) is the fifth largest U.S. city as well as the birthplace of our nation. It was founded at the junction of the Delaware and Schuylkill Rivers by William Penn in 1682, and given its name from the Greek word for "brotherly love." It was the first U.S. city laid out on a grid system, the core being a 22 by 8 block grid with four public parks. Philadelphia reached a population of 10,000 by 1720 and became the most important commercial and political center in North America. Today, Philadelphia is known for its rich ethnic heritage, its historical sites, and its industry, trade, and commerce, and some of the East Coast's finest restaurants.

PITTSBURGH (369,879) is America's busiest inland river port. In 1753, upon the recommendation of George Washington, settlers began building a fort on the site to protect against the French and the Indians. The French drove off the English and completed Fort Duquesne in 1756. The British recaptured the area in 1761, calling it Fort Pitt, in honor of British prime minister William Pitt. In 1792, the first blast furnace was built, leading to the development of the iron and steel industry in the mid-1800s. Primary metals are still the foundation of this major manufacturing region today.

HARRISBURG (52,376) was founded as a ferry terminal and fur trading post by James Harris in 1719. The name was changed to Louisburg (for King Louis XVI) in 1785, but became Harrisburg in 1791. Harrisburg was named the state capital in 1812. The building of the Pennsylvania Canal in 1834 and the arrival of the railroad in 1836 led to the city's growth as a trade center.

PENNSYLVANIA'S PEOPLE

With 12,048,271 residents, Pennsylvania is the nation's fifth most populous state. The state is well known as the home of the Pennsylvania Dutch, who include the Amish, whose religion prohibits the use of most modern machines and appliances. The state has a long history of religious tolerance, which has attracted a very diverse population.

PENNSYLVANIA'S ECONOMY

Much of Pennsylvania's economy has long been based on what comes out of the ground. The state has large bituminous coal reserves as well as the only anthracite (hard) coal fields in the U.S. The very first U.S. oil well was drilled in Pennsylvania in 1859. These energy sources fueled the growth of the iron and steel industry, which continues to be important today. Pennsylvania also has one of the nation's most diverse industrial economies, with the most important being industrial machinery, printed materials, food products, electrical equipment, glass, chemicals, transportation equipment, clothing and textiles, and paper. Manufacturing in turn spawned large financial and service industries.

Agriculture provides a relatively small part of the state's income. However, Pennsylvania leads the nation in mushroom production. About 70% of farm income comes from livestock and livestock products, especially dairy and poultry products. Hay and corn are major crops.

Tourism brings Pennsylvania about $9 billion per year. Among the major tourist centers are Philadelphia, Pennsylvania Dutch country, and Hershey, home of the chocolate company.

Places to Visit

Philadelphia (800-537-7676) is the home of the Liberty Bell, Independence Hall, the Philadelphia Museum of Art, and lots of other attractions. Nearby Bucks County (215-345-4552) is known for antiques, covered bridges, and country inns. Pennsylvanian Dutch Country (717-299-8901) can take you back to a different world. Within a 90 minute drive are Hershey (717-534-3005), which features Chocolate World and Hershey park, and the Gettysburg National Military Park (717-334-1124). Tourist information: Pennsylvania Office of Travel Marketing, 453 Forum Bldg., Harrisburg, PA 17120 (800-847-4872/717-787-5453)

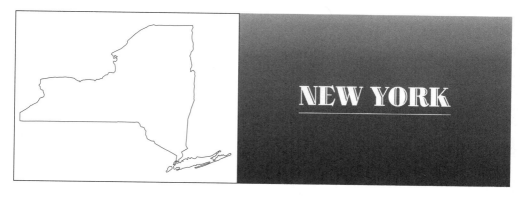

NEW YORK

Although New York stretches north as far as any New England state except Maine, it's topography and history make it more appropriate to classify it as a Middle Atlantic state. New York is bordered on the north by the Canadian provinces of Ontario and Québec, on the east by Vermont, Massachusetts, and Connecticut, on the southeast by the Atlantic Ocean, on the south by New Jersey and Pennsylvania, and on the west by Pennsylvania and Ontario. The state is shaped roughly like a right triangle with a "heel" extending down to New York City and with Long Island jutting out into the Atlantic. Lake Champlain and the Poultney River form part of the northeastern border, the Hudson and Delaware Rivers form part of the southeaster, Lake Erie and the Niagara River are part of the western border, and Lake Ontario and the St. Lawrence Seaway are part of the northern border. New York has 127 miles of coastline. It's extreme dimensions are 330 miles east-west and 280 miles north-south.

New York City is 222 miles southwest of Boston, 101 miles northeast of Philadelphia, and 203 miles northeast of Washington, D.C. on I-95. Buffalo, the second largest city, is 393 miles northwest of New York City and 191 miles northeast of Cleveland on I-90.

THE LAY OF THE LAND

New York has seven different geographic areas. The St. Lawrence Lowlands are gently rolling areas adjacent to the Seaway on the north and Lake Champlain on the east. The Eastern Great Lakes Lowlands begin at the edge of Lakes Erie and Ontario. Flat land gives way to rolling hills, many of which are egg-shaped "drumlins," which were carved by glaciers.

The one-quarter of the state north of a line running from Albany west to Utica is the Adirondack Mountains, which consist of very old (1 billion years +) land pushed upward by the pressure of magma below the earth's crust. This wild, sparsely populated range is geographically much different from the other mountain ranges of the eastern U.S.

South of the Adirondacks are the Hudson-Mohawk Lowlands, which consist of a broad river valley that runs up the Hudson River from New York City to Albany, then west along the Mohawk River. To the east of the Hudson River is part of the New England Uplands, which have three separate regions—the Taconic Mountains, the Hudson Highlands, and the Manhattan Hills.

The large swath of land that runs west from the Hudson River and south of the Mohawk Valley and Great Lakes is part of the Appalachian Plateau. The eastern part of this region consists of Rip Van Winkle's Catskill Mountains. In the south central part of the state are long, narrow valleys carved by glaciers which formed the famous Finger Lakes. The western portion is a plateau with broader valleys.

Finally, Long Island and Staten Island are part of the flat Atlantic Coastal Plain.

New York borders on three of North America's largest lakes—Lake Erie, Lake Ontario, and Lake Champlain. The state has numerous other lakes, including the Finger Lakes, Lake Placid, and Lake George. New York City and Buffalo are connected by water because of the Hudson River and the Erie Canal. Canals also connect the Hudson River to Lake Champlain to Canada.

HOW NEW YORK BECAME A STATE

On the basis of Henry Hudson's third voyage in 1609, the Dutch West India Co. claimed a wide area of the Atlantic Coast. It established settlements at Fort Orange (in the area of Albany) in 1624 and on the southern tip of Manhattan Island in 1626. New Amsterdam grew under the leadership of Peter Stuyvesant, who became governor in 1647.

However, in 1664, the English sent a force to evict the Dutch, and New Amsterdam became New York, in honor of James, Duke of York.

The colony grew when it acquired Long Island from Connecticut. When James became King James II in 1685, he combined New England, New York, and New Jersey into the Dominion of New England. However, James was deposed by the Glorious Revolution of 1688. Jacob Leisler rebelled against royal rule, declaring himself governor. But royal rule was restored and Leisler was hanged for treason 1691.

The French, with the aid of their Indian allies, controlled much of upstate New York into the mid-18th century. A force led by Lord Jeffrey Amhearst ousted the French from Fort Ticonderoga in 1755, and the French withdrew completely after the conclusion of the French and Indian War in 1763.

By this time enmity between many New Yorkers and the British was fermenting. New York convened the Stamp Act Congress to protest restriction of trade in 1765, and in 1767, the British disbanded the legislature after it refused to provide housing and supplies for British troops. Many areas of the state were battlegrounds in the Revolutionary War; the surrender of Gen. John Burgoyne to the Americans at Saratoga in 1777 prevented the British from sending troops from Canada to fight the Americans. New York joined the Union on July 26, 1788 as the eleventh state.

NEW YORK'S CITIES

NEW YORK CITY (7,322,564), one of the world's great cities, was founded by the Dutch in 1626 on the lower end of Manhattan Island, near the present day Wall Street. A number of small communities arose in the area, which is why so many streets twist and turn today south of 14th Street in Manhattan. North of 14th street, the island is laid out in a grid system. The other four boroughs that make up the city—Queens, Brooklyn, the Bronx, and Staten Island—grew as collections of hamlets that eventually merged into one administrative unit. Trade fueled the growth of the city, especially after the completion of the Erie Canal provided a water link to the Great Lakes in 1804. New York's Ellis Island became the gateway to America for tens of millions of immigrants, who contributed to the cosmopolitan aura of the city. Today, New York City is the nation's cultural center, the home of the major financial markets, and the headquarters of more major corporations than any other city.

BUFFALO (328,175), located at the junction of Lake Erie and the Niagara and Buffalo Rivers, was settled by the Seneca Indians in 1780. The site was purchased in 1790 by the Holland Land Co. and was laid out in 1803 using the new nation's capital, Washington, D.C., as a model. The name of the city comes from Buffalo Creek, which was not derived from the animal, but from a corruption of the French phrase "beau fleur," or "beautiful flower." After the completion of the Erie Canal, Buffalo became one of America's leading flour-milling centers. It is one of the nation's largest rail junctions, and it's importance as a shipping center was enhanced with the completion of the St. Lawrence Seaway in 1959. I find myself mentioning Buffalo a lot during my weather reports, because moist air moving inland from Lake Erie gives it one of the nation's highest annual snowfall totals.

ALBANY (100,031), the state capital, was the site of Fort Nassau (1614), then Fort Orange (1624). It became the town of Beuerwyck until the British gained control in 1664. The name was changed to honor James, Duke of York and Albany. In 1754, the city hosted the Albany Congress, at which Benjamin Franklin presented his Plan of Union, the forerunner to the Declaration of Independence. It became the capital of New York in 1797. Today, Albany is the center of state government as well as a major transportation, banking, medical, and education center.

NEW YORK'S PEOPLE

It's 18,197,154 people make00000 New York the nation's second largest state. It is also a racial and ethnic melting pot. New York has more African-Americans than any other state and more Jews than the nation of Israel. The New York City area, which is home to half of the state's residents, has ethnic conclaves that give visitors the impression that they've been whisked off to a foreign land. Even the upstate cities tend to be cosmopolitan and culturally diverse.

NEW YORK'S ECONOMY

New York was an important agricultural state in its early history. Today, dairy farming is the number one farm industry, with beef cattle, chickens, turkeys, and ducks other important livestock products. The Finger Lakes region is famous for its grapes and wine, and the state is one of the leading producers of apples and cherries.

New York City is the foremost cultural center in the United States, attracting the world's finest artists to its theaters and galleries. The city is one of the world centers of finance and international trade. The city is also home to the broadcasting, publishing and advertising industries.

New York State manufacturers employ 1.2 million people and generate $80 billion in revenue. The major industries are printed materials, precision instruments, electrical equipment, industrial machinery, clothing, aircraft, chemicals, pharmaceuticals, and motor vehicle parts. The state is the headquarters of many of the nation's largest companies.

Tourism brings in another $19 billion, a significant portion from foreign visitors.

Places to Visit

After two decades, I still love New York City (212-397-8222), where you can spot nearly every famous person in the world from time to time—even me. Hundreds of attractions from Broadway to the Bronz Zoo to the Statue of Liberty have made the Big Apple the world's number one tourist destination. All the way on the other end of the state, Niagara Falls (716-278-1701) is a a must-see even if you're not on your honeymoon. In between are the numerous resorts of the historic Catskill Mountains (800-542-2414), the six million acre Adirondack Park (518-327-3000) with its 2,500 lakes and 1,000 miles of rivers, the National Baseball Hall of Fame in Copperstown (607-547-9988), the 11 Finger Lakes in a region filled with gorges and waterfalls (800-548-4386), and a whole lot more. Tourist information: New York State Division of Tourism, 1 Commerce Plaza, Albany, NY 12245 (800-225-5697/518-474-4116)

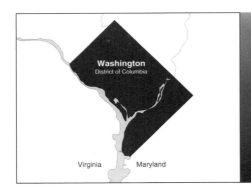

Washington
District of Columbia

Virginia   Maryland

DISTRICT OF
COLUMBIA

WASHINGTON, D.C (606,900), our nation's capital, used to represent an enigma for me—I always wondered why it had two names. I never could figure out what the difference was between Washington and the District of Columbia. Fortunately, doing this book spurred me to find out.

And what's the answer? Today, , there is no difference—they're exactly the same area. But when the capital was created, there was more than one city in the District. You see, in 1783, the U.S. Congress decided the new country needed a brand new capital. George Washington selected the site at the junction of the Potomac and Anacostia Rivers, on land donated by both Virginia and Maryland. Washington, who had been a mason, laid the cornerstone for the U.S. Capitol in 1793. The district was named for Christopher Columbus and the capital city, first called Federal City, was named in honor of the first President. However, it included the communities of Alexandria in Virginia and Georgetown in Maryland.

Washington was laid out by Pierre L'Enfant in a grid system that had four quadrants: northwest, northeast, southwest, and southeast. At the center was a large open space called The Mall.

The White House, the home of every U.S. President except George Washington, was completed in 1800. Much of the city, including the White House, was burned by the British in 1814.

The Nation's Capital experienced relatively slow growth. In 1847, the U.S. Government gave Alexandria back to Virginia and the District of Columbia shrunk from 100 square miles to its present 68 square miles. But the population began to rise after the Civil War, largely due to black migration from the South.

In this century, Washington has become the focus of a large and growing metropolitan area. In 1840, 70% of the area's population lived in the District; by 1990, that figure had shrunk to just 16% . About two-thirds of the District's residents are black.

The Federal Government is the focus of the city's economy. One-third of all workers are U.S. employees. Washington is home to thousands of organizations, embassies, law firms, banks, and other service companies. With 18 million people visiting every year, tourism is a major industry.

Politically, Washington, D.C. is a strange animal. The city's residents are the only U.S. citizens with representation in Congress—an amendment to give the city seats in the Senate and House was passed by Congress but not ratified by enough states to become law. The city is administered by a locally elected mayor and Council under the supervision of the Federal Government

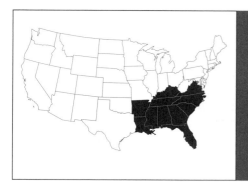

THE SOUTH

Comparing the Southern States

STATE National rankings in parenthesis	POPULATION (1993 U.S. Bureau of the Census)	SQUARE MILES (U.S. Geological Survey)	PER CAPITA INCOME (U.S. Bureau of Labor Statistics)
Florida	13,678,914 (4)	65,758 (22)	$14,698 (18)
North Carolina	6,945,180 (10)	53,821 (28)	$12,885 (33)
Georgia	6,917,140 (11)	59,441 (24)	$13,631 (22)
Virginia	6,490,634 (12)	42,769 (35)	$15,713 (12)
Tennessee	5.098,798 (17)	42,146 (36)	$12,255 (37)
Louisiana	4,295,477 (21)	51,843 (31)	$10,635 (47)
Alabama	4,186,086 (22)	52,423 (30)	$11,486 (40)
Kentucky	3,788,088 (24)	40,411 (37)	$11,153 (44)
South Carolina	3,642,718 (25)	32,007 (40)	$11,897 (38)
Mississippi	2,642,748 (31)	48,434 (32)	$9,648 (50)
Arkansas	2,424,418 (33)	53,182 (29)	$10,530 (49)

I won't even attempt to mask my bias when talking about the South, because I am so proud to be a southerner. This may come as a surprise to some, because I am black; and, having grown up during the 1950s and 1960s, I vividly recall the ugliness of racial segregation in the Old South and the harsh and hateful treatment of blacks during the civil rights struggle. However, in spite of those painful memories, I also recognize that the South has made the most rapid and dramatic social progress of any region in this country. After the initial resistance to change in the '50s and '60s, the South's longtime tradition of civility began to prevail, and people—black and white—began to get on with the business of living together and learning together, working together and worshipping together. Are there still racist attitudes in the South? Without doubt! But, racism is a fundamental flaw of the human condition which knows no geographic boundaries. After having traveled all over this great land, it is my firm belief that the most troubling pockets of racial hostility and polarization today exist outside the South.

I am also proud of the South for its tremendous economic growth over the last few decades. Let us not forget that the Civil War was largely an *economic* conflict between the industrialized North and the agrarian South. The devastation suffered by the Confederate States was total and terrible, and recovery took more than half a century. Eventually the mild climate, a willing and relatively inexpensive work force, low cost of living, and cooperative local governments attracted more and more manufacturing to areas that had previously depended primarily on agriculture. Today the New South has a diverse and rapidly expanding economy which can match that of any region of the country.

The South also has a rich history that dates back to the earliest explorations of this land. The first Europeans were the Spanish. Juan Ponce de León, Spanish governor of what is now Puerto Rico, discovered Florida in 1513. Pánfilo de Narváez and Alvar Nunez Cabeza de Vaca explored parts of Florida and the northern Gulf Coast between 1528 and 1536. Then Hernando de Soto, who had seized great wealth in the Spanish conquest of Peru, spent three years looking for gold in the Southeast, visiting parts what is now Alabama, Mississippi, Tennessee, Arkansas, Georgia and Louisiana. However, the Spanish were unable to establish permanent colonies.

The English had more success. Sir Walter Raleigh established short-lived settlements off the coast of North Carolina as early as 1585. The first permanent settlement in the New World was founded at Jamestown, in what is now Virginia, in 1607. The British went on to establish control of the Atlantic coastal regions from Virginia to Georgia.

Much of the rest of the region was discovered by French explorers coming down from the north. Jacques Marquette and Louis Jolliet traveled down the Mississippi River as far as northern Mississippi in 1663, and Robert Cavelier, sieur de La Salle, traveled all the way down to the Gulf of Mexico in 1682. He claimed the entire Mississippi Valley for France, naming it Louisiana after King Louis XIV.

Control of many areas passed several times between nations until the fledgling United States was born and expanded west. But the multi-nation heritage left its remnants in a rich Southern history and culture that is fascinating.

Finally, I love the South for one feature that hasn't changed over the centuries—the breathtaking natural beauty of the region. From the scenic vistas of ranges like the Blue Ridge Mountains to the rolling, tree covered hills of the Cumberland Plateau to the fertile farmland of the Atlantic and Gulf Plains to the pristine beaches, the South has terrain for every taste.

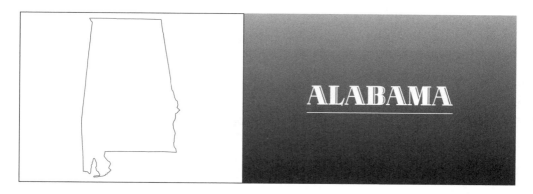

ALABAMA

You'll find Alabama in the center of America's south (guess why it's nicknamed "The Heart of Dixie), sandwiched between Georgia on the east (the Chattahoochee River makes up the southern half of this border) and Mississippi on the west. Tennessee is the entire northern border, while Florida is its southern neighbor. Alabama is roughly rectangular in shape, except for a small southern section of the state, shaped like the heel of a boot, that extends to the Gulf of Mexico As a result, Alabama has 53 miles of coastline.

Birmingham, Alabama's largest city, is on I-65, 189 miles south of Nashville and 663 miles south of Chicago. Atlanta is 148 miles east of Birmingham on I-20, while Dallas is 648 miles to the west. Mobile, Alabama's busy Gulf port, is 303 miles west of Jacksonville, Florida is on I-10 and 148 miles northeast of New Orleans.

THE LAY OF THE LAND

Many people picture the deep south as an endless vista of flat cotton plantations, But they might be surprised to find that there was a tremendous variety of terrain. Two-thirds of Alabama's 52,423 square miles are covered with timber, making it the nation's 5th most heavily forested state. The northern half of the state, part of the ancient Appalachian Mountain highlands that stretches north all the way into Canada, consists of hills, valleys, plateaus, caverns, and broken terrain. These scenic highlands are rich in such minerals as iron ore, coal, and limestone— Alabama is the only state with all the raw materials necessary to make iron and steel.

You can find vintage plantation country in the band of very fertile dark soil, known as "The Black Belt," that stretches across the central portion of the state. This rich agricultural region and the rest of southern Alabama are part of the flat Gulf Coastal Plain that stretch from southern Texas through Georgia and Florida. The Mobile River delta right on the Gulf of Mexico includes swamps and bayous that remind me of nearby Louisiana. Much of the state's beachfront is on Pleasure Island and Dauphin Island, two beautiful vacation areas.

The broad Tennessee River cuts through the northern quarter of the state flowing east to west on its way to the Ohio River. The remaining portion of the state is drained by a number of south-flowing rivers (including the Alabama and the Tombigbee) which become the Mobile River shortly before flowing into the Gulf of Mexico.

HOW ALABAMA BECAME A STATE

Having a great appetite for history, I have found the saga of Alabama a truly rich feast. Prehistoric peoples found their way to the area very early—Russell Cave in northeast Alabama has been continuously occupied for at least 9,000 years. At the time of Columbus, the region had one of the largest Native American populations in the New World, including Creeks, Choctaws, Cherokees, and Chickasaws.

The French established the first permanent settlement near the present day site of Mobile in 1702—and celebrated the first Mardi Gras in the New World (Sorry, New Orleans). However, France was forced to give up all of its territories east of the Mississippi to Britain 1763 after losing the French and Indian War.

Here's how the U.S. ended up with what is now Alabama:

Victory in the Revolutionary War gave the fledgling U.S. the Mississippi Territory.

Before the War of 1812. U.S. General Andrew Jackson seized West Florida from the Spanish, an area that included the Gulf Coast of present day Alabama and Mississippi.

In 1817, the Mississippi Territory was divided into the state of Mississippi and the Alabama Territory (the part of West Florida around Mobile was tacked on to give the territory access to the Gulf of Mexico). The name Alabama came from the name of a tribe in the Creek confederacy.

Alabama was admitted to the Union on December 14, 1819, as the 22nd U.S. state. It was probably named for an Indian tribe whose name means "vegetation gatherers" or "thicket clearers."

ALABAMA'S CITIES

BIRMINGHAM (265,968) was incorporated in 1871 at the junction of two railroad lines. With all the natural resources necessary to make iron and steel available in the surrounding hills, Birmingham grew rapidly to become the "Pittsburgh of the South."—I have found it to be a very charming and vibrant city. Appropriately, the city was named for the center of the English iron and steel industry, Birmingham. The Birmingham Museum of Art is the Southeast's largest municipal museum.

MONTGOMERY (187,106), in the center of the state on the Alabama River, became a major cotton and livestock trading city in the early 1800s. It became the capital of Alabama in 1847 and the first capital of the Confederate States in 1961 (the White House of the Confederacy is now a popular museum). The city was named for American Revolutionary War General Richard Montgomery.

MOBILE (196,278) has been a major Gulf Coast port since 1711. Among the area's many beautiful ante-bellum mansions is Bleeingrath Gardens and Home, which includes magnificent azalea gardens. "Mobile" is a French corruption of a Native American tribal name.

ALABAMA'S PEOPLE

Most of Alabama's 4,186,806 residents were born there—only 23% were born in other states, and only 1% were born in foreign countries. Most Alabamans are descended from settlers who moved to the region from the North and East and from Africans who were forcibly brought to America as slaves. At the time of the Civil War, the population of Alabama was 55% white and 45% black. Although many African-Americans migrated to the North after the Civil War, they still make up 25% of the present population, the sixth highest percentage among U.S. states.

Alabama's Economy

Cotton was the foundation of Alabama's economy in its early years of statehood, but the Civil War left the plantations in ruins and the state's economy in a shambles. Although the plantations slowly recovered, it was manufacturing stimulated by the Birmingham-centered iron and steel industry that helped get the state back on its feet.

Alabama's twentieth-century economic story is dominated by two great scientists and one bug.

First, the bug. In 1910, the boll weevil worm devastated Alabama's huge cotton crop. It was nearly a decade before the insect was brought under control.

One of my favorite historical figures was the great, black Alabama scientist George Washington Carver, who discovered an astounding 300 uses for the peanut and 150 uses for soybeans. His work prompted many farmers throughout to switch to raising soybeans and peanuts instead of cotton.

In 1950, Dr. Werner von Braun brought 120 German rocket scientists to Huntsville, which later became the "Rocket Capital of the World" and home to the famous Space Camp. Rocket research in turn drew other high-tech industries to Alabama.

Today, industries such as primary metals, aerospace, paper, textiles, and food processing dominate Alabama's economy. The chief agricultural products are soybeans, peanuts, cotton, and poultry. Although Alabama was one of the most rural U.S. states in 1920, 60% of the population now lives in urban areas.

Places to Visit

Birmingham (800-962-6453) is home to the Southeast's largest municipal museum, the Birmingham Museum of Art, as well as the Jazz Hall of Fame and the new Birmingham Civil Rights Institute. Forty miles northwest of Birmingham is DeSoto Caves (205-378-7252), a network of vast onyx chambers that was used as a Native American burying ground. Montgomery (205-262-0013) has many lovely ante-bellum mansions and the White House of the Confederacy. Gulf Shores (205-968-7511) is the focal point of 50 miles of pristine white beaches. Tourist information: Alabama Bureau of Tourism and Travel, 401 Adams Ave., Box 4309, Montgomery, AL 36103 (800-252-2262/205-242-4169)

MISSISSIPPI

It's not hard to find Mississippi on a map, even if the outlines of the states aren't supplied. Just find the Mississippi River, the focal point of the vast system that drains half the United States and bisects America's heartland. Run your finger down the black line to the point where the river empties into the center of the Gulf of Mexico shoreline. The last 350 miles of the river are the western border of the state of Mississippi, a roughly rectangular area that extends 180 miles to the east. Mississippi is bordered on the east by Alabama, on the north by Tennessee, on the west by Arkansas and on the west and south by Louisiana. The state has 44 miles of coastline, and offshore are the glittering white sand beaches of the Gulf Islands National Seashore.

Jackson, the capital and largest city, is 180 miles north of New Orleans and 213 miles south of Memphis on I-55. Atlanta is 383 miles to the northeast and Dallas is 409 miles to the west on I-20.

THE LAY OF THE LAND

When I think of Mississippi, I picture trees—not only the breathtaking flowering magnolias (the state tree and the reason for its nickname, "The Magnolia State"), but vast expanses of forest. More than half the state is covered by trees, the result of one of the nation's most extensive reforestation programs restoring natural beauty to areas plundered by uncontrolled logging.

Mississippi lies entirely within the Gulf Coastal Plain. The trees adorn rolling hills, but no mountains—the highest elevation in the state is just 806 feet above sea level. Mississippi slopes gently from north to south. Mississippi was blessed with generous areas of unusually fertile soil. The Black Prairie, which has rich, dark soil, covers a broad section of the state. A band stretching eastward about 65 miles from the Mississippi is delta, a combination of silt and organic material deposited by periodic flooding of the great river known as "Big Muddy." The Gulf Coast area has a thin band of sandy soil as well as marshland.

The western portion of the state is drained by two major tributaries of the Mississippi River, the Yazoo and the Big Black. The Pearl River drains the central portion of the state, while the Pascagoula River drains southeastern Mississippi. The state's natural lakes are primarily "oxbow" lakes, bodies of water created when the Mississippi River changed course. Man-made reservoirs dot many of the rivers, especially the Pearl and the Yazoo.

HOW MISSISSIPPI BECAME A STATE

Rivers were the highways for the early European explorers. Because the Mississippi River was the most important river in the New World, it's not surprising that several nations vied for control of the waterway and its banks. At various times, the flags of France, Britain and Spain flew over the present-day state.

The French established the first permanent settlement, Fort Maurepas, on the site of the present-day Ocean Springs near what is now Biloxi in 1699. In 1716, the French established Fort Rosalie on the present-day site of Natchez.

Here's how the U.S. ended up with what is now Mississippi:

Britain won the area from France in the French and Indian War that ended in 1763.

Britain ceded control to the fledgling U.S. after the Revolutionary War, but Spain had taken advantage of the conflict to seize physical control of the area.

The U.S. gained control of most of the land between Georgia and the Mississippi River through negotiation, and in 1798 the U.S. Congress created the Mississippi Territory.

US General Andrew Jackson seized the territories near the Gulf Coast from Spain during the War of 1812.

The Mississippi Territory was divided by the U.S. Congress in 1817, and the western portion was admitted to the Union as the 20th state on December 10 of that year. The eastern portion became the Alabama Territory. The state was named after the river, which comes from an Indian word for "father of waters" or big river."

MISSISSIPPI'S CITIES

JACKSON (196,637), named for Andrew Jackson, was incorporated shortly after statehood and became the capitol in 1821. The thriving city was burned to the ground by Union forces during the Civil War (it was nicknamed Chimneyville because chimneys were all that were left standing.) The city didn't recover until gas wells were discovered nearby in the 1930s. Jackson is now the state's manufacturing, financial and medical center and is home to the Mississippi Museums of Art and Natural History. In 1993, I had the great pleasure of being a guest in the historic Governor's mansion, which has been continuously inhabited since 1841.

BILOXI (46,319), Mississippi's Gulf port, is a major tourism and fishing center. The French built a fort on the site in 1716, and a town grew up around it that was named for the Biloxi Indians. Excursion boats to the beaches and wilderness areas of the Gulf Islands leave from Biloxi.

NATCHEZ (19,460), located on the Mississippi and named for the Natchez Indians, was a thriving commercial center under French and British rule and was the capitol when the Mississippi Territory was created by Congress in 1789. Natchez was the southern end of the Natchez Trace, an important 19th century road that followed long-used Indian paths that began in Nashville. Natchez boasts more than 500 pre-Civil War mansions.

MISSISSIPPI'S PEOPLE

Over 35% of Mississippi's 2,642,748 people are African-Americans, the highest percentage of any state. Most Mississippians are descended from English settlers who moved to the region from the north and east and from Africans who were forcibly brought to America as slaves. Blacks outnumbered whites until 1940, when migration to northern cities began. 53% of Mississippians live in rural areas, making it one of the least urbanized U.S. states.

MISSISSIPPI'S ECONOMY

Cotton plantations made Mississippi one of the most prosperous pre-Civil War states. However, more than 60,000 state residents were killed in the war (nearly 1 out of every 10 residents) and the economy was shattered. Cotton made a slow comeback, but many Mississippi farmers diversified when the boll weevil devastated the cotton crop in the early part of this century. Today, Mississippi's most valuable agricultural product is still cotton (it's the third leading cotton producing state.) Corn, soybeans, broiler chickens and cattle are also important.

Mississippi's forests provided the raw material for a growing lumber and wood products industry after 1900. Today, the apparel and transportation equipment industries also employ large numbers of state residents. However, Mississippi residents have the lowest per capita and family incomes of any state.

Places to Visit

Because Natchez (800-647-6724) was virtually untouched by the Civil War, it is the site of more surviving opulent plantation homes than any other southern city. Jackson's(800-354-7695) Governor's Mansion is the oldest continually occupied state residence, and the Mississippi Museum of Art is also located there. The Gulf Islands National Seashore (601-875-9057) has unspoiled beaches and Vicksburg (800-221-3536) is the site of the beautiful and historic National Military Park commemorating the Civil War Battle. Tourist information: Mississippi Division of Tourist Development, Box 22825, Jackson, MS 39205 (800-647-2290/601-359-3297)

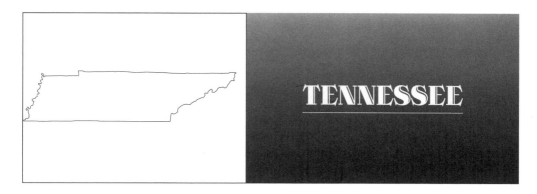

TENNESSEE

If you straddled the Tennessee border as you circled the state clockwise, your left foot would touch eight other states—a distinction Tennessee shares only with Missouri. Tennessee is bordered by Kentucky and Virginia on the north, North Carolina on the east, Georgia, Alabama, and Mississippi to the South, and Arkansas and Missouri on the west. The Mississippi River makes up the entire western border. The state is shaped like an upside-down anvil, with an east-west width of 440 miles, but a north-south width of just 120 miles.

Memphis, the state's largest city, is 210 miles southwest of Nashville and 137 miles northeast of Little Rock on I-40 and 393 miles north of New Orleans on I-55. Nashville is 175 miles south of Louisville and 243 miles northwest of Atlanta on I-24. Knoxville is 255 miles south of Cincinnati on I-75 and 256 miles west of Charlotte on I-40.

THE LAY OF THE LAND

Tennessee slopes from the Blue Ridge Mountains in the east to the Mississippi River on the west. The highest peaks (including Clingman's Dome, at 6,643 ft. the second-highest in the entire Appalachian Chain) are in the Unaka Mountains in the southeastern corner. West is a region of fertile valleys separated by rocky, tree-covered ridges. The center of the state is part of the flat Cumberland Plateau, which gives way to the rolling hills of the Gulf Coast Plain, then a narrow band of soil enriched by Mississippi River flood waters.

The Tennessee River, which drains 39,000 square miles, begins near Knoxville at the junction of the Holston and French Broad rivers, then flows in a great U-shape, southward into Alabama, then back northward through western Tennessee and Kentucky before emptying into the Ohio River. The Tennessee often flooded and was difficult to navigate before the Tennessee Valley Authority (TVA) built 33 dams along its length. A by product of the TVA's work has been the creation of several large lakes. Tennessee's only large natural lake is Reelfoot (located in the northwest corner), which was created by the violent earthquakes centered around New Madrid, Missouri in 1811 and 1812.

The Mississippi is the other large river that drains a significant portion of Tennessee. An area along the river is part of the Mississippi Flyway, a path for tens of millions of migrating birds.

HOW TENNESSEE BECAME A STATE

The area was inhabited for at least 2,500 years before Columbus by native peoples known as the Mound Builders. When Robert Cavelier, sieur de La Salle, claimed the area for France in 1682, it was a rich hunting ground for the Chickasaw, Creek, and Cherokee tribes. The French had little impact on the area before they ceded it to Great Britain in 1763.

Hunters from the coastal colonies first ventured into Tennessee in the 1760s, and settlers followed. Far from any existing government, these settlers formed their own independent government, the Wantauga Association, in 1772 to protect themselves. Four years later, when the Revolutionary War broke out, the Wantaugans feared attacks from British-controlled Indians. So they asked North Carolina to annex them for protection.

Here's the somewhat tangled process by which this new North Carolina county became a separate state:

In 1783, North Carolina ceded the area to the U.S. government.

In 1784, the people of the area declared themselves a new state, which they named Franklin. The State of Franklin petitioned to join the United States.

North Carolina promptly voided the cessation of the territory. Franklin managed a precarious existence until 1789, when North Carolina finally managed to regain control.

Under pressure from the territory and Washington, North Carolina once again ceded the area to the U.S. In 1790, Congress declared it the Territory of the United States South of the River. After the 1795 Census showed the area had the required 60,000 population, it was admitted to the Union as the 16th state on June 1, 1796. It was named Tennessee, after a Cherokee Indian village.

TENNESSEE'S CITIES

MEMPHIS (610,337), the largest city, was founded on the site of former French and British forts by General Andrew Jackson in 1819. It was named for the capitol of the ancient Egyptian empire, which also lay on one of the world's great rivers. The location of Memphis has made it a thriving transportation hub, cotton market, and livestock center. As the home of the blues and the late Elvis Presley, Memphis has also made a major contribution to American culture.

NASHVILLE (516,880), the state capitol, began in 1779 as Fort Nashborough, named for Revolutionary War General Francis Nash. Located on the Cumberland River and at the northern end of the Natchez Trace, Nashville became a thriving commercial and distribution center. But today Nashville is best known as the home of country music—even I've joined millions of other Americans in making a pilgrimage to the Grand Ole Opry where I once broadcast live for *Good Morning America*.

KNOXVILLE (165,039), named for General Henry Knox, the first U.S. secretary of war, was founded in 1786 and soon became a supply center for settlers moving westward. Knoxville is the location of the University of Tennessee and the headquarters of the Tennessee Valley Authority (TVA).

TENNESSEE'S PEOPLE

Tennessee's 5,098,798 residents are 83% white, 17% Black, and less than 1% other minorities. Traditionally an agricultural state, Tennessee experienced very slow population growth between 1830 and 1930, and a majority of the population still lived in rural areas as late as 1950. But rapid industrialization has led to equally rapid growth of metropolitan areas. Today, two-thirds of Tennessee's people are urban dwellers and half live in the four largest metropolitan areas.

TENNESSEE'S ECONOMY

A dramatic change in the economy of Tennessee began in 1933, when Congress created a federal corporation to build dams to control flooding on the Tennessee River, to build power plants for local use as well as national defense, and to develop new types of fertilizers. The Tennessee Valley Authority made the Tennessee River fully navigable, created jobs and recreational areas, and, most importantly, provided low-cost electricity to an 80,000 square mile area that was very attractive to industry. Today, Tennessee has a diverse manufacturing base that includes chemicals, clothing and textiles, industrial machinery, and transportation equipment.

Livestock such as cattle, calves, hogs, and chickens contribute more than half of the state's agricultural income, while soybeans and tobacco are the two most important crops. Tourism and the music industry also make significant contributions to the economy.

Places to Visit

Nashville (615-259-4700) is the home of country music and a mecca for visitors to the Grand Ole Opry, Opryland, and dozens of other attractions. Memphis (800-447-8278), on the other hand, is home of the blues and was the residence of Elvis Presley, the "King of Rock and Roll." Graceland (800-238-2000), Presley's estate, is one of America's most visited homes. Nestled in the mountains of eastern Tennessee is Dollywood (800-365-5996), Dolly Parton's theme park. Tourist information: Tennessee Department. of Tourist Development, Box 23170, Nashville, TN 37202 (615-741-2158)

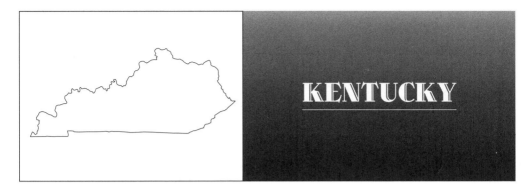

KENTUCKY

Kentucky, which is shaped like a high-top sneaker pointing westward, has wiggley borders on three sides that have largely been drawn by rivers. Its northern border is the Ohio River, which separates the Bluegrass State from Ohio, Indiana and Illinois. Part of the eastern border with West Virginia and Virginia is formed by the Big Sandy and Tug Fork rivers. Tennessee lies to the south of a straight line border drawn on a map, while the Mississippi River forms the short western border with Missouri. Kentucky has a maximum east-west length of 425 miles, and is 175 miles from north to south at its widest point.

Louisville, the largest city, is 299 miles southeast of Chicago on I-65, 103 miles southeast of Cincinnati on I-71, and 258 miles east of St. Louis on I-70. Lexington is 74 miles east of Louisville on I-70, 173 miles northwest of Knoxville on I-75, and 174 miles east of Charleston, West Virginia on I-64.

THE LAY OF THE LAND

When I think of Kentucky, the first thing that pops into my mind is bluegrass country and the magnificent thoroughbreds that compete in the Kentucky Derby every spring. This relatively flat grassland region, called the Lexington Plain, covers much of the center of the state. It's surrounded by the Pennyroyal Plateau, an uplands area. Much of the Pennyroyal lies on top of limestone that has been eroded away over the years to produce natural wonders like the Mammoth Cave system. The northwestern part contains the rich Western Coal Fields.

The eastern quarter of Kentucky is part of the mountainous Cumberland Plateau, which contains the Eastern Coal Fields. Bisecting the rugged peaks is the Cumberland Gap, through which so many settlers came west. The southwestern corner of the state is a fertile part of the Gulf Coast Plain known as the Purchase.

Most of Kentucky's significant lakes were formed by damming rivers. The state does contain some spectacular waterfalls, including the Cumberland Falls.

HOW KENTUCKY BECAME A STATE

The most famous name in early Kentucky history was the frontiersman Daniel Boone. Boone and five companions crossed the Cumberland Gap and explored eastern Kentucky in 1767. But others were reluctant to follow because of the hostility of the Indians. In 1774, James Harrod and 40 other Virginia residents formed Harrodsborough, the first permanent settlement and one year later, Boone founded Boonesboro.

Daniel Boone alleviated some of the Indian problems by negotiating the sale of 17 million acres of land, 65% of present-day Kentucky, to a group called the Transylvania Company, which was headed by Richard Henderson. The price: £10,000 pounds sterling. Unfortunately for the Transylvania Company, Virginia refused to recognize the legitimacy of the sale.

Henderson continued to work towards independence. In 1776, Virginia organized the entire region as Kentucky County (the name comes from an Iroquois Indian word meaning "meadowlands"), with a county seat at Harrodsborough and separate delegates to the Virginia legislature. On June 1, 1792, Kentucky was admitted to the Union as the 15th state and the first carved from the western wilderness.

KENTUCKY'S CITIES

LOUISVILLE (269,555), the largest city, was founded on the Ohio River in 1778 by George Rogers Clark. The settlement was first located on Corn Island, but was soon moved to the shore. In 1780, the settlement was named for King

Louis XVI in gratitude for France's assistance during the American Revolution. Louisville soon became a major river port and later a major rail center. Louisville is also the home of historic Churchill Downs, where the Kentucky Derby is run each May. I have had the great pleasure of broadcasting live from Churchill Downs on the eve of the Derby each year except one, since 1986. Don't miss the Kentucky Derby Museum, the Kentucky Museum for the Arts, and the Museum of Science and Industry as well.

LEXINGTON (225,366) was selected as a site for settlement in 1775 by Robert Patterson and Simon Kentun, who had just received word that the colonists had dealt a blow to the British at the Battle of Lexington. The site was settled in 1779, and the first state legislature met at Lexington in 1792. Located in the middle of the fertile Bluegrass Region, the city became a major tobacco market and the center of the thoroughbred horse breeding industry. Major attractions are the Kentucky Horse Park, the John Hunt Morgan House, and the Mary Todd Lincoln House (owned by the parents of the President's wife).

FRANKFORT (26,535), the state capitol, was founded in 1786 at Frank's Ford, named in honor of Stephen Frank, who died fighting Indians. Located halfway between Louisville and Lexington, the city became the center of state government.

KENTUCKY'S PEOPLE

The state's 3,788,088 residents include only 8% minorities, one of the smallest of any southern state. One reason is that the pre-Civil War economy was not dependent on slave labor—in fact, Kentucky outlawed slavery in 1833. Although the ban was repealed in 1850, the issue remained so politically contentious that Kentucky did not join the Confederacy. Today, Kentucky remains predominately a rural state, with just 46% of the population living in urban areas.

KENTUCKY'S ECONOMY

Although Kentucky has developed a diverse modern economy, it's known primarily for coal, tobacco, whiskey and thoroughbred horses. The state is the nation's leading producer of bituminous coal, which accounts for 90% of its income from minerals. Kentucky produces the nation's second largest tobacco crop and has the world's largest loose-leaf tobacco market. The state has the distinction of being the world's leading producer of whiskey, and its farms also raise significant numbers of cattle in addition to horses. It's not surprising to learn that Kentucky sells more fine grass seed than any other state. Kentucky's not called the Bluegrass State for nothing.

Manufacturing produces an increasing amount of the state's income. The leading industries are non-electrical machinery, food products, electric and electrical products, and printing and publishing.

Places to visit

Kentucky is known for its racehorses, and Lexington (800-845-3959) is the home of the Kentucky Horse Park and the focus of the 400 horse farms of Bluegrass Country. Louisville (800-626-2646) hosts the Kentucky Derby at Churchill Downs and is home to the Kentucky Center for the Arts and the Museum of History and Science. Mammoth Caves National Park (502-758-2328) contains 350 miles of spectacular underground passages. Tourist information: Kentucky Department. of Travel Development, 2200 Capital Plaza Tower, Frankfort, KY 40601 (800-225-8747/502-564-4930)

LOUISIANA

Mention Louisiana, and a smile forms on my lips—the state's unique culture and hospitable people make it a favorite tourist destination and a personal favorite of mine. But the state that surrounds the mouth of the Mississippi River has also played a long and crucial role in American history. Louisiana, an L-shaped state that "faces" east, is bordered on the north by Arkansas, the east by Mississippi, the south by the Gulf of Mexico, and the west by Texas. The Mississippi and Pearl Rivers form much of the eastern boundary, while the Sabine River separates Louisiana and Texas along two-thirds of the boundary. Louisiana has 397 miles of coastline on the Gulf of Mexico.

New Orléans is 352 miles east of Houston and 143 miles west of Mobile on I-10. The home of Mardi Gras is 393 miles south of Memphis and 183 miles south of Jackson on I-55. Baton Rouge is 80 miles northwest of New Orleans and 226 miles southeast of Shreveport.

THE LAY OF THE LAND

Louisiana's physical geography is dominated by the Mississippi River, which forms the eastern border of the top of the L, then plunges through the middle of the lower portion of the state before emptying into the Gulf at New Orleans. The river has created a 15,000 square mile delta of built-up silt as well as broad flood plain on either side of its course. Because soil deposited over many centuries has made the banks higher than the flat land beyond, frequent flooding was a major problem until an extensive series of levees was built.

The western portion of the state, part of the Gulf Coast plain, is gently rolling prairie land that slopes from the north to the south. Near the Gulf is a region of marshes that give way to sandy beaches. In the eastern portion of the state is another small plain area between Lake Pontchartrain and the Mississippi River. Louisiana has a mean elevation of just 100 feet, making it the second lowest state after Delaware.

The brackish Lake Pontchartrain, which is located north of New Orleans, is the state's largest inland body of water. The meanderings of the Mississippi and its tributaries have created numerous oxbow lakes that dot the state. Louisiana's waterways include many bayous, slow-moving streams through marshland.

HOW LOUISIANA BECAME A STATE

Robert Cavelier, sieur de La Salle, named the future state as well as the entire Mississippi basin when he claimed the area for King Louis XIV after reaching the mouth of the river in 1682. In 1712, Antoine Crozat, a French merchant, was granted a charter to operate the colony, and the first permanent settlement was founded at Natchitoches. In 1718, Crozat's charter was turned over to a colorful Scottish gambler and businessman named John Law. That same year New Orleans was founded, and it become the capital of the territory in 1722. During this time, a large number of Germans immigrated to Louisiana.

But Law's charter was revoked in 1731, and the French government took over. France ceded the Louisiana Territory to Spain in 1763. After a short-lived rebellion by French settlers, Spanish governors brought stability and growth to New Orleans. In 1800, Spain gave the territory back to France. However, the importance of the Mississippi River became apparent to President Thomas Jefferson and other leaders. When the government of New Orleans banned American trading goods from the port, many residents of Tennessee, Mississippi and other river states called for war. Jefferson sent James Monroe to France to negotiate with Napoleon Bonaparte. In a brilliant stroke of diplomacy, Monroe arranged to buy the entire Louisiana Territory for $15 million dollars. In 1804, the lower part of the Purchase was organized as the Territory of Orleans, and it was admitted to the union as the State of Louisiana on April 30, 1812. But U.S. control of the important port was not secure until the end of the War of 1812, when Gen. Andrew Jackson defeated the British at the Battle of New Orleans in 1815.

LOUISIANA'S CITIES

NEW ORLEANS (496,938) was founded in 1718 by Jean Baptiste le Moyne, sieur de Beinville. He named it Nouvelle Orleans, in honor of the duc d'Orleans, regent of France. The city prospered under several different governments before the U.S. secured control after the War of 1812. New Orleans' oldest district, the French Quarter, is a world famous tourist attraction and Bourbon Street, the main drag, is the home of Dixieland Jazz. The city is most famous for its celebration of Mardi Gras ("Fat Tuesday"), a wild and crazy carnival that now lasts for the week before Ash Wednesday, the first day of Lent.

BATON ROUGE (219,513), the furthest point of deep water navigation of the Mississippi, was founded by the French in 1719. The city's name means "red stick," supposedly derived from a red cypress tree that stood in the middle of an Indian village. Baton Rouge was an important shipping center that became the state capitol in 1849.

SHREVEPORT (198,525) was named for its co-founder Henry R. Shreve, a trader and steamboat builder who decided to build a settlement on the spot while helping to clear a huge hundred-mile-long log jam on the Red River. Shreveport, an important trading center for cotton and other agricultural products, was the Confederate capitol of Louisiana. The discovery of petroleum nearby spurred the growth of the city over the last four decades.

LOUISIANA'S PEOPLE

Louisiana's 4,295,477 people represent one of the most unusual ethnic mixes of any state, a mix that resulted from the pattern of settlement of the area. The French were the first settlers, but after 1718 came an influx of so many German and Swiss immigrants that for a time they outnumbered the French. In 1755, the British deported a group of French Canadians known as Arcadians from Nova Scotia. More than 2,500 moved to Louisiana, where they settled in the bayou country. Their descendants are known as Cajuns. Spanish rule brought Spanish-speaking Canary Islanders (sometimes called Islenos), slaves imported from Africa, and whites and free blacks from the West Indies.

The result of this immigration is that a variety of languages and dialects are still spoken in Louisiana. Some generations-old families still speak French, the Cajuns use a French-based patois, some descendants of the Canary Islanders speak Spanish, and some blacks use a mixed dialect of African origin called "Gumbo." New Orleans has an architecture and government influenced by the Spanish, a French-based culture, and a tradition of jazz music that has African-American origins.

LOUISIANA'S ECONOMY

The new state of Louisiana had a vigorous economy based on shipping and agriculture. The inland regions were plantation country, with cotton the dominant crop in the north and sugarcane in the southern portion of the state. Today, Louisiana agriculture is still primarily crop-based—it's number one in sweet potatoes, number two in sugarcane and rice, and number three in soybeans. Cotton is still an important crop.

Steamboat traffic on the Mississippi made New Orleans the nation's second leading port by 1840 and the third largest city by 1852. I was surprised to find out that today the Port of South Louisiana near New Orleans is America's busiest port, handling 200 million tons per year, while Baton Rouge and New Orleans rank fifth and sixth.

The discovery of huge oil reserves in Louisiana early in this century gave the state another source of income. After Texas, Louisiana is second in the nation in income derived from mineral wealth, providing 15% of America's petroleum and 28% of our natural gas. The availability of petroleum and ease of shipping also attracted industry to the state, especially the chemical industry.

I'm not so sure that anyone has estimated with accuracy what Louisiana's restaurant industry contributes to the state's economy, but I can say from extensive personal experience that no state has a higher concentration of great eating establishments than Louisiana.

Places to Visit

New Orleans (504-566-5031), site of Mardi Gras, America's most famous street festival, might also be the most fun-loving U.S. city. Don't miss the French Quarter, Bourbon Street, or the world-famous creole and cajun cuisine. Southwest Louisiana (800-456-7952) is Cajun country, which was settled in the seventeenth century by French-speaking settlers who were forced out of Canada—with 40% of the residents still speaking Cajun French, you'll think you're in a different country. Tourist information: Louisiana Office of Tourism, Box 94291, Baton Rouge, LA 70804 (800-334-8626)

GEORGIA

Georgia, a state of great natural diversity, is the southernmost of the original 13 colonies. It's bordered by Tennessee and North Carolina to the north, South Carolina and the Atlantic Ocean to the east, Florida to the South, and Alabama to the West. Georgia is roughly rectangular in shape, with South Carolina chopping off the northeast corner. The state stretches 320 miles north to south, and 255 miles east to west. The Savannah River forms the entire eastern border with South Carolina, while the Chatahoochee River is the southern two-thirds of the western border with Alabama. The state has 100 miles of coastline on the Atlantic.

Atlanta, the "New York of the South," is 246 miles southeast of Nashville and 256 miles northwest of Savannah on I-24. On I-85, Atlanta is 240 miles southwest of Charlotte and 157 miles northeast of Montgomery. Savannah, on the Atlantic coast, is 140 miles north of Jacksonville, 499 miles north of Miami, and 243 miles south of Raleigh on I-95.

THE LAY OF THE LAND

The Blue Ridge Mountains extend into the northern and northeast portions of this geographically diverse state. North of Atlanta, the mountains give way to the Piedmont Plateau, a broad swath of rolling hills that cover the center of the state. In southern Georgia, the Piedmont drops off abruptly to lowlands that are part of the Atlantic Coastal Plain in the east and the Gulf Coastal Plain in the west. The Atlantic Coastal Plain, which has the more fertile soil, gives way to marshes and swamps as it nears the ocean. Much of the Gulf Coast Plain is on top of vast limestone deposits, so it's marked by many sinkholes.

Right off the coast is part of a chain of islands (including St. Simons, Cumberland, and St. Catherines) that extend north to South Carolina and south into Florida. Georgia has no large natural lakes, but damming of rivers has produced several of the man-made variety. On the southern border with Florida is the 160 mile Okefenokee Swamp, an area of islands, lakes, brush vines, and cypress trees that is drained by the St. Marys and Suwannochee rivers.

HOW GEORGIA BECAME A STATE

The Spanish built a mission on St. Catherines Island in 1566, and later added a number of other forts and missions on the coast. However, French and British pirates plagued the Spanish, especially the famous Blackbeard, who made his headquarters on Blackbeard's Island in 1716.

The first permanent settlement was created by a British philanthropist named James Oglethorpe. Moved by the plight of debtors languishing in British jails, Oglethorpe petitioned King George II for a charter for a New World colony in which these unfortunates could be given a second chance. Eager to have a buffer between Spanish Florida and South Carolina, the Crown granted Oglethorpe in 1732 a charter for an area he named Georgia in honor of the King. In 1733, he built the first settlement at Savannah.

Oglethorpe proved the King wise when a force at his command defeated the Spanish at the Battle of Bloody Marsh in 1742, ending their attempts to move northward. The colony prospered, especially after James Wright became governor in 1753.

A significant portion of Georgia residents were royalists when the American Revolution broke up, but the state did join the Continental Congress and sign the Declaration of Independence. In late 1778, the British did capture Savannah, but guerrilla forces kept them from moving inland. The British evacuated the state in 1782.

Georgia became the fourth of the original thirteen states on January 2, 1788, and it was one of only three states that unanimously approved the new constitution.

GEORGIA'S CITIES

ATLANTA (394,017) is the state capitol and the commercial, financial, and transportation center of the South. The site was ceded to Georgia by the Creek Indians in 1821, and it was settled in 1837 at the southern end of a new rail line to Chattanooga. The city was first named Terminus, then Marthasville before the residents settled on Atlanta, the feminine form of Atlantic. By the time of the Civil War, Atlanta was a major rail hub. Although it was burned by General Sherman on November 15, 1864, the city quickly recovered and became Georgia's capitol in 1868. On my many visits to Atlanta, I've been impressed by the vibrancy and energy of the city, and it's easy to understand why it was chosen to host the 1996 Summer Olympics. Southern hospitality is more tahn a cliché here.

SAVANNAH (137,812) is one of the oldest and most beautiful cities in the South. James Oglethorpe designed the city so that each family's plot fronted on a public square. Named either for the Sawana Indians or a Shawnee Indian word for Savannah River, the city soon became a thriving port that was the state's capitol from 1754 to 1780. Beginning in 1895, the Savannah Cotton Exchange set world cotton prices. Today, Savannah is a frequent tourist destination and a major shipping and shipbuilding center.

AUGUSTA (44,707) was founded in 1735 at a fur trading post and named for the mother of King George III. It became the capitol of the state from 1786 to 1795 and was the commercial center for a large area of Georgia and South Carolina. Nearby is Fort Gordon, the headquarters of the U.S. Army Signal Corps., which contributes a great deal to the area's economy. Augusta is best known as the home of the Masters golf tournament, which is held at the Augusta National Golf Club every April.

GEORGIA'S PEOPLE

Georgia's population of 6,917,140 makes it the eleventh largest state, and it is growing rapidly (four of the ten fastest growing counties in the U.S. are in Georgia). A little over one-quarter of the state's residents are black, the fifth highest percentage of any state. Atlanta's Dr. Martin Luther King, Jr. was a towering figure in the struggle for racial equality that helped make Atlanta the progressive and cosmopolitan city it is today.

Although Georgia was largely rural for much of its history, today two-thirds of its residents live in metropolitan areas.

GEORGIA'S ECONOMY

Agricultural fueled Georgia's early growth, and it remains important today. Jimmy Carter's election focused attention on the fact that the state is America's leading peanut producer, and it's also number one in pecans and peaches. Cotton and soybeans are also important crops. More than half the agricultural income comes from livestock and livestock products, especially broiler chickens, cattle, and eggs.

Georgia is a major producer of apparel and textiles, including carpets. Other major industries in a very diverse economy are transportation equipment, processed foods, and paper and paper products.

Atlanta is also the South's primary financial, communications, transportation, and service industry center—I love the view of its dramatic skyline as I approach downtown. Its airport is the nation's fourth busiest, and it is also a major tourist center.

Places to Visit

As host to the 1996 Summer Olympics, Atlanta ((404) 222-6688) will be showcased to the world. When you visit, don't miss the King Center, focal point of the Martin Luther King Jr. Historical District. North of the capital is Stone Mountain Park (404-498-5690), which features the world's largest sculpture, museums, and recreational activities.   The historic district of Savannah (800-444-2427) is the nation's largest urban landmark with its 1,000 restored homes. Calloway Gardens (800-282-8181) in Pine Mountain, 70 miles southwest of Atlanta, is a 2,500 acre horticultural fantasyland. Tourist information: Georgia Department. of Trade and Tourism, Box 1776, Atlanta, GA (404)656-3590)

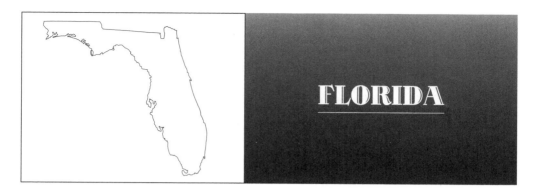

FLORIDA

Florida, the southernmost of the contiguous 48 states (Hawaii is nearer the equator), is primarily a large peninsula about the size of the Korean peninsula. Florida's Panhandle region (class, can you guess what shape this region is?) extends westward on the north side of the Gulf of Mexico. Florida is bordered by Alabama and Georgia on the north, the Atlantic Ocean on the east, the Straits of Florida (which separate it from Cuba) on the south, and the Gulf of Mexico and Alabama on the west. The state is 450 miles from north to south (almost twice the distance between New York and Washington DC.), but only an average of 130 miles east-to-west. The state includes several groups of islands, including the Sea Islands, the Florida Keys, the Marquesas Keys, the Dry Tortugas, and the Ten Thousand Islands. The state's 1,350 miles of coastline (the most of any state except Alaska) has numerous sandbars and coral reefs.

THE LAY OF THE LAND

With the exception of the Panhandle, Florida is part of the Atlantic Coastal Plain. With a mean elevation of 100 feet, it's also one of the flattest states. Most of the land in the central and northern portions of the state has sand and clay on top of limestone beds. Water eating away at the limestone has created numerous sinkholes and lakes. The Everglades and the Big Cypress Swamp dominate most of southern Florida. The Panhandle, part of the Gulf Coast Plain, consists of low, swampy land giving way to red clay hills.

Florida has few rivers but a large number of lakes. Lake Okeechobee north of Miami is the third largest lake completely within U.S. borders. The Everglades, a vast, freshwater marsh covering 5,000 square miles, was created by water overflowing from Lake Okefenokee. One of the wildest and least accessible areas in the U.S., the region is covered by sawgrasses, cypress trees, mangos, oaks and pines. The Everglades is one of America's most important wildlife refuges.

HOW FLORIDA BECAME A STATE

Florida was first explored in 1513 by Juan Ponce de León, who searched in vain for the Fountain of Youth. He named the state after the Spanish word for flower, partly because he arrived on Easter, which is "pascua florida" in Spanish. In 1562, Jean Ribeaut led a group of French Huguenots (Protestants) to Florida and claimed it for France. However, the French were soon driven off by the Spanish, who built the first permanent settlement at St. Augustine in 1567.

The Spanish efforts to colonize the area were seriously hampered by attacks by the Indians and the French. In 1763, Spain traded Florida to Great Britain in exchange for Cuba. Britain divided the territory into East Florida (the peninsula) and West Florida, which stretched all the way to the Mississippi. Not waiting for a British defeat, Spain invaded Florida in 1779 and got the territory back in the 1783 Treaty of Versailles by giving Britain the Bahamas and Gibraltar.

The new United States was eager to acquire this Spanish property, and by 1815 had managed to gain possession of West Florida. In 1818, General Andrew Jackson invaded East Florida to retrieve runaway slaves and captured Pensacola. Under pressure, Spain ceded its remaining territory to the U.S. in the Adams-Onis Treaty of 1819.

However, the fighting was far from over. The Seminole Indians resisted giving up their lands and moving west, but after the Second Seminole War (1835-1842), almost all were dead or relocated. Although the U.S. Congress was reluctant to add another slave state, Florida was admitted to the Union as the 27th state on March 13, 1845.

FLORIDA'S CITIES

MIAMI (358,548) is the focus of a 25 community metropolitan area known as Florida's "Gold Coast." The U.S. Army established Fort Dallas on the site in 1835, but it did not become a settlement until the 1870s. It's the population didn't begin to grow until 1896, when Henry M. Flagler built a railway line to the area. Miami, which received its name from an Indian word meaning "Big water" (probably because of Lake Okefenokee), soon experienced a land boom and became a major tourism and retirement center. Cuban immigrants fleeing their homeland after Castro took over revitalized the area's economy and gave the city a distinctive Latin flavor.

JACKSONVILLE (672,971), the state's largest city, was the site of Fort Caroline, established in 1765. The civilian settlement was founded in 1816, and in 1832 the city was named for General Andrew Jackson, the first territorial governor of Florida. Jacksonville soon became the center of finance and commerce in northern Florida as well as important deep-water port. At 774 square miles, it encompasses one of the largest areas of any U.S. city.

TAMPA (280,015) originated as a plantation in 1823 and was incorporated 1851. Growth began after three developments occurred: phosphates used to make fertilizer were discovered nearby in 1883; the railroad reached the area in 1885; and the cigar-making industry began in 1886. Tampa (the name is from an Indian word meaning "sticks of fire") became a major commercial center and port. Today, as the site of Busch Gardens, it is a major tourist center and the home of the NFL's Tampa Bay Buccaneers.

FLORIDA'S PEOPLE

Florida, with 13,678,914, is the nation's fourth largest and fastest growing state. It also has one of the lowest percentages of residents who were born in the state. One reason is immigration—Florida's population is 12% Hispanic, and it has received large numbers of Haitian refugees. The second reason is retirement—18% of the state's population is age 60 or older, the highest of any state. More than 85% of Florida residents live in urban areas.

FLORIDA'S ECONOMY

Tourism is a cornerstone of Florida's economy, with 39 million visitors spending $28 billion in an industry employing 650,000 state residents. Orlando has more hotel rooms than any U.S. city except New York, and Disneyworld is the nation's leading single tourist attraction.

It's not surprising that contraction would be a major industry in the fastest-growing state. Service industries and government are two other areas that expand with the population.

Agriculture was the foundation of the early Florida economy and it still generates $6 billion in income. Florida is the nation's leading producer of citrus fruits, the second leading vegetable grower, and a major producer of flowers and nursery plants, cattle, sugarcane, chickens, and eggs.

Manufacturing has also been attracted by the climate. Major industries are electronics equipment, transportation equipment, precision instruments, and processed foods.

Places to Visit

The construction of Walt Disney World (407-824-4321) began the development of Orlando (407-363-5871) as America's number one tourist destination. Tampa/St. Petersburg (800-826-8358) are meccas for beachlovers as well as families who want to visit Busch Gardens and the Florida Aquarium. Greater Miami (800-283-2707) is the focal point of south Florida's Gold Coast. Among my favorite spots are the Florida Keys (800-352-5397), Everglades National Park (305-242-7700), and St. Augustine (904-825-1000), the oldest city in the U.S. Tourist information: Florida Division of Tourism, 126 Van Buren St., Tallahassee, FL 32301 (904-487-1462)

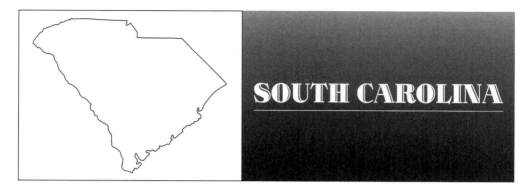

SOUTH CAROLINA

South Carolina, a small state that played a crucial role in early U.S. history, is a triangular-shaped state with 187 miles of coastline on the Atlantic Ocean. The state is bordered by North Carolina on the north, the Atlantic Ocean on the east, and Georgia on the south and west. The Savannah and Chattanooga rivers form most of the Georgia border. Numerous islands called the Sea Islands line the coast, including such famous resorts as Hilton Head and Kiawah Island.

Charleston, one of the nation's oldest cities, is 525 miles south of Washington DC. and 580 miles north of Miami on I-95. It is 320 miles east of Atlanta on I-26 and I-20. Columbia, the capitol, is 91 miles south of Charlotte on I-77 and 215 miles east of Atlanta on I-20.

THE LAY OF THE LAND

South Carolina has three distinct geographic regions. About half the state is part of the Atlantic Coastal Plain. On the shoreline are some of the nation's most beautiful beaches and a wide stretch of swamps and marshes. Further inland is an area known as the Pine Barrens, which gives way to rolling hills. A line that runs roughly through Columbia marks the beginnings of the Piedmont Plateau, an area colonists called the "upper country." In the northwestern part of the state the Plateau rises into the Blue Ridge Mountains.

The Savannah, the Santee, and the Pee Dee are the state's three major rivers, all of which flow into the Atlantic. South Carolina has no major natural lakes, but does have beautiful man-made bodies such as Lake Moultrie and Lake Marion.

HOW SOUTH CAROLINA BECAME A STATE

The very first European settlement on the Atlantic Coast was San Miguel de Guadalupe, founded by the Spanish in 1526. However, it lasted only a few months. In 1562, a group of French Huguenots settled on Parris Island (now the home of the Marine training base), the first Protestant settlement in America.

But the British soon dominated the area. In 1619, Charles I of England granted the area that is now North and South Carolina to Sir Robert Heath (Carolinus is the Latin word for Charles), and the charter later passed to a group known as the Lord Proprietors. The first permanent settlement was founded in 1670 and named Charles Towne. That same year slaves were first imported, and the area soon began to thrive. North Carolina became a separately governed colony in 1712. After the residents revolted against the Proprietors in 1719, South Carolina became a royal colony.

Despite the close ties to the crown, South Carolinians despised the high tariffs imposed by the British and enthusiastically joined the Continental Congress. More Revolutionary War battles (137) were fought in South Carolina than any other state. The heroic leadership of a group of guerrilla fighters by Francis Marion earned him the nickname "The Swamp Fox." The Battle of King's Mountain (October 7, 1780) and the Battle of the Cowpens (January 17, 1781) were critically important to the American victory. South Carolina became the eight state on May 23, 1788.

SOUTH CAROLINA'S CITIES

CHARLESTON (79,925) was founded on a peninsula between the Ashley and Cooper Rivers in 1670. The bounty of the state's plantations made it one of the wealthiest cities in the colonies. It attracted immigrants from all over the world, including the largest Jewish population in the New World. It was captured by the British in the

Revolutionary War, but afterwards the name was shortened to Charleston to lessen the connection to the British crown. The first shots of the Civil War were fired on Fort Sumter in Charleston Harbor on April 12, 1861. Today, Charleston attracts many tourists to its beautiful historic homes and gardens and the nearby resort islands. I've always enjoyed the city and I've been pleased to see how quickly the area rebounded from the destruction caused by Hurricane Hugo in 1989.

COLUMBIA (98,052) was laid out as a new state capitol by the legislature in 1786 in an attempt to ease a growing hostility between residents of coastal areas and those of the upper country area. The city was named for Christopher Columbus. The city was largely burned during the Civil War, but it rebounded to become a major center for the industry that moved to South Carolina in the last half-century.

GREENVILLE (58,282) and **SPARTENBURG** (43,467), located just 15 miles apart in the northwest corner of the state, have become a major metropolitan area. Greenville, named for Isaac Green, a local textile operator, was settled in 1776 and saw its first factory built in the 1820s. Spartanburg, named for the Spartan regiment that fought in the Revolutionary War, was founded in 1785 and incorporated as a city in 1831.

SOUTH CAROLINA'S PEOPLE

South Carolina's 3,642,718 people ranks it 25th among U.S. states. However, that figure represents an 11% increase over the last decade, making South Carolina one of the fastest growing states. The state's population is 30% black, the fourth highest percentage of any state. Its new popularity as a retirement area produced the seventh largest increase in numbers of residents age 60 and over, according to the 1990 Census. About 60% of residents live in urban areas.

SOUTH CAROLINA'S ECONOMY

Agriculture built South Carolina, and the state is still an important producer of tobacco, soybeans, and corn. Livestock and livestock products include broiler chickens, cattle, eggs, and hogs. Most of the farms are in the Atlantic Coastal Plain.

Manufacturing companies have been increasingly attracted to the Piedmont Plateau region by the low cost of labor. The leading industries are clothing and textiles, chemicals, industrial equipment, rubber and plastics, electrical equipment, and paper. The growth in manufacturing is demonstrated by the fact that Charleston went from the nation's 59th to 14th busiest port over just two decades of this century.

Tourism is the other major factor in the South Carolina economy. Myrtle Beach, the center of a stretch of magnificent beach known as The Grand Strand, is one of the nation's top ten tourist destinations. Hilton Head, off the southern shore, attracts many famous visitors, including U.S. presidents.

Places to Visit

Charleston (803-853-8000) is one of the best preserved and most beautiful cities in the south, and has attractions ranging from plantations to Fort Sumter to the aircraft carrier Yorktown. Myrtle Beach (800-6356-3016), the heart of the 60 miles swath of beautiful beaches known as the Grand Strand, is one of the top ten family vacation destinations in the country. Hilton Head Island (803-785-3673) is a playground for golfers, tennis players, and U.S. presidents. Tourist information: South Carolina Division of Tourism, 1205 Pendleton St., Box 71, Columbia, SC 29202 (800-346-3634/803-734-0235)

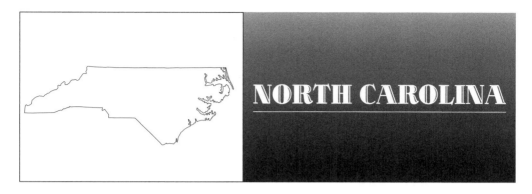

NORTH CAROLINA

North Carolina, one of the most geographically and economically diverse states in the South, is a roughly rectangular state with a "heel" that thrusts southward into South Carolina along the Atlantic Coast. It's bordered by Virginia on the north, the Atlantic Ocean to the east, South Carolina and Georgia on the south, and Tennessee on the west. The state covers an impressive 500 miles east-to-west, but just 150 miles north-to-south. North Carolina has 301 miles of coastline, off of which are the famous barrier islands known as the Outer Banks, including Cape Hattaras and Cape Fear.

Charlotte, the state's largest city, is 198 miles south of Roanoke and 91 miles north of Columbia, SC. on I-77, and 240 miles northeast of Atlanta on I-85. Raleigh, the state capital, is 253 miles south of Washington DC. on I-95.

THE LAY OF THE LAND

North Carolina is a state with starkly contrasting geographic areas. The eastern two-thirds of the state is part of the Atlantic Coastal Plain. The first 50 miles from the coastline are low-lying land marked by many swamps and marshes, including the Great Dismal Swamp. The land gradually elevates until it rises more abruptly at the Piedmont Plateau, which also covers about two-fifths of the state. The Plateau, covered with fertile, rolling land, slowly rises until it reaches the Blue Ridge Mountains that cover the westernmost 20% of the state. The Blue Ridge Mountains, which includes the Great Smoky Mountains, contains 40 peaks with elevations over 6,000 feet, including Mt. Mitchell (6684 feet), the highest elevation in the Appalachians.

Rivers in most of the state flow toward the Atlantic, while the Blue Ridge Mountains drain westward. North Carolina has few large natural lakes—Mattamuskeet, the largest, is in the middle of the Great Dismal Swamp. The Blue Ridge Mountains contain many beautiful waterfalls.

HOW NORTH CAROLINA BECAME A STATE

On July 4, 1584, two English explorers commissioned by Sir Walter Raleigh dropped anchor off the North Carolina coast; on August 17, 1585, 108 English settlers landed on Roanoke Island. Short of supplies, they returned to England the next year in ships commanded by Sir Francis Drake. In 1587, John White led another group of 121 settlers that formed a colony, and a month later, White's granddaughter, Virginia Dare, became the first European baby born in the New World. However, White went back to England to fetch supplies, and when he returned in 1590, the colony had vanished without a trace—one of the great mysteries of U.S. history.

In 1629, Sir Robert Heath was given a charter for what is now North and South Carolina, but he did little with it. So in 1663, the King sold the charter to eight men known as the Lord Proprietors. Under the proprietorship, South Carolina grew rapidly. But North Carolina lagged behind, hampered by Indian wars, pirates (until Blackbeard was killed in 1718) and competition from Virginia to the north. In 1712, North Carolina got a separate governor, and in 1728, the British government repurchased the charter from all but one of the Proprietors.

The settlers in North Carolina proved to be fiercely independent people who vigorously protected their rights. There was constant tension between the governors and the population. In 1765, settlers formed a group called the Regulators to protest onerous taxes and regulations. The Regulators were defeated in a 1771 battle that some consider the first of the Revolutionary War.

Given its desire for self-rule, it's not surprising that North Carolina was the first state to authorize its delegates to sign the Declaration of Independence. And it's also not surprising that North Carolina refused to ratify the Constitution after the war because the legislators believed it infringed on states' rights. North Carolina did not participate in the first presidential election, and it joined the Union on November 19, 1789, after the Bill of Rights was passed.

NORTH CAROLINA'S CITIES

CHARLOTTE (395,934) was settled in 1750 and named for Charlotte-Sophia of Mecklenburg-Strelitz, the wife of King George II of England. The city became an important commercial center and led North Carolina's vigorous support of the Confederate cause. It was also the site of the last Confederate cabinet meeting of the Civil War. Manufacturing began to dominate the economy in this century, with printing, chemicals, micro-electronics, textiles, and machinery becoming the leading industries. One of its prime tourist attractions is the Discovery Place, an award-winning hands-on science center.

RALEIGH (207,951) was founded after the American Revolution, when the legislature decided to build a capital city in the center of the state. The new city, which was named for Sir Walter Raleigh, was laid out in 1792. It became a center of the tobacco industry and the hub of a major manufacturing region. Among its attractions are the North Carolina Museums of Art and Natural History.

GREENSBORO (183,521) was founded in 1808 and named for General Nathaniel Greene, who led the U.S. forces at the Battle of Guilford Courthouse. It became the center of the cotton and textile industries. North Carolina's Zoological Park is south of the city.

NORTH CAROLINA'S PEOPLE

North Carolina has a population of 6,945,180, and the population is growing as high-tech industries move into the state. However, North Carolina is still one of the nation's least urbanized states, with 50% of the people living in rural areas. North Carolina also has more than 79,000 Native Americans, the largest population east of the Mississippi. About 22% of the population is Black.

NORTH CAROLINA'S ECONOMY

Agriculture was the foundation of the economy into the 20th century, and North Carolina's farms still produce $5 billion a year in revenue. The state is the leading tobacco grower, with sweet potatoes, peanuts, corn and soybeans the other important crops. Chief livestock produced are broiler chickens, hogs, and turkeys.

Manufacturing has become the star of the state's economy. North Carolina is the nation's leading producer of cloth and textiles, furniture, and tobacco products. Other leading industries include pharmaceuticals, fabricated metal products, rubber and plastics, and paper. High-tech industries have begun to thrive in the Raleigh-Durham-Chapel Hill "Research Triangle," which includes the University of North Carolina and Duke University.

Tourism is a $7 billion industry. The Outer Banks are a popular seaside destination, while Great Smoky Mountain National Park is one of the busiest of all national parks.

Places to Visit

Great Smoky Mountains National Park (615-436-5615), with more than 500,000 acres of natural grandeur in western North Carolina, is the most visited of all the national parks. At the opposite end of the state, the Outer Banks provided a haven for eighteenth century pirates and are now a prime vacations spots. Cape Hattaras National Seashore (919-473-2111) and Cape Lookout National Seashore (919-728-2250) are two of my favorite seaside locations. Raleigh (800-849-8499) has museums of history and natural sciences and is the gateway to the Triangle Area. Tourist information: North Carolina Division of Travel and Tourism, 430 N. Salisbury St., Raleigh, NC 27611 (800-847-4862/919-733-4171)

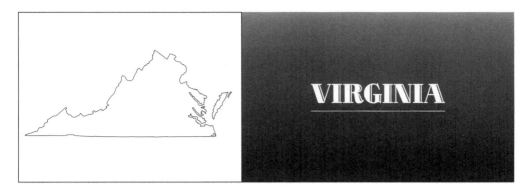

VIRGINIA

Right off the bat, I have to admit I'm biased—I'm extraordinarily proud of Virginia's place as the focal point of so much U.S. history. My native state is the northernmost of the southern states. Virginia is bordered by West Virginia, Maryland, and Washington, DC. on the north, Maryland and the Atlantic Ocean on the east, North Carolina and Tennessee on the south, and Kentucky on the west. The Potomac River makes up the state's northern boundary. The majestic Chesapeake Bay separates the main body of Virginia from the Eastern Shore, and the state has 112 miles of coastline. The state stretches 440 miles east-to-west and 200 miles north-to-south.

Richmond is 105 miles south of Washington, DC. and 148 miles north of Raleigh on I-95, and 301 miles northeast of Charlotte on I-85. Roanoke is 283 miles south of Harrisburg, Pennsylvania and 257 miles northeast of Knoxville on I-81.

THE LAY OF THE LAND

Virginia has five different geographic regions. The eastern fifth of the state is part of the Atlantic Coastal Plain. This region, which includes the small Eastern Shore area, is flat and covered by relatively infertile soil. The coastal areas are dotted with marshes and swamps. The Piedmont Plateau, gently rolling terrain, covers the central third of the state.

The Blue Ridge Mountains form a narrow band in northern Virginia that widens as it reaches Roanoke. On the other side of the mountains is a Valley and Ridge region that includes the Valley of Virginia or the Great Valley, of which the storied Shenandoah Valley is a part. The valley soil is extremely fertile. This region gives way to the Cumberland Mountains, an area of valleys dotted with higher peaks.

The most impressive body of water in the state is Chesapeake Bay, 1,500 square miles of it are part of Virginia. The largest lake, Lake Drummond, is in the Great Dismal Swamp. Virginia has some beautiful beaches, especially in the Virginia Beach area.

HOW VIRGINIA BECAME A STATE

Virginia was the site of an Indian empire headed by the powerful Chief Powhattan. In 1570, the Indians drove off Spanish settlers. In 1607, three English ships anchored in the James River and founded James Towne, (which later became Jamestown) the first permanent English settlement. In 1609, James Towne's leader, John Smith, made peace with Chief Powhattan, perhaps with the help of the chief's daughter Pocahontas. By 1619, the settlers were organized enough to elect representatives to the House of Burgesses, the first legislative body in America. It was named Virginia in honor of Elizabeth I, the Virgin Queen.

In 1670, the influx of full scale slavery ushered in the era of the large plantation. That same year, voting rights were restricted to property owners. The House of Burgesses grew very powerful and assertive. The property owners encouraged westward exploration, and at times the state claimed portions of what are now Illinois, Indiana, Ohio, Michigan, Wisconsin, Minnesota, Kentucky, and Tennessee. In 1738, the House of Burgesses created Augusta County, which stretched westward all the way to the Mississippi.

They also quarreled with the British. Nathaniel Bacon led a revolt against the governor after proclaiming what he called "America's declaration of independence," but he died and 20 of his followers were hanged. More organized resistance formed when Patrick Henry joined the House of Burgesses in 1765. In 1776, the Virginia Convention declared that the colony was free and independent.

Curiously, considering its location, Virginia was the site of very little fighting in the Revolutionary War. Perhaps because of the stability at home, the state was able to lend such leaders as George Washington, Robert Henry Lee,

Patrick Henry, and Thomas Jefferson to the struggle to create a new nation. However, George Rogers Clark did lead a force that kept the British at bay in the western portion of the state. And the American and French forces lured the British troops into a trap at Yorktown, which led to the British surrender on October 18, 1781.

Virginia became the tenth state on June 25, 1788, and it contributed four of the first five American presidents, and eight altogether—more than any other state.

VIRGINIA'S CITIES

RICHMOND (203,056), the state capital, was in an area on the James River visited by John Smith in 1607. The site was settled in 1637 and named for Richmond on Thames, England. It became an important seaport and commercial center , and was named capital of the state in 1779. It became the capital of the Confederacy during the Civil War, and as such it was the focus of much Union attention during the war. Today Richmond is a vibrant, modern city that is an important commercial, financial and medical center as well as a seaport. Notable attractions are the Virginia Museum of Art, the Museum of the Confederacy, the State Capitol Building, and the Medical College of Virginia, part of Va. Commonwealth University and one of the nation's foremost medical schools. Richmond is also the place where I began my TV career back in 1971. I still think of it as home.

NORFOLK (261,2500), named for Norfolk, England, was founded in 1682 and became a major center for the West Indies trade. The city became the leading grain shipping port on the east coast, but it was virtually destroyed in the Civil War. The local economy received a tremendous boost from a shipbuilding industry that began during World War II. Neighboring Virginia Beach is the state's largest city and major tourist center.

WILLIAMSBURG (11,530) was founded in 1633 as Middle Plantation. It became the state capital in 1699 and was named in honor of King William III of Great Britain. The city diminished in importance after Richmond became the capital following the American Revolution. But in 1926, under the sponsorship of John D. Rockefeller II, 120 colonial buildings in the center of the city were restored as Colonial Williamsburg. The city is now a major tourist center, and such attractions as Busch Gardens have been built nearby.

VIRGINIA'S PEOPLE

Virginia's 6,490,634 people make it the twelfth most populous state. The state has an increasingly diverse ethnic mix that includes one of the nation's largest Asian populations. Nearly 70% of Virginia residents live in metropolitan areas, but that population is spread among to many medium-sized cities rather than compressed into huge urban centers. The state has been a leader in breaking away from old southern stereotypes, and, in 1989 Virginia made L. Douglas Wilder the first black elected governor in U.S. history.

VIRGINIA'S ECONOMY

Agriculture built Virginia, and it is still important today. Livestock accounts for two-thirds of farm income, with Smithfield ham (which comes from hogs specially raised and fed peanuts) as one of the distinctive products. Tobacco, hay, peanuts, soybeans, and apples are major crops.

Manufacturing is the most significant sector of the economy. The leading industries are shipbuilding, transportation equipment, clothing and textiles, chemicals, printing, electrical equipment and processed foods. The western portion of the state is one of the nation's leading producers of bituminous coal.

Given its historical significance (great beaches don't hurt either), Virginia generates $8.2 billion in tourist income annually. Virginia Beach, Williamsburg, and the areas adjacent to Washington, DC. are the prime destinations. The Federal government and military installations also generate a significant number of jobs and opportunities for service businesses.

Places to Visit

In northern Virginia near Washington, D.C. are countless of historical attractions from George Washington's Mount Vernon (703-780-2000), the most visited house museum in the country, and Arlington National Cemetery (703-692-0931), site of the Kennedy graves to the Tomb of the Unknown Soldiers. Richmond, my home town, has many historic houses and nearby is Colonial Williamsburg (800-368-6511), a must see for school children and history buffs. In western Virginia is the beautiful Shenandoah Valley (703-740-3132) and Monticello (804-984-9800), Thomas Jefferson's home. Tourist information: Virginia Division of Tourism, 1021 E. Cary St., Richmond, VA 23219 (800-932-5827/804-786-2051)

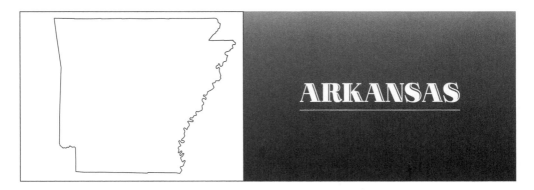

ARKANSAS

Arkansas, which has the most optimistic nickname of any state ("The Land of Opportunity"), is the northwestern most of the Southern states. It is bordered on the north and northeast by Missouri, on the east by Tennessee and Mississippi, on the south by Louisiana, on the southwest by Texas, and on the west by Oklahoma. Named for the Arkansas Indians, this state is roughly square, with both a north-south and east-west width of 220 miles. The Mississippi River constitutes the entire eastern border.

Little Rock, the state capital, is 127 miles west of Memphis and 340 miles east of Oklahoma City on I-40, and is 326 miles northeast of Dallas on I-30. Fort Smith is 291 miles west of Memphis and 181 miles east of Oklahoma City on I-40.

THE LAY OF THE LAND

Northeast Arkansas is part of the Ozark Plateau that extends northward into Missouri. It's an area of rugged hills and valleys with relatively fertile soil. The Arkansas River valley, through which I-40 runs, separates the Ozark Plateau from the Ouachita Mountains, which occupy the west central part of the state.

Southwestern Arkansas, part of the Gulf Coast Plain, is an area of rolling hills, many of which are heavily forested. The eastern third of the state is part of the Mississippi River flood plain. Until the late 1800s, much of this area consisted of swamps because of river flooding. However, an extensive system of levees turned much of this flat expanse into fertile farmland. Rising out of the floodplain in a north-south line is Crowley's Ridge, a rock formation up to 500 feet high that varies in width from one to twelve miles.

Eastern Arkansas has many oxbow lakes formed by changes in course of the Mississippi. Lake Chicut, in southeastern Arkansas, is the largest. Other large lakes include Lake Oachita, Beaver Lake, and Nimrod Lake. The Arkansas River, which flows diagonally through the center of the state from northwest to southeast, is a major transportation thoroughfare.

HOW ARKANSAS BECAME A STATE

Arkansas was home to the Osage, Caddo, Cherokee, Chictaw, and Quapaw Indians when the Spanish first explored the area in the early 1500s. But it was a Frenchman, Henri de Tonty, who established a trading post (the Arkansas Post) at the junction of the Arkansas and Mississippi Rivers in 1686, and for more than 80 years, it was the only trading post in a huge part of the Louisiana Territory. Arkansas state and river were named for the Arlcansa Indians. Arkansas was part of the Louisiana Territory that was claimed by France, ceded to Spain in 1763, returned to France in 1800, then sold to the U.S. in the 1803 Louisiana Purchase.

In 1812, Arkansas became a county in the newly created Missouri Territory. In 1819, it was granted territorial status on its own, and the capital was moved from Arkansas Post to Little Rock in 1821. The population in 1820 was just 14,273, but cotton farmers, with slaves, began to pour into the state in such numbers that the population doubled to 30,000 by 1830, and reached 50,000 when the territory became the 35th state on June 15, 1836.

ARKANSAS'S CITIES

LITTLE ROCK (175,795), the capital and largest city, is in the center of the state on the Arkansas River. In 1722, Bernard de la Harpe built a trading post on the site near the smaller of two rock outcroppings. In 1812, William Lewis built his home near that same "little rock," and the subsequent settlement took that name. The territory capital was moved to Little Rock in 1821, and remained there when Arkansas became a state in 1836. Little Rock began to grow when the railroad reached it in 1880, and the first manufacturing began in 1890. When I visited Little Rock for the first time, I was struck by how modern the city seemed.

FORT SMITH (72,798) is located at the junction of the Arkansas and Poteau Rivers on the Oklahoma border. In 1817, a fort was built on the site to keep peace between the Cherokee and Osage Indians, and it was named for General Thomas A. Smith, commander of the troops that built the fort. The Army moved out in 1871, but Fort Smith became the site of a Federal court in 1875. Today, the city is a huge livestock market and is home to agricultural processing and electrical appliance industries.

PINE BLUFF (57,140) is a port on the Arkansas River southeast of Little Rock. The town was founded as Mount Marie in 1819, and in 1832 was renamed for a forest of pines overlooking the river. Pine Bluff is commercial center for the cotton, rice and soybean growers, and has a diverse industrial base that includes paper, fabricated metals, and transformers. The Federal Toxology Research Center is located nearby.

ARKANSAS'S PEOPLE

Most of Arkansas's 2,424,418 people are descended from the Anglo-Saxon farmers who came into the state after 1815 and the slaves they brought with them. More than three-quarters of the current residents were born in Arkansas, and less than 1% are foreign-born. Although the population was more than 90% rural early in this century, 54% now live in urban areas. The population is 83% white and 16% black.

ARKANSAS'S ECONOMY

Although originally a cotton state, livestock accounts for the majority of the farm income that makes Arkansas one of the top ten agricultural states. It is the number one producer of broiler chickens, and cattle, hogs, and turkeys are also raised. Rice, soybeans, cotton, wheat, hay, spinach, and tomatoes are the major crops.

Arkansas is the source of 90% of our nation's bauxite, the raw material in the making of aluminum. The state also has significant petroleum and natural gas reserves. Murfreesboro is the site of the only diamond mine on the North American continent.

Most of Arkansas's manufacturing uses the raw materials produced by the state. Little Rock is the center of such industries as food products, petrochemicals, paper and paper products, clothing and textiles, and electrical equipment.

Places to Visit

Little Rock's (800-844-4781) history is displayed in the restored 19th century Quapaw Quarter and the Arkansas Territorial Restoration. The state's most famous tourist destination is Hot Springs (800-772-2489), which features thermal waters nestled between five bodies of water called "diamond lakes" because of their purity. Fort Smith (800-637-1477) celebrates its frontier heritage with the preserved old Fort, the courtroom of the "hanging " Judge Parker, and the only former bordello on the National Register of Historic Places. Tourist information: Arkansas Department. of Parks and Tourism, 1 Capital Mall, Little Rock, AK 72201 (800-872-1259/501-682-7777)

The Other United States

The Virgin Islands of the United States, a favorite Caribbean vacation spot, consist of 3 islands and about 50 islets. The three islands are Saint Thomas (32 square miles), Saint John (20 square miles), and Saint Croix (80 square miles). The islands begin about 70 miles east of Puerto Rico, and extend to the east—on a map, it looks as if the Virgin Island's are Puerto Rico's tail. The Caribbean Sea is to the south and the Atlantic Ocean is to the north. The capital of the Virgin Islands is Charlotte Amalie on St. Thomas. The only other larger communities are Christiansted and Frederiksted, both on St. Croix.

The terrain on the three main islands is basically rugged. St. Thomas has a series of ridges running east to west, with little space for agriculture. St. John also has lofty hills and steep, narrow valleys. St. Croix is hilly in the north, but has flatter and wetter land in the south.

Sailing for Spain, Christopher Columbus discovered the Virgin Islands on his second voyage in 1493 . He named them for St. Ursula, a virgin martyr. Spain colonized the islands in 1555, and warfare, disease, and oppressive treatment wiped out the entire native population, and Spain abandoned the Islands. In 1662, the Danish West Indies Company that colonized St. Thomas and controlled the islands until 1755, when the King of Denmark bought them. They became a slave trading market and a major producer of sugar cane. In actions related to the Nepoleanic Wars in Europe, Great Britain twice occupied the islands—once, briefly, in 1801, and from 1807 to 1815. But Denmark regained control.

However, after slavery was outlawed in 1848, sugar production declined and the territory became less profitable for its owner. The U.S. began negotiating to buy the Virgin Islands in 1867, but it wasn't until 1917,. fifty years later, that the sale was completed for $25 million.

The U.S. Navy controlled the islands until 1931, when the responsibility passed to the Interior Department. A governor was appointed until 1968, when Congress passed legislation permitting gubernatorial elections. Virgin Island residents have been U.S. citizens since 1927, and they elect a non-voting delegate to the U.S. House of Representatives. The vast majority of the Virgin Island population are descended from the slaves imported by Denmark.

Tourism is the most important industry in the Virgin Islands. Over 2 million visitors spend over $1 billion. The Islands also produce rum, watches, textiles, and pharmaceuticals.

The U.S. originally purchased the Virgin Islands for defense purposes, and there is an air base on St. Thomas and an airfield on St. Croix today.

The Commonwealth of Puerto Rico (Estado Libre Asociado de Puerto Rico) is composed of one large island and several small islands located 1,000 miles southeast of Miami. Puerto Rico is the easternmost of the group of islands called the Greater Antilles, which includes Cuba, Hispaniola, and Jamaica. The island is only 500 miles north of Caracas, Venezuela. Puerto Rico is bordered on the north by the Atlantic Ocean, on the east by the Virgin Passage (which separates it from the Virgin Islands), on the south by the Caribbean Sea, and on the west by the Mona Passage (which separates it from the Dominican Republic). The island is roughly rectangular, stretching 110 miles east-west and 40 miles north-south. With an area of 3,515 square miles, it is larger than Rhode Island and Delaware.

About three-quarters of the land area consists of mountain ridges running east-west across the state. The highest point is Cerro del Punta (4,389 ft.) There is a narrow coastal plain averaging 12 miles wide on the Atlantic Coast (north) and another averaging 8 miles in width on the Caribbean coast (south). San Juan (437,7450) , the capital and largest city, is on the northeast coast.

On his second voyage in 1493, Christopher Columbus reached a large island he called San Juan Bautista. Juan Ponce de Leon conquered the island in 1508 and became the first governor. The Spanish enslaved the Borinqueno Indians, the island's natives. Within a decade, the original inhabitants had been decimated by cruel treatment and disease. They were replaced by black African slaves who worked the plantations and sugar mills. In 1521, the city of San Juan was founded and the island was Puerto Rico, or "rich port."

Spanish control of Puerto Rico was challenged by the British and Dutch over the course of more than a century. In 1595, Sir Francis Drake and Sir John Hawkins tried to oust the Spanish, but were driven off. The Dutch burned San Juan in 1625, and British raids lasted into the eighteenth century. In the nineteenth century, tension grew between the Spanish government and the Puerto Rican population, and several armed conflicts occurred. Span finally abolished slavery in 1873 and made the island self-governing in 1897.

However, in the very next year Spain was forced to cede Puerto Rico to the United States in the Treaty of Paris that ended the Spanish-American War. Puerto Ricans were made U.S. citizens in 1917. In the early 1940s, a large scale development program called Operation Bootstrap helped create a diversified economy. In 1948, the first democratically elected governor, Luis Munoz Marin, took office. In 1951, Puerto Ricans approved a U.S. law that gave them the right to create their own constitution, and in July, 1952, Puerto Rico became a Commonwealth. However, there was opposition from people who wanted Puerto Rico to be independent. These nationalists had tried to assassinate President Harry Truman in 1950, and four wounded five Congressmen when they fired shots into the House Chamber in March, 1954. Periodic violence from Puerto Rican nationalists occurs today. However, Puerto Rican voters decided to remain a Commonwealth in 1967. Despite a growing statehood movement, Puerto Rico retains its commonwealth status.

Puerto Rico's 3,522,307 people speak Spanish. About 80% are Roman Catholic and 71% live in urban areas. After San Juan, the largest cities are Bayam¢n, Carolina, and Ponce

The majority of Puerto Rican farms are small and produce enough to sustain one family. The large farms produce coffee, sugarcane, bananas, pineapples, and vegetables. Dairy and beef cattle are the main livestock, followed by poultry. Puerto Rico has developed a much more diversified economy over the last three decades. Clothing is the leading product, followed by electronic goods, processed foods, and chemicals. The Commonwealth also benefits from a strong U.S. military presence. Puerto Rico's beaches, golf courses, and casinos attract more than four million annual visitors and bring in over $2 billion.

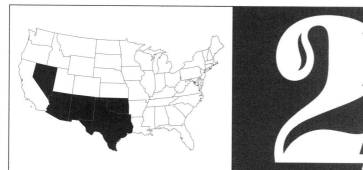

THE SOUTHWEST

Comparing The Southwestern States

STATE National rankings in parenthesis	POPULATION (1993 U.S. Bureau of the Census)	SQUARE MILES (U.S. Geological Survey)	PER CAPITA INCOME (U.S. Bureau of Labor Statistics)
Texas	18,031,484 (3)	268,601 (2)	$12,904 (32)
Oklahoma	3,231,464 (28)	69,903 (20)	$11,893 (39)
Arizona	3,936,142 (23)	114,006 (6)	$13,461 (25)
New Mexico	1,616,483 (36)	121,598 (5)	$11,246 (42)
Nevada	1,388,910 (38)	110,567 (7)	$15,214 (13)

Two things come to mind immediately when I think of the Southwestern States. The first are all those unforgettable images from western movies—cattle being driven through endless prairie, gunfights in mining boom towns, ambushes in steep-walled canyons, posses chasing bad guys through plateaus dotted with breathtakingly beautiful mesas and buttes. The vast expanses of the Southwest (4 of the 5 states are among America's 10 largest) contain geographic diversity that ranges from snow-capped mountains to colorful deserts, from spectacular gorges carved by rivers to immense dams built by man. The region includes some of the world's most famous natural wonders—the Grand Canyon, the Painted Desert, Carlsbad Caverns—as well as some of the most desolate wastelands.

The second thing that comes to mind is the unique culture of the Southwest provided by its Spanish and Indian heritage. Archaeologists have uncovered evidence of sophisticated Native American civilizations that date back more than 12,000 years. Today, the states in this region have the largest Native American populations, the largest reservations, and the largest tribal structures.

Four of these five states were under Spanish control for nearly three centuries. Settlers from the United States came to this region late in our history, and three of these states weren't admitted to the Union until the twentieth century. From place names to architecture, the Southwest has a wonderful atmosphere all its own.

Of course, the Southwest states haven't lingered in the past. This region is America's fastest growing, for reasons that range from rich mineral wealth to wonderful climates. It's dynamic economy, friendly people, fascinating history, and dramatic landscapes make the Southwest one of my favorite destinations.

EARLY EXPLORATION

Of all the Spanish who explored the Southwest and laid claim to the area, by far the most famous is Francisco Vasquez de Coronado. Coronado was born in Spain in 1510, and shortly after coming to the New World was appointed governor of four Mexican provinces. When reports of a fabulously wealthy Indian civilization that lived in the Seven Cities of Cibola reached Mexico, Coronado was given a force of 200 soldiers and a larger number of Indians to search for the gold and jewels. His quest, which began in 1539, lasted four years and carried him through the present states of Arizona, New Mexico, Texas, and Oklahoma. By the time he returned to Mexico, he had seen many natural wonders (including the Grand Canyon) but had found no wealth. His displeased superiors soon relieved him of his position.

Most of the Spaniards who followed Coronado were looking for souls rather than gold. Franciscans and Jesuits established many missions in the area to convert the natives. However, their methods tended to be of un-Christian, and often incited the Indians to rebellion. Because the Spanish wouldn't allow settlers from any other nations into the territory, the Southwest was very sparsely populated into the early 1800s.

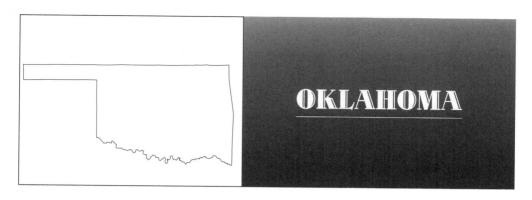

OKLAHOMA

The stirring Rogers and Hammerstein score of one of the greatest of all musicals, "Oklahoma," aptly celebrates a state blessed with many natural resources. Oklahoma is bordered on the north by Colorado and Kansas, on the east by Missouri and Arkansas, on the south by Texas, and on the west by Texas and New Mexico. The Red River forms the entire southern border of the state shaped like a rectangle with a panhandle on the western side. The panhandle gives Oklahoma a 400 miles east-west length, while the north-south width is 220 miles.

Oklahoma City, the largest city, is 208 miles north of Dallas and 157 miles south of Wichita on I-35, and 470 miles west of Memphis and 573 miles east of Albuquerque on I-40. Tulsa is 105 miles northeast of Oklahoma City and 393 miles southwest of St. Louis on I-44.

THE LAY OF THE LAND

Oklahoma has a surprisingly varied landscape which features five separate regions. The Panhandle is part of the Great Plains, elevated terrain that rises from east to west. The central two-thirds of the state is part of the Osage Plains, a large, gently rolling region that's interrupted by isolated mountain ranges.

The northeastern section of the state is part of the Ozark Plateau that stretches southwestward from Missouri and Arkansas. The land is hilly with valleys cut by streams. South of the Ozark Plateau, running diagonally from north central to the southeast border, is the fertile Arkansas River Valley, a rich agricultural region. In the southeastern corner are the Ouachite Mountains, Oklahoma's most rugged terrain.

The state is drained primarily by the Arkansas and Red Rivers. There are few natural lakes, but more than 200 artificial lakes created by damming. Forests cover just 16% of the state.

HOW OKLAHOMA BECAME A STATE

The Spanish explorer Coronado was the first European to traverse the area, but it was part of the vast area claimed for France by Robert Cavelier, sieur de la Salle, in 1682. French trappers and traders worked the area in the seventeenth and eighteenth centuries, but there were no permanent settlements.

In 1803, the U.S. acquired Oklahoma in the Louisiana Purchase. In 1817, the government began relocating to the area five eastern Indian tribes (known as the Five Civilized Tribes)—the Creek, Cherokees, Chicsaw, Choctaw, and Seminoles. The journey of these tribes to their new home is called the Trail of Tears, because it involved so much death and suffering. In 1834, Oklahoma was designated the Indian Territory by Congress. The Indians, some of whom owned slaves, angered the government by siding with the Confederacy in the Civil War. Between 1866 and 1883, the Five Tribes were forced to cede western Oklahoma to other tribes.

Pressure also grew to open parts of the state to white settlement. Although some moved into the state illegally (they were called "Sooners"), the first land rush began on April 22, 1889. Over 50,000 people flooded into Oklahoma on the very first day, staking out farms and building tent cities overnight. More parcels of land were subsequently opened in additional land rushes.

In 1890, Congress combined the territory controlled by the Indians, the white settlements, and the Panhandle into the Territory of Oklahoma (the name comes from the Choctaw Indian word for "red people"). And on November 16, 1907, Oklahoma became the 46th state.

OKLAHOMA'S CITIES

OKLAHOMA CITY (444,719), at the junction of the Canadian and North Canadian Rivers, covers an area of 621 square miles, making it one of the largest U.S. cities. The city became a metropolis on April 22, 1889—between noon and midnight, 10,000 people settled on the site. In 1910, Oklahoma City became the capital. It has since become a major financial, commercial and transportation center with a large petroleum production and refining industry. Oklahoma City is home to the National Cowboy Hall of Fame and the National Softball Hall of fame.

TULSA (367,302) is located on the Arkansas River in the eastern portion of the state. A settlement on the site was founded by Creek Indians from Alabama who called it Tulsee Town, from "tullahasee," the Creek word for "old town." The town began to add population with the arrival of the railroad in 1882, but it was the discovery of oil in the area in 1901 that made Tulsa into a major city. The city gained an added impetus with the completion of the McClellan-Kerr Arkansas River Navigation System, which gave Tulsa water access to the Gulf of Mexico and the Great Lakes. The city produces aerospace equipment, oil-field supplied, and refined petroleum.

LAWTON (80,561) is the commercial center for a large wheat-growing and cattle-raising region. It was opened to white settlement in 1901 and named for General Henry W. Lawton, who was killed in the Philippines during the Spanish-American War. Today, the city of Lawton produces tires, jewelry, cosmetics, clothing, and processed foods. Fort Sill, the Army artillery training center, is nearby.

OKLAHOMA'S PEOPLE

Oklahoma's 3,231,464 people include 252,087 Native Americans, the largest number of any state. These Native Americans belong to many different tribes. Most of the state's people live in the eastern half of the state. A little over two-thirds of the population lives in urban areas.

OKLAHOMA'S ECONOMY

The homesteaders who poured into Oklahoma after 1889 were overwhelmingly farmers and ranchers. Today, beef cattle are the state's leading agricultural product. Oklahoma's farms also produce wheat, hay, cotton, peanuts, sorghum, and soybeans.

The Oklahoma economy really heated up after large reserves of natural gas and petroleum were discovered early in this century. Oklahoma is a major oil producer and refiner, and Tulsa is one of the capitals of the petroleum industry. Other significant mineral products are gypsum and iodine.

The supply of oil spurred a growth in manufacturing. The leading industries are petrochemicals, mobile home manufacturing, plastics, aerospace equipment, industrial machinery, transportation equipment, fabricated metals, electronic equipment, and processed foods.

Tourism brings about $2.5 billion per year to the state.

Places to Visit

Oklahoma City (800-225-5652) celebrates the Old West with the National Cowboy Hall of Fame, the Western Heritage Museum, and the Harn Homestead and 1889er Museum. North of Tulsa is the Woolaroc Museum (918-336-0307), which includes a drive-through wildlife preserve, and Will Rogers Birthplace (918-275-4201). In southwestern Oklahoma is the Wichita Mountains Wildlife Refuge (405-429-3222), which contains some of the most spectacular scenery in the state. Tourist information: Oklahoma Tourism and Recreation Department, Box 60789, Oklahoma City, OK 73146 (800-652-6552/405-521-2409)

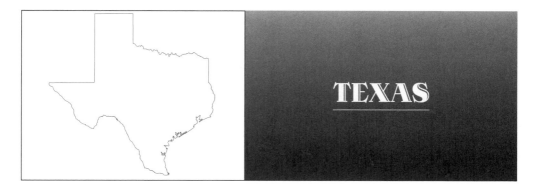

TEXAS

Even people who don't know anything about geography know that Texas is big—in fact, it's by far the biggest of the 48 contiguous U.S. states (60% larger than California), although it covers less than half the area of Alaska. Texas is bordered on the north by Oklahoma, the northeast by Arkansas, the east by Louisiana, the southeast by the Gulf of Mexico, the south by Mexico, and the west by New Mexico. The Rio Grande River forms all of the border with Mexico, while the Red River makes up part of the northern border and the Sabine River part of the eastern border. Texas runs an impressive 800 miles north-south and 775 miles east-west. It has 367 miles of coastline on the Gulf of Mexico.

Houston, the largest city, is 272 miles west of Baton Rouge and 199 miles east of San Antonio on I-10 and 246 miles southeast of Dallas on I-45. Dallas is 209 miles south of Oklahoma City and 195 miles north of Austin on I-35, and 325 miles southwest of Little Rock on I-30. San Antonio is 555 miles southeast of El Paso on I-10 and 79 miles south of Austin on I-35.

THE LAY OF THE LAND

Despite its size, Texas has just four different regions. In the southwest corner of the state, between the Pecos and Rio Grande Rivers, is a Basin and Range area consisting of plains broken by isolated mountains and deep canyons. Most of central Texas and the Panhandle are part of the southern Great Plains, which are dry and treeless before giving way to the Texas hills. Irrigation is necessary for agriculture on the Plains.

South central Texas is part of the Osage Plains, rolling country that has stretches of black, fertile soil. The eastern two-thirds of Texas are part of the Gulf Coastal Plain. In contrast to the Great Plains, this region receives ample rainfall and has fertile sandy soil.

Texas has few natural lakes but many artificial ones. The Gulf Intercoastal Waterway, is a key shipping link to the other Gulf states.

HOW TEXAS BECAME A STATE

The Spanish were the first explorers—Pineda sailed along the coast in 1519 and Alva Nunex Cabeza de Vaca was shipwrecked and captured by Indians in 1528. He was held in captivity for many years before before escaping. Coronado conducted a fruitless search for gold between 1539-1541, and at the same time Luis Moscoso explored eastern Texas. The very first Spanish settlement was established near El Paso in 1682. In 1687, the French established Fort Saint Louis on Matagurdo Bay and claimed the area, but the Spanish reasserted their claims by establishing a number of new missions by 1690.

The Spanish refused to allow settlers from any other nation into Texas, and in 1820, there were only 2,000 white settlers in the entire area. But after Mexico won its independence in 1821, it began to allow Americans to enter. Among the first was Stephen Austin, who led a group of "empresarios," or colonists. More settlers followed over the next decade, and the U.S. made offers to Mexico to buy the area. Some settlers tried to establish the independent Republic of Fredonia in East Texas, but were squashed by the Mexican army.

Finally, in 1835, Texans revolted when Santa Anna overthrew democracy in Mexico and became dictator. They captured San Antonio and drove the Mexicans south of the Rio Grande. However, Santa Anna led an Army back into the state in the spring of 1836. In the process of recapturing San Antonio, Santa Anna's troops killed every single defender of a fort called the Alamo, including the famous frontiersmen Davy Crockett and James Bowie.

The Texans then appointed Sam Houston as commander-in-chief of their forces. They defeated the Mexicans at the Battle of San Jacinto on April 21, 1836, capturing Santa Anna. Houston was named president of the Republic of Texas, which was an independent nation for the next nine years. Texas was admitted to the Union December 29, 1845, with one unique provision—if it chooses, Texas can divide itself into five different states.

Mexican anger at the admission of Texas gave President Polk justification for launching the Mexican War. In 1848 the Treaty of Guadalupe Hidalgo, set the Texas border at the Rio Grande.

TEXAS'S CITIES

HOUSTON (1,630,553), America's fourth largest city, is the center of our nation's petroleum industry. The site was settled as Harrisburg by John Harris in 1826, but it was destroyed by Santa Anna. Later that same year, Augustus and John Allen laid out another town at the same location. It was the capital of the new Texas Republic from 1837-1839 and was named for Sam Houston, the first president of Texas. The city grew slowly until oil was discovered in the region in 1901. The completion of the Houston Ship Canal in 1914 gave the city a port just two miles from downtown. Houston today is a leading producer of petrochemicals, insecticides, fertilizers, oil field supplies, paper products, and electrical equipment. It is also home to the Lyndon B. Johnson Space Center.

DALLAS-FORT WORTH (1,453,892) is a metropolitan area consisting of two cities just 28 miles apart. Dallas, which has twice the population of Fort Worth, was founded in 1841 by John Neely Bryan, whose original log cabin survives. It was named in 1846 for U.S. Vice President George Dallas. It became a center of cotton-growing and oil-producing in the 1920s. Today, it is a major financial and banking center. Fort Worth was founded in 1849 as a frontier army post and named for Mexican War General William Jenkins Worth. It thrived as a stop on the Chisholm Trail, along which cattle were driven to Kansas. I've had the privilege of touring Fort Worth's famous stockyards. The city is also a grain-processing and shipping center.

SAN ANTONIO (935,933) is famous for its Spanish, Mexican, and Indian cultures and the beautiful Pasco del Rio walkway surrounding the San Antonio River in the downtown area. It was settled as a mission and presidio (fort) by the Spanish in 1718. San Antonio had primarily a Spanish population until the Texas Revolution in 1835. The Alamo, the site of the infamous massacre, still stands as a major tourist attraction. The city began to grow rapidly when the railroad arrived in 1877.

AUSTIN (465,622), the state capital, was founded as a Franciscan mission on the Colorado River in 1730. In 1838, it was settled and named Waterloo. It became the Texas capital in 1839 and was named in honor of Stephen F. Austin. It grew as a stop on the Chisholm Trail and a transportation center on the Colorado River. It's the home of the University of Texas and it has become a high-tech research center.

EL PASO (515,342), which is located in westernmost Texas on the Rio Grande River is an important point of entry and international trading center. In 1520, Juan de Onato had given the name El Pasa del Rio del Norte to a ford on the Rio Grande River. A mission was built on the site in 1659 and a presidio in 1780. The first houses were built in 1827, and El Paso later became a stop on the Santa Fe Trail that linked St. Louis and San Francisco. The arrival of the railroad in 1881 and the building of the Elephant Butte Dam in 1915 were two major factor's in El Paso's growth. The city is the center of a major mining and agricultural region.

TEXAS'S PEOPLE

With 18,031,484 people, Texas is the third most populous state, and it will soon pass New York to move into the second position behind California. One-quarter of the state's residents are Hispanic and 12% are black. Four-fifths of all Texans live in urban areas, which is why the state has so many large cities. Texas has an astounding 254 counties and 1175 cities and towns.

TEXAS'S ECONOMY

The original settlers came to Texas to farm and ranch, and today the state has the second highest agricultural income. Livestock, especially beef cattle, sheep, and chickens, account for two-thirds of farm revenue. Texas is also the nation's leading cotton growing state, and its other major crops include wheat, corn, sorghum, watermelon, and cabbage. Texas has the nation's largest number of tree farms. The state has a significant shrimp-fishing industry.

You can't think of Texas without thinking of oil. It's petroleum and natural gas reserves have created many millionaires and billionaires. Texas has significant coal reserves and it is number one in sulphur production.

Texas also receives the fourth largest income from manufacturing. It is the leading petroleum refiner and is home to a huge petrochemical industry. Other leading industries are transportation equipment, industrial machinery, electrical products, fabricated metals, processed foods, and clothing. Texas has five of the top 26 U.S. ports, including number two, Houston.

Tourism accounts for $20 billion in annual revenue for Texas.

Places to Visit

Big Bend National Park (915-837-2326), the state's premier natural attraction, offers almost one million acres of canyons, desert, and mountains. San Antonio (800-447-3372), with the historic Alamo and the beautiful River Walk, has Texas's most inviting downtown area. Many people visit Dallas (800-232-5527) to see the site of JFK's assassination, but end up admiring its many cultural institutions. If you go, don't miss Fort Worth's (800-433-5747) Stockyards Historic District just 30 miles away. Houston (800-365-7575) is a brash, sprawling city that has some wonderful museums. Tourist information: Department of Texas Tourism, Box 5064, Austin, TX 78763 (800-888-8839)

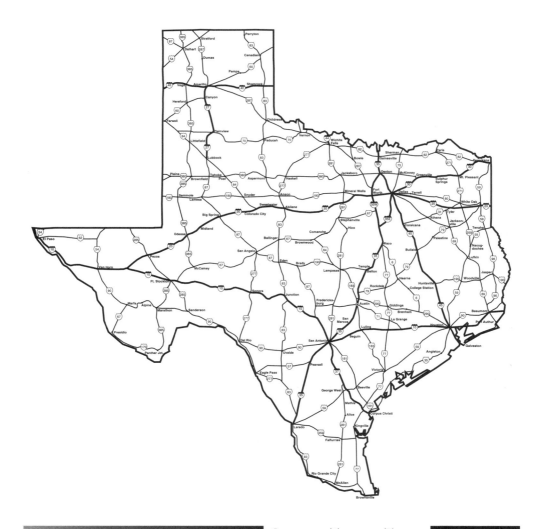

NEW MEXICO

New Mexico's rich Spanish and Indian heritage gives the state a unique atmosphere that draws tourists from all over the United States. The state is bordered on the north by Colorado, on the east by Oklahoma and Texas, on the south by Texas and Mexico, and on the west by Arizona. New Mexico is nearly square, with a 390 mile north-south width and a 350 mile east-west length. The northwestern most point is part of the famous Four Corners, where visitors can lie down on the ground and touch Colorado, Utah, Arizona, and New Mexico at the same time.

Albuquerque is 267 miles north of El Paso on I-25, and 288 miles west of Amarillo, Texas and 327 miles east of Flagstaff, Arizona on I-40. Santa Fe, the capital, is 58 miles north of Albuquerque on I-25.

THE LAY OF THE LAND

New Mexico has a geographic diversity that ranges from towering, snow-capped peaks to parched desert basins. Overall, New Mexico has the fourth highest mean elevation of the U.S. states, and 85% of its area is over 4,000 feet. The eastern third of the state is part of the Great Plains that stretch from Texas to the Canadian border. The northern part of these section is dotted with numerous mesas and buttes, while the southern section, the "Llano Estaendo" or "staked plains," is some of the flattest land in the U.S. Limestone underlies some of this area, allowing erosion to carve the magnificent Carlsbad Caverns near the Mexican border.

In the central third of the state is the southernmost part of the Rocky Mountains, boasting Mt. Wheeler (13,161 ft.), the highest point in the state. This rugged country is divided almost in half by the Rio Grande River. Many crops are grown in this fertile valley.

Most of the western third is made up of the high plateau known as the Basin and Range region. This area includes mountain ranges interspersed with desert valleys. In the northwest corner is part of the Colorado Plateau, valleys and plains cut by deep canyons.

New Mexico has no large natural lakes, but dams have created several huge reservoirs.

HOW NEW MEXICO BECAME A STATE

New Mexico has a rich pre-Columbian history that dates back more than 10,000 years. The Pueblo Indians developed a sophisticated culture that dominated the area between 700 A.D. and 1300 A.D.

New Mexico was crossed by Marcus de Niza in 1539 and Coronado in 1540. In the 1560s, Spanish explorers gave the territory the name "Nueva Mexico," or New Mexico. In 1598, Don Juan de Onato founded San Juan, the first European settlement in New Mexico and the second in the entire New World. Santa Fe was established in 1610, and by 1626 the Spaniards had more than 40 missions. In 1821, Mexico gained independence from Spain and took control of New Mexico. The Mexicans allowed Americans to enter the territory, and in 1822, William Becknell arrived with the first wagon loads of goods from St. Louis. They forged the Santa Fe Trail which was to become a major trade route.

After the U.S. declared war on Mexico in 1846, General Stephen Watts Kearney marched into the state with a force that captured Santa Fe on August 18. The U.S. was given most of New Mexico in the Treaty of Guadalupe Hidalgo which ended the Mexican War. Congress created the Territory of New Mexico, which included present-day Arizona, as part of the Compromise of 1850 that led to California's admission to the Union. A slice of land near the present Mexican border became part of New Mexico as a result of the Gadsden Purchase of 1853.

In 1862, Congress separated Arizona from New Mexico. But New Mexico remained a territory for another half century. One reason was that state residents weren't eager for the higher taxes that would result from joining the Union. Another was that Congress worried that the Spanish culture wasn't conducive to democracy. However, New Mexico schools began teaching English in 1898, and New Mexico was finally admitted as the 47th state on January 6, 1912.

NEW MEXICO'S CITIES

ALBUQUERQUE (384,730) was founded in 1706 on the Chihuahua Trail, a trading route between Santa Fe and Mexico City. It was named for the Duke of Albuquerque (somehow, the r was dropped). Albuquerque was a U.S. military post from 1846 to 1870. In 1880, a new town was laid out to meet the newly built railroad, and eventually the it grew to encompass the old town. Today, Albuquerque is a commercial, financial, and tourism center as well as the focus of a manufacturing economy that includes electronic and nuclear research, electronic equipment, processed foods, aerospace equipment, textiles, clothing, and printed materials.

SANTA FE (55,859)the capital was founded in 1609 as the capital of Nueva Mexico, and is the oldest state capital in the U.S. The Spanish were expelled by the Pueblo Indians in 1680, but regained control in 1693. After the Mexican War, Santa Fe became capital of the New Mexico Territory, and it was briefly held by the Confederacy during the Civil War. Today, it is a year-around tourist mecca that is famous for its Indian and Spanish pottery, jewelry and art. Santa Fe has retained the flavor of the old Southwest in its architecture and lifestyle and is home to a large colony of artists and filmmakers. Santa Fe is also one of my personal favorite vacation spots.

LOS CRUCES (62,360) is on the Rio Grande River in the central southern part of the state 46 miles north of El Paso. Located in the part of the state obtained in the Gadsden Purchase, it's name is Spanish for "the crosses," a reference the graves of 40 travelers massacred by Apache Indians in 1830. The city is the commercial center for the southern agricultural region.

NEW MEXICO'S PEOPLE

Of New Mexico's 1,616,483 people, 39% are Hispanic, the highest percentage of any state. A language other than English is spoken in 35% of New Mexico homes, again the highest percentage of any state. The Spanish influence in the state's culture is still strong and many residents speak both Spanish and English. New Mexico has 154,355 Native Americans, 9% of the population, the second highest percentage of any state.

NEW MEXICO'S ECONOMY

For much of its long history, New Mexico has been a self-sufficient farming and ranching state. The state saw bitter battles over grazing land between sheep and cattle ranchers, including the famous Livingston County War of 1870 in which Billy the Kid played a leading role. Today, cattle and sheep provide the largest share of the state's agricultural income. Irrigation has increased the amount of crops grown, with hay, cotton, and vegetables providing the most revenue.

The discovery of gold and silver brought miners to the state in droves in the late 1800s. Today, New Mexico is the leading producer of uranium and potash and the second leading producer of copper. The state also has significant natural gas, petroleum, and coal reserves.

The nuclear age arrived to a great extent because of research at New Mexico's Los Alamos Nuclear Research Center. Other major Federal research centers in the state include Sandia Laboratories and the Air Force Special Weapons Center. New Mexico is famous for its handicrafts and works of art, and has a diverse manufacturing base that includes electronic equipment, precision instruments, transportation equipment, printed materials, and industrial machinery.

Tourism brings more than $2 billion to the state's economy each year.

Places to Visit

Santa Fe's (505-983-7317) blend of Spanish and Native American culture make it an exotic treat for visitors. Taos, 60 miles north of Santa Fe, is an enchanting art and literary center as well as a ski resort. Carlsbad Caves National Park (505-785-2232) contains the 77 caves that make up one of world's largest cave systems. The Albuquerque (800-284-2282) area has numerous scenic and historic sites. Tourist information: New Mexico Department of Tourism, 491 Old Santa Fe Trail, Santa Fe, NM 87503 (800-545-2070/595-827-7400)

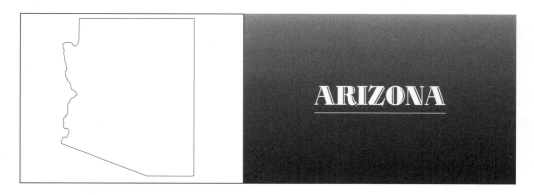

ARIZONA

Arizona is the home of the Grand Canyon, perhaps our greatest natural wonder. The state is bordered on the north by Utah, on the east by New Mexico, on the south by Mexico, and on the west by California and Nevada. The Colorado River forms most of the western border. Arizona is roughly rectangular, with a north-south width of 395 miles and an east-west length of 345 miles. The northeastern point of the state is part of the Four Corners, where you can lay down and touch the states of Arizona, New Mexico, Utah, and Colorado. The Federal government owns 42% of the land in Arizona.

Phoenix, the largest city, is 117 miles northwest of Tucson and 369 miles east of Los Angeles on I-10. Tucson is 320 miles west of El Paso on I-10 and 410 miles east of San Diego on I-8.

THE LAY OF THE LAND

The northern third of Arizona is part of the Colorado Plateau. Level areas of land are cut by deep gorges and marked by mesas and buttes. In the western part of the plateau is the Grand Canyon, a spectacular 1 mile deep gorge cut by the Colorado River as internal forces caused the land around it to rise. The Painted Desert cuts a swath across the southern and eastern sections of this region—in the east is the famous Petrified Forests, six separate stands of trees that were buried when the area was covered with marshland and turned to stone.

The Mexican Highlands slashes across the center of the state from northwest to southeast. In the San Francisco Mountains at the northern end of the Highlands is Humphrey's Peak (12,633 ft.), the state's highest point. In southwestern Arizona is the Sonoran Desert (sometimes called the Gila Desert), an area of arid valleys marked by isolated mountain ranges.

The Colorado River, which cuts through northern Arizona before heading south to form much of the border with California, is the most important natural body of water. Arizona has few natural lakes, and many rivers flow only in the rainy season. The state borders on some of the largest man-made lakes in the nation, such as Lake Mead, Lake Powell, Lake Mojave, and Lake Havasu.

HOW ARIZONA BECAME A STATE

The highly civilized Pueblo Indians created a culture that dominated present day Arizona and New Mexico between 700 A.D. and 1300 A.D. Both Marcus de Niza and Coronado crossed the area in a futile search for the fabled Seven Cities of Cibola. The area became part of New Spain in 1598, but was ignored until the Jesuit missionary Eusebio Francisco Kino established missions beginning in 1692. The first permanent settlement was built in the Santa Cruz Valley in 1752, and the Spanish built a presidio (fort) at what is now Tucson in 1776.

Mexico took control of the area in 1821, but Indian attacks discouraged settlement. The U.S. acquired Arizona north of the Gila River in the Treaty of Guadalupe Hidalgo that ended the Mexican War in 1848. In 1849, more than 60,000 people crossed the state to join California's Gold Rush. In 1853, the U.S. purchased a strip of southern New Mexico and the lower quarter of present-day Arizona from Mexico in the Gadsden Purchase. In 1858, gold was discovered near Yuma, but the deposits quickly ran out and the Arizona Gold Rush soon fizzled. Arizona was separated as a self-governing unit from New Mexico in 1862 (the name comes from a Pima Indian word for "place of the small spring"). But frequent Indian raids by Chief Geronimo discouraged settlement, and the population in 1870 was only 9,000.

Geronimo was captured in 1887, and irrigation began to make more farmland available. The population began to grow, and Arizona finally became the last of the 48 contiguous states to join the Union on February 14, 1912.

ARIZONA'S CITIES

PHOENIX (983,403), the largest city and capital, was built on the site of Indian ruins in the 1860s (the name comes from the mythical bird that rises from its own ashes). The city's ability to survive in this arid desert was possible because of the restoration of ancient irrigation ditches built by the Hohokam Indians five hundred years previously. Phoenix became the capital of the territory in 1889, but its major growth stems from 1911, when the building of the Theodore Roosevelt Dam on the Salt River provided a consistent source of water. Phoenix has recently become one of the fastest growing cities in the U.S. It is a tourist and retirement mecca, and it has developed a diverse industrial base that includes aerospace equipment, processed foods, metal products, cosmetics, sporting goods, and paper products.

TUCSON (405,390) was founded as the San Xevaierde Bac Indian Mission in 1700. A fort was built on the site in 1776, and in 1853, the Gadsden Purchase made it part of the U.S. Tucson (the name comes from an Indian word for "foot of the mountain") was the territory capital from 1867 to 1877. The city began to grow when the railroad arrived in 1880. Today, Tucson is the center of a copper-mining area and is known for the manufacture of electronic and electrical equipment.

FLAGSTAFF (45,853) is located at the foot of the San Francisco Mountains in northern Arizona. It was founded in 1871 and named for a flagpole erected during the Centennial celebration in 1876. Flagstaff is a tourist resort and the home to logging companies that process cuts from the area's extensive pine forests. Since 1894, it has been home to Lowell Observatory, from which astronomers from all over the world continue to make significant discoveries about our universe.

ARIZONA'S PEOPLE

Arizona's current population of 3,936,142 is an incredible eight times higher than it was in 1940. The new arrivals have flocked to the cities, which is why 88% of the population lives in urban areas. Arizona has 203,009 Native Americans, the third largest population of any state. The Navajos, the nation's largest tribe, live on the largest reservation in the United States. Arizona's population is 19% Hispanic.

ARIZONA'S ECONOMY

Irrigation is responsible for the state's considerable agricultural output, and because of the expenses involved, Arizona has the largest average farm size of any state. Major products include cattle, cotton, dairy products, vegetables, citrus fruits, hay, and lettuce.

Arizona leads the nation in the production of non-fuel minerals. It's the number one in copper ore, and it has significant amounts of gold, silver, and molybdenum.

The processing of copper is one of the state's dominant industries. Others are electronic equipment, transportation equipment, precision instruments, industrial machinery, primary metals, and fabricated metals.

Tourism is a rapidly growing industry that brings nearly $7 billion in annual revenue.

Places to Visit

The 277 mile long Grand Canyon i(602-638-7888) is one of the most unforgettable places in the world—everyone should have the privilege of experiencing this natural wonder. I am drawn back to the canyon again and again. But Arizona has a lot more to see. The Petrified Forest National Park (602-524-6228) includes a portion of the breathtaking Painted Desert. If you're a fan of westerns, you'll recognize the rock formations of Monument Valley Navajo Tribal Park (801-727-3287). Glen Canyon Dam ((602-645-2511) is an awe-inspiring man-made structure that created the hauntingly beautiful Lake Powell. Tourist information: Arizona Office of Tourism, 1100 W. Washington St., Phoenix, AZ 85007 (602-542-8687)

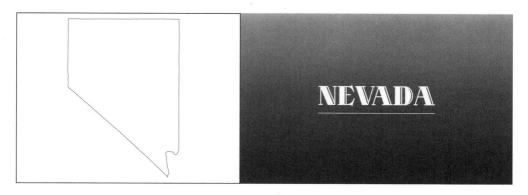

NEVADA

Gambling made Nevada famous, but all the jokes about going to the state to visit your money are true for another reason—78.9% of Nevada is owned by the Federal government, which makes it our land. Nevada is bordered on the north by Oregon and Idaho, on the east by Utah and Arizona, and on the south and southwest by California. The Colorado River makes up part of the eastern border. The state is shaped like a rectangle with the lower left corner cut off to a point. The seventh largest U.S. state, Nevada stretches 485 miles north-south and 320 miles east-west.

Las Vegas, the largest city, is 417 miles southwest of Salt Lake City and 275 miles northeast of Los Angeles on I-15. Reno is 535 miles west of Salt Lake City and 228 miles east of San Francisco on I-80.

THE LAY OF THE LAND

Most of Nevada lies in the Great Basin, an area of land between the Rocky Mountains and the coastal ranges. Much of the land is covered with thick layers of rock spewed forth from ancient volcanoes. The movement of the earth's crust has tilted huge blocks of rock, elevating them above other blocks. The result is hundreds of mountain ranges separated by broad, mostly desert plains.

A small spur of the Sierra Nevada Mountains juts into the state around the Lake Tahoe area, and this region includes Boundary Peak (13,143 ft.), the state's highest point. Along the Idaho border in the northeast is a small section of the Columbia Plateau. The southern point of Nevada is part of the Sonoran Desert.

The only major natural lakes in the state are in the Sierra Nevada region. They include the famous resort destination, Lake Tahoe, and Lakes Pyramid and Walker. However, Nevada borders on two of the largest man-make lakes in the U.S., Lake Mead and Lake Mojave.

HOW NEVADA BECAME A STATE

Although archaeological discoveries indicate Nevada was inhabited more than 12,000 years ago, it was the last area of the contiguous United States to be explored. Franciscan missionaries may have entered the area in the late 1700s seeking a shorter route from New Mexico to California, but no further Spanish or Mexican exploration or settlement was recorded. In 1825, Peter Skene Ogden of the Hudson Bay Company led the first documented exploration, crossing the northern border to trap along the Humboldt River. In 1826-1827, Jedediah Smith led a party from Salt Lake City through Nevada to California, then returned. In 1833, Joseph Walker explored a route across the state that later became known as the California Trail.

The U.S. acquired Nevada from Mexico in the Treaty of Guadalupe Hidelgo. When gold was discovered in California in 1849, the California Trail became a conduit for thousands of fortune hunters. The first permanent settlement in Nevada was a trading post established by the Mormans on the California Trail in 1850. That same year Nevada became part of the Utah Territory.

The state's history changed dramatically when huge silver deposits known as the Comstock Lode were discovered in 1859. Almost overnight, a settlement called Virginia City sprang up, and it soon became one of the wealthiest in the west. When the Civil War broke out, the United States government was anxious to secure Nevada's mineral wealth. Nevada (from the Spanish for "snow-covered") became a separate territory in 1861 and was admitted to the Union on October 31, 1864—less than five years after the first significant settlement was founded.

NEVADA'S CITIES

LAS VEGAS (258,295) was founded by Mormans as a trading post in 1855. It's name comes from the Spanish for "grasslands," which sprang up around spring-fed desert streams. Fort Baker was built on the site as an outpost for soldiers fighting Indian uprisings in 1884. The railroad arrived in 1905, bringing the first significant settlement. Growth was further spurred by completion of the nearby Hoover Dam in 1930. But the major factor in the rise of Las Vegas was the legalization of gambling in 1931. Las Vegas has become one of the nation's top ten tourist destinations and one of America's fastest growing cities.

RENO (133,850) was founded as Lake's Crossing in 1868 by the Central Pacific Railroad at the beginning of a pass through the Sierra Nevada Mountains. It was renamed for American Civil War General Jesse Lee Reno. The settlement grew as a supply center for Virginia City, and when the Comstock Lode petered out, the railroad insured its survival. Gambling made Reno a tourist mecca. The city also has significant building materials, electronic equipment, and metal and wood manufacturing industries.

CARSON CITY (40,443), the state capital, was founded in 1851. In 1858, it was named for the Carson River. It became the capital of the Territory of Nevada in 1861. A U.S. branch mint operated in Carson City until 1893. Today, Carson City derives its income from tourism as well as being the center of state government.

NEVADA'S PEOPLE

Nevada has a population of 1,388,910 people. The population jumped 50% between 1980-1990 and another 16% since, making it America's fastest growing state. Almost all the growth took place in the metropolitan areas. Nevada's population is 10% Hispanic. The state also leads the nation in the number of marriages per 1,000 residents and the number of divorces per 1,000 residents.

NEVADA'S ECONOMY

Mining built Nevada, and even today it produces gold, silver, copper, gypsum, mercury, magnetite, sand and gravel, and petroleum. The lack of arable land limits agricultural production. The state does produce cattle, sheep, hay, potatoes, wheat, barley, and vegetables.

Nevada's recent growth was the result of legislation passed in 1931, when the state was suffering from the effects of the Depression. That legislation legalized casino gambling, eliminated the waiting period to get married and decreased the waiting period for divorces to six weeks. After World War II , gamblers, lovers, and couples on the rocks flocked to Las Vegas and Reno. Gambling and tourism are the state's major industries.

Most early manufacturing in the state involved processing minerals. Today, the state also produces electronic equipment, gaming devices, precision instruments, and stone, clay and glass products.

Places to Visit

Some people call Las Vegas ((702-892-0711) a "Disneyland for Adults."—words fail me when I try to sum up the unique atmosphere of this desert town. Everyone should visit Los Vegas at least once in a lifetime. Reno (800-367-7366) offers gambling in a less crowded and less overwhelming atmosphere. Lake Tahoe (800-468-2463), the largest Alpine lake in the U.S., is a ski resort in the winter and a boating, fishing, and hiking paradise in the summer. Tourist information: Nevada Commission on Tourism, Capitol Complex, Carson City, NV 89710 (800-237-0774)

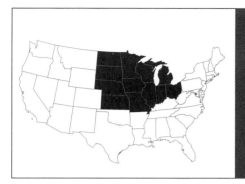

THE MIDWEST

Comparing The Midwestern States

STATE	POPULATION	SQUARE MILES	PER CAPITA INCOME
National rankings in parenthesis	(1993 U.S. Bureau of the Census)	(U.S. Geological Survey)	(U.S. Bureau of Labor Statistics)
Illinois	11,697,336 (6)	57,918 (25)	$15,201 (14)
Ohio	11,091,301 (7)	44,828 (34)	$13,461 (24)
Michigan	9,477,545 (8)	96,705 (11)	$14,154 (20)
Indiana	5,712,799 (14)	36,420 (38)	$13,149 (29)
Missouri	5,233,649 (16)	69,709 (21)	$12,989 (30)
Wisconsin	5,037,928 (18)	65,499 (23)	$13,276 (28)
Minnesota	4,517,416 (20)	86,943 (12)	$14,389 (19)
Iowa	2,814,064 (30)	56,276 (26)	$12,422 (35)
Kansas	2,530,746 (32)	82,282 (15)	$13,300 (27)
Nebraska	1,607,199 (37)	77,358 (16)	$12,452 (34)
South Dakota	715,392 (45)	77,121 (17)	$10,661 (47)
North Dakota	634,935 (47)	70,704 (19)	$11,051 (45)

Although we tend to think of oil, gold, or silver when we think about natural resources, many experts believe that the most important natural resource in the next century will be farmland. Food in America is so abundantly available and so inexpensive (the average family spends about 12% of its income on food) that we take it for granted. But with a world population approaching 6 billion people, food is going to become in increasingly short supply.

That's why the Midwestern States are so important to our economy. The heartland of our country contains more fertile soil that any other area in the world. On farms from Ohio to Kansas, from Missouri to North Dakota, our under appreciated farms raise crops and livestock that not only fill our tables, but produce a surplus that can be sold overseas to reduce our chronic balance of payments deficit. That's why agriculture is the characteristic that comes to mind first when I think of the Midwest.

The second is manufacturing. Included in this vast area are the Great Lakes and the Mississippi River, two systems that served as highways for the transportation of the agricultural products, timber, and minerals produced by the region. It was practical that factories would be built at transportation hubs to process these natural resources. That's why cities such as Chicago, St. Louis, Cleveland, Detroit, Indianapolis, Milwaukee, Minneapolis-St. Paul, and Kansas City became major industrial centers.

Weather also comes to mind when I think of the Midwest. Because the middle of the United States has no mountain ranges to get in the way, it serves as an alley for warm air masses from the south and Arctic air masses moving southward from the north. The results: thunderstorms, tornadoes, and dramatic temperature swings from season to season as well as from day to day. If you don't think Midwestern weather is especially tempestuous, ponder this: almost three-quarters of the world's tornadoes occur in the United States, and most of those touch down in the Midwest. The residents of this area also have to deal with more violent thunderstorms, more hail, and more flooding then any other region. No wonder the Midwest is known for the determination and hard work of its people.

EARLY EXPLORATION OF THE MIDWEST

Almost all of this area was first explored by the French moving down from Canada. The list of explorers include Robert Cavalier, sieur de La Salle, Louis Jolliet, Father Jacques Marquette, Jean Nicolet, Pierre Esprit Raddison, Medard Chouart, Father Louis Hennepin, and Daniel Greysolon, sieur Duluth. These brave men claimed the area for France and established a handful of mission or trading posts. The French, however, had little lasting impact on the area. Instead, British trappers and traders had the most influence even after the U.S. acquired title to the land east of the Mississippi in 1783 and the land west of the Mississippi in the Louisiana Purchase of 1803.

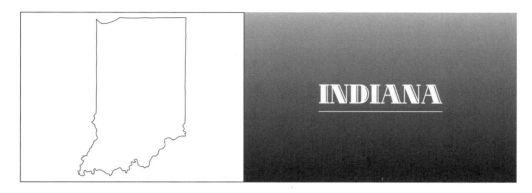

INDIANA

Indiana's state motto is "The Crossroads State," for good reason—not only do four major interstate highways cross in Indianapolis, but the state also has access to the two most important water systems in the eastern U.S. Indiana is bordered on the north by Lake Michigan and Michigan, on the east by Kentucky and Ohio, on the south by Kentucky, and on the east by Illinois. The Ohio River, part of the Mississippi River system, forms all of the southern boundary. The Wabash River forms part of the western border with Illinois. The state is roughly rectangular, stretching 275 miles north-south and 170 miles east-west.

Indianapolis, the largest city, is 185 miles southeast of Chicago and 114 miles northwest of Louisville on I-65, is 246 miles northeast of St. Louis and 178 miles southwest of Columbus, Ohio on I-70, and is 110 miles northwest of Cincinnati on I-74. Fort Wayne is 132 miles north of Indianapolis on I-69.

THE LAY OF THE LAND

Moving from north to south, you'll cross three different regions in Indiana. The third of the state nearest Lake Michigan is part of the Eastern Great Lakes Lowlands. This primarily flat area contains lots of lakes and bogs, with sand dunes on parts of the lake front. Drained swampland in the northwest makes great farmland.

The central Till Plain (till means materials deposited by glaciers) is Indiana's main agricultural region. The terrain goes from flat to low hills and shallow valleys. The southern third of the state is part of the Interior Low Plateau. This region wasn't smoothed by glaciers, so it features higher ridges and valleys with steeper sides. Much of the area sits on limestone deposits which have sinkholes, underground rivers, caves, and mineral springs. Wyandotte and Marengo are two of America's most beautiful caves, while French Lick, West Baden, and Martinsville are famous for their springs.

Indiana has many natural lakes, of which Lake Wawasee is the largest. Two-thirds of the state is drained by the Wabash River.

HOW INDIANA BECAME A STATE

French trappers from Canada were the first Europeans to venture into Indiana. In 1677, Robert Cavalier, sieur de La Salle, explored the area and claimed it for France. Between 1732-35, the French built three forts in the Maumee-Wabash Valley—Fort Vincennes, Fort Miami, and Fort Ouiatenon. But the area was very sparsely settled when the British took the area from France in the Treaty of Paris that ended the French and Indian War in 1763. The Indians resented British rule and in an uprising known as Pontiac's War, destroyed Fort Miami and Fort Ouiatenon before they were defeated in 1767.

Partly to appease the French settlers, the British made the area part of the Canadian province of Quebec in 1774, but this move angered the English colonies. In 1779, George Rogers Clark captured Fort Vincennes, one of a series of vital victories that secured the west and contributed to the success of the American Revolution.

With its victory, the new United States acquired title to all lands east of the Mississippi. Virginia and other states that had claimed part of Indiana ceded that land to the Federal government, and in 1787, Indiana became part of the newly created Northwest Territory. In 1800, Ohio became a separate territory and William Henry Harrison became governor of the remaining area, which was known as the Indiana territory (the name is Latin for "land of the Indians"). The Indians remained hostile until 1811, when they were defeated by forces led by Harrison at the Battle of Tippecanoe (a battle so famous that Harrison's slogan in his successful 1840 presidential election was "Tippecanoe and Tyler, too").

Settlers poured into the state, and it was admitted to the Union as the 19th state on December 11, 1816.

INDIANA'S CITIES

INDIANAPOLIS (731,327), was built on a site on the White River in the center of the state that was selected to be the state capital. It was named in 1821 and designed by Alexander Ralston, who had helped Pierre L'Enfant lay out Washington, D.C. The capital moved to Indianapolis in 1825, and in 1830, the first railroad arrived. The city soon became a major commercial and transportation hub for a rich agricultural district. Today, Indianapolis has a diverse manufacturing economy that is famous for the annual Indianapolis 500 auto race held each Memorial Day at the Indianapolis Motor Speedway.

FORT WAYNE (173,072) sits on the Maumee River where it is created by the junction of the St. Marys and St. Joseph Rivers. The chief village of the Miami Indians was located here, and the French built Fort Miami on the site between 1732-35. The British captured the Fort in the French and Indian War, but it was destroyed by the Indians in Pontiac's War. General "Mad" Anthony Wayne finally defeated the Miami Indians in 1794 and built a fort that bore his name. Economic growth began in 1840 with the completion of the Wabash-Erie Canal, and was further stimulated with the arrival of the railroad in 1850. Fort Wayne today is a manufacturing city that produces motor vehicles, magnet wire, military electronic items, and frozen desserts.

EVANSVILLE (126,272) is a port city on the Ohio River. It was founded in 1812 and named for Robert Evans, who had mapped the area. The key factor in its growth came when it was reached by the Wabash-Erie Canal in 1853. Today, Evansville produces electric appliances, plastics, aluminum, and processed foods.

INDIANA'S PEOPLE

Indiana's 5,712,799 people make it the 14th most populated state. The residents proudly refer to themselves as Hoosiers, although the derivation of the name is unknown. Indiana's population is 91% white and 8% black. About two-thirds of the residents live in urban areas.

INDIANA'S ECONOMY

Agriculture was the foundation of Indiana's growth, and as part of the Corn Belt, it is still has the 10th highest farm income. Three-fifths of the agricultural income comes from crops, especially corn, soybeans, hay, wheat, oats, potatoes, and vegetables. The major livestock products are hogs, beef cattle, dairy cattle, and turkeys. Indiana is part of one of America's major coal producing regions, and also produces masonry cement and building limestone.

Indiana is classified as a major manufacturing state. The northwest region of the state around Gary, which is close to Chicago, is an iron and steel production center. Other major products are aluminum, appliances, telephones, motor vehicles and motor vehicle parts, aerospace equipment, recreational vehicles, oil refining, chemicals, pharmaceuticals and rubber and plastic items.

Tourism brings about $4.5 billion in annual revenues.

Places to Visit

In addition to the Indianapolis Speedway Hall of Fame Museum, Indianapolis (800-323-4639) offers the 19 story Hoosier Dome, the National Track and Field Hall of Fame, and the Indiana State Museum. Southern Indiana (800-767-7752) features historic communities such as 300 year old Vincennes, New Harmony, and Corydon, the state's first capital. South Bend (800-392-0051) is the home of Notre Dame. Tourist information: Indiana Division of Tourism, 1 N. Capital Ave., Indianapolis, IN (800-289-6646/317-232-8860)

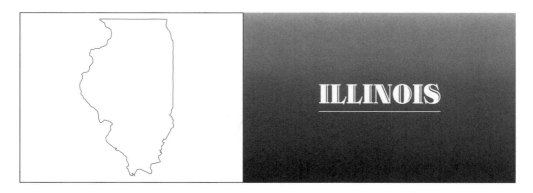

Illinois, the Land of Lincoln and the home of Chicago, the largest metropolitan area between New York and Los Angeles, is bordered on the north by Wisconsin, on the east by Lake Michigan and Indiana, on the south by Kentucky and Missouri, and on the west by Missouri and Iowa. The Wabash River makes up part of the eastern border, the Ohio River forms the southern border, and the Mississippi River is the state's entire western border. The state, which is roughly rectangular, stretches 380 miles north-south and 210 miles east-west.

Chicago, the largest city, is 90 miles south of Milwaukee on I-43, 409 miles southeast of Minneapolis and 360 miles west of Cleveland on I-90, and 185 miles northwest of Indianapolis on I-65. Springfield, the capital, is 202 miles south of Chicago and 97 miles north of St. Louis on I-55.

THE LAY OF THE LAND

About 90% of the state was at one time covered by glaciers. One pocket that escaped was the far northwestern corner, which is known as the Driftless Region. This area of wooded hills includes the state's highest point, Charles Mound (1,235 feet). Near Lake Michigan is flat land that is part of the Eastern Great Lakes Lowlands. This area is not well-drained, and contains marshes and bogs.

Central Illinois is part of the Till Plain, flat to gently rolling land covered by mineral-rich soil deposited by the glaciers. This area, which is divided into several broad river valleys, is fertile farm country. The southernmost section of the state, part of the Interior Low Plateau, also escaped the glaciers and has steep valleys and narrow ridges. Finally, at the extreme southern tip is an area of the Gulf Coastal Plain that is known as Little Egypt because of its similarity to the area around Cairo—it includes Cairo, Illinois, where the Ohio River meets the Mississippi.

Illinois has many small lakes, including Chain O'Lakes in the northwest, which are popular with vacationers. However, the most important waterways are Lake Michigan, the Mississippi River, and the Illinois Waterway that connects the two.

HOW ILLINOIS BECAME A STATE

The first important European explorers were Louis Jolliet and Father Jacques Marquette, who traveled down and up the Mississippi in 1673. France claimed the area and built a fort at the foot of Peoria Lake in 1680. In 1699, French priests built the first permanent settlement at Cahokia, and in 1720, the settlement of Kaskaskia was established at the foot of Peoria Lake. The state had fewer than 3,000 white residents when the British were ceded the area after the French and Indian War in 1763. However, the British did not actually control the area until they defeated Pontiac, chief of the Ottawa Indians, in 1765.

Beginning in 1769, settlers from Virginia began to move in and Virginia claimed large areas of the state. In 1774, Britain attached Illinois to Quebec, angering the Virginians. In 1779, George Rogers Clark ended British control with his victories at Chokia and Kaskaskia.

Virginia attempted to take control of Illinois after the Revolution, but ceded its claims to the U.S. in 1784. Illinois became part of the Northwest Territory in 1787 and the Indiana Territory in 1800. In 1809, Congress created the Illinois Territory, which also consisted of Wisconsin and part of present-day Minnesota (the name Illinois comes from the French version of an Algonquin word meaning "men"). Even though the population was well below the 60,000 minimum established by the Ordinance of 1787, Illinois was admitted to the Union as the 21st state on December 3, 1818.

ILLINOIS'S CITIES

CHICAGO (2,783,726), the nation's third largest city, has 29 miles of shoreline on Lake Michigan at the mouths of the Chicago and Calumet Rivers. Indians who lived in this swampy area called it "Checagou," because it smelled like wild onions. Jolliet and Marquette passed through the site in 1673, but the first house was built by a black French trapper named Jean Batiste Point du Sable in 1773. In 1803, the British built Fort Dearborn, but the current city started to grow with the start of construction of the Illinois and Michigan Canal in 1837. The arrival of the railroad in the 1840s brought huge waves of European immigrants to a city that had become the major commercial center on the Great Lakes. The Great Chicago Fire of 1871 destroyed one-third of the city, but it was soon rebuilt. Today, Chicago has the second-highest number of manufacturing employees of any city, and is the most important rail and trucking center in America. Among the city's jewels are its museums, orchestra, and magnificent shoreline on Lake Michigan.

SPRINGFIELD (105,227) was founded on the Sangamon River in 1818 by poor farmers who moved west to find their own land. One of them was an Indiana man named Tom Lincoln, who arrived in 1830. His son, Abraham, fought in the last Indian war in 1832. Lincoln practiced law in Springfield from 1837 until he became president in 1861. Springfield became the state capital in 1837. The city has several Lincoln-related attractions, including the Lincoln Home National Historic Site.

PEORIA (113,504) is a port on the Illinois River that was founded as Kaskaskia in 1720. The city was burned by U.S. troops in the War of 1812, but rebuilt as Fort Clark in 1813. The city was settled and named for the Peoria Indians in 1826. Today, it is a manufacturing center that produces earth-moving and mining equipment, off-road vehicles, steel, and wire.

ILLINOIS'S PEOPLE

With 11,697,336 people, Illinois is the sixth largest U.S. state. Chicago has been the mecca for waves of immigrants. In the 1800s, it became the new home for large numbers of Germans, Scandinavians, and Polish immigrants, and those ethnic communities are still very large today. In the twentieth century, it has attracted blacks fleeing the poverty of the rural South. The population is 78% white and 15% black, and 85% of the people live in urban areas.

ILLINOIS'S ECONOMY

Illinois was the leading farm state in 1870, and in 1990 this Corn Belt state still had the fifth highest farm income. Illinois is the leading soybean producer and, after Iowa, is the number two producer of corn. Other important crops are wheat, hay, oats, and commercial flowers. Illinois is second in the nation in production of hogs.

Much of Illinois is underlain with coal, which is a significant source of income along with petroleum, natural gas, sand and gravel, and fluorite.

More than 1,000,000 Illinois residents work in one of the nation's leading manufacturing industries. Major products are farm and construction equipment, steel products, printed materials, electronic goods, foods, chemicals, and rubber and plastic items.

Tourism brings in $15 billion per year.

Places to Visit

Chicago's (800) 487-2446) Loop area is a living museum of skyscraper architecture that includes the Sears Tower, the world's tallest building. Chicago is also home to the famous Art Institute, Field Museum of Natural History, and the Lincoln Park Zoo. Springfield (800-545-7300), the capital, has the Lincoln Home National Historic Site and other attractions related to the famous president. Galena (800-747-9377), which overlooks the Mississipi River, features beautifully preserved pre-Civil War architecture. Tourist information: Illinois Bureau of Tourism, 310 S. Michigan Ave., Chicago, IL 60604 (800-223-0121)

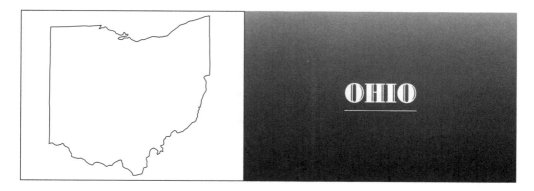

OHIO

Ohio, the easternmost of the Midwestern States and the first to be admitted to the Union, is bordered on the north by Michigan and Lake Erie, on the east by Pennsylvania and West Virginia, on the south by West Virginia and Kentucky, and on the west by Indiana. The Ohio River makes up most of the eastern and all of the southern border. The state, which is shaped like a square with the lower corners rounded off, stretches 225 miles east-west and 215 miles north-south. Ohio has 310 miles of shoreline on Lake Erie.

Cleveland, center of the largest metropolitan area, is 157 miles northwest of Pittsburgh on I-70, 360 miles east of Chicago on I-90, and 178 miles southeast of Detroit on I-90/75. Columbus, the largest city, is 142 miles southwest of Cleveland and 111 miles northeast of Cincinnati on I-71 and is 178 miles east of Indianapolis on I-70.

THE LAY OF THE LAND

Ohio is the gateway to the Corn Belt. The Eastern Great Lake Lowlands, flat and frequently wet land, rings Lake Erie. South of the Lowlands in the western part of the state is the fertile Till Plains, gently rolling land covered with rich soil deposited by the glaciers that stretches westward into Indiana and Illinois. This is Ohio's most productive agricultural region.

In the south is an area of the Interior Low Plateaus, hilly land with high bluffs along the rivers. Eastern Ohio is part of the Appalachian Plateau that stretches into Pennsylvania, West Virginia and Kentucky. This land has steep hills and narrow valley floors. Although the north is smoother than the south, this area generally has thin soil.

Ohio has many small lakes, but Lake Erie and its rivers are by far the most important bodies of water.

HOW OHIO BECAME A STATE

The first European explorer in Ohio was Robert Cavalier, sieur de La Salle, who claimed the region for France in 1669. However, British settlers were interested in this land, too. In the 1830s, traders from Virginia trekked through Ohio. In 1849, the British Government granted a charter for the area to the Ohio Company, which was owned by British merchants and Virginia plantation owners. To counter the British claims, Canadian governor Marquis de La Galissoniérie sent Pierre de Bienville to buy six lead markers on the Ohio River to reinforce France's claim (part of one of these markers was found by small boys and is now in a museum).

Conflicts between two nations and the Native American allies of the French was one cause of the outbreak of the French and Indian War in 1754. The British gained the French territories east of the Mississippi in the Treaty of Paris ending the war in 1763, and it 1774, Ohio was made part of Quebec. Except for some fights with the Indians, few Revolutionary War battles were fought in the state. In 1788, Rufus Putnam Shaw of the Ohio Group of Associates established a settlement at Marietta; Cincinnati was also founded the same year. In just twelve months, 10,000 settlers entered the state after floating down the Ohio River. The Indians were hostile, but they were moved out of the state after 1795.

The path to statehood was now blocked only by land claims of other states. Connecticut controlled the Western Reserve, and area that included Cleveland; Virginia claimed the Virginia Military Survey. Connecticut gave up the Western Reserve in 1800, which was sufficient for Ohio become the first state in the Northwest Territory to join the Union when it became the 17th state on March 1, 1803. By the way, Virginia didn't officially give up its claim until 1852.

OHIO'S CITIES

COLUMBUS (632,910), the largest city and capital, is located in the center of the state on the Scioto and Olentangy Rivers. The site for a capital was selected across the river from an existing town called Franklinton. The new city, named for Christopher Columbus, absorbed Franklinton shortly after becoming the capital in 1827. In 1831, a feeder canal to the Ohio and Erie Canal was completed, and in 1850, the railroad arrived. Columbus became a major commercial and shipping center, and it was famous in the 19th century for the manufacture of horse-drawn carriages. Today, it is a government, insurance, and medical center.

CLEVELAND (505,616) is an important manufacturing and port city on the Lake Erie, halfway between New York and Chicago. The area, part of the Western Reserve claimed by Connecticut, was surveyed in 1796 by Moses Cleaveland (the a was dropped in 1832 to fit a newspaper masthead). In 1832, the Ohio Canal provided a water link to the Erie Canal, which went all the way to New York City. Cleveland's industry took iron ore from the Great Lakes and coal from nearby Kentucky and West Virginia and developed a large steel industry. The population soared from 17,034 in 1850 to 560,663 in 1910. The population dropped 42% between 1960 and 1990, but the city remains the center of culture and commerce for the area and it's new baseball stadium, opened in 1993, is a symbol for renewal.

CINCINNATI (364,040) was founded on two plateaus on the Ohio River in 1788. Originally named Losantiville, it was the site of Fort Washington. The name was changed in honor of the Society of Cincinnati, an organization of Revolutionary War veterans. In 1811, steamboat traffic began on the Ohio River, and in 1827, the Miami Canal was opened. Cincinnati became the "Queen City of the West," a gateway and supply center for the people seeking to settle in what is now Indiana and Illinois. Cincinnati is a major manufacturing and transportation center today.

OHIO'S PEOPLE

Ohio has a population of 11,091,301 people, ranking it seventh among U.S. states. That population is 88% white and 11% black. Nearly three-quarters live in 12 metropolitan areas, the most of any U.S. state.

OHIO'S ECONOMY

Western Ohio is part of the Corn Belt, and in addition to its namesake grain, that region produces soybeans, wheat, dairy products, hogs, beef cattle, fruits, and vegetables. Ohio has significant coal reserves, and also receives income from national gas and petroleum.

Ohio has more well-known manufacturing cities than any other state. In addition to Cleveland, Columbus and Cincinnati, these cities include Akron, Toledo, Dayton, Canton, Youngstown, and Lorain-Elyria. More than 1.1 million residents are employed in manufacturing, and the state ranks fourth among all states in manufacturing income. Ohio produces motor vehicles, iron and steel, machinery, machine tools, soap, office machines, aircraft, refrigerators, and tires and other rubber products.

Places to Visit

One of my favorite Ohio destinations is Canton (800-533-4302), site of the Pro Football Hall of Fame and the National Afro-American Museum and Cultural Center. Cleveland (800-321-1001) has great museums, a great symphony, and a great zoo (and Sea World is nearby). Cincinnati (800-344-3445) is a friendly metropolis on the Ohio River. The Mound City Group National Monument (614-774-1125) south of Columbus is a prehistoric burial ground that dates back to 200 B.C. Tourist information: Ohio Division of Travel and Tourism, Box 1001, Columbus, OH43266 (800-282-5393)

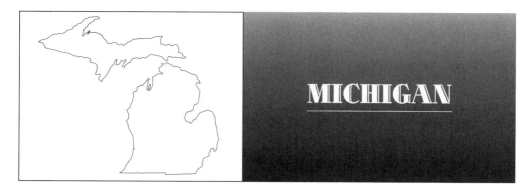

MICHIGAN

Michigan consists of two separate peninsulas that have shores on four of the five Great Lakes. The Upper Peninsula is bordered by Lake Superior on the north, Lake Michigan on the south, and Wisconsin on the southwest. The Lower Peninsula, separated from the Upper by the Straits of Mackinac on the north, is bordered on the east by Lake Huron, Lake Erie, and the Canadian province of Ontario, on the south by Ohio and Indiana, and on the west by Lake Michigan. The Lower Peninsula, which is shaped like a left-handed mitten, stretches 285 miles north-south and 195 miles east-west; the Upper Peninsula extends 325 miles east-west and 175 miles north-south. Michigan has more than 3,000 miles of shoreline, the most of any inland state. Lansing is the state capital.

Detroit, the largest city, is 279 miles east of Chicago on I-94/90, 272 miles north of Cincinnati on I-75, and 188 miles northwest of Cleveland on I-90/75. Ann Arbor, the capital, is 41 miles west of Detroit on I-94.

THE LAY OF THE LAND

The western half of the Upper Peninsula is part of the Superior Uplands, an extension of the Canadian Shield that consists of the oldest rock in North America. Because the harder rock was more resistant to the scouring of the glaciers, the terrain is more rugged, consisting of ridges and valleys.

The eastern half of the Upper Peninsula and the entire Lower Peninsula are part of the Western Great Lakes Lowlands. The eastern Upper Peninsula is level and has many swampy areas. The northern half of the Lower Peninsula is primarily rolling hills covered with a type of thin, sandy glacial soil that has limited fertility. The southern Lower Peninsula, on the other hand, is more level and was left by the ice packs with a richer loam that makes it an excellent agricultural area.

Glaciers scraped out numerous depressions in the state, and their retreat left Michigan with over 11,000 lakes. The largest is Houghton Lake. The state has several important islands, including Isle Royale and Mackinac Island.

HOW MICHIGAN BECAME A STATE

Étienne Brûlé discovered Lake Superior in 1620, and Jean Nicolet was the first European to sail into Lake Michigan in 1634. The Jesuits preached to the Indians as early as 1641, and in 1669, Father Jacques Marquette established the first permanent settlement, a mission, at Sault Ste. Marie. Fur traders soon followed, and in 1701, Antoine de La Mothe, sieur de Cadillac, founded Fort Pontchartrain on the present site of Detroit.

The British won control of Michigan in the French and Indian War. Although the Treaty of Versailles required the British to give Michigan to the U.S. in 1783, they didn't leave the area. Their Indian allies fought the U.S. until defeated by General Anthony Wayne in 1794. Finally, after the signing of Jay's Treaty, the British evacuated Detroit in 1796.

Michigan was part of the Northwest Territory between 1796—1800. The name of the state comes from Lake Michigan, taken from an Algonquin Indian term meaning "big water." It was split between the Northwest and Indiana Territories from 1800-1803, was consolidated into the Indiana Territory in 1803 and became a separate territory in 1805. At the start of the War of 1812, Governor William Hull led an invasion of Canada but was forced to surrender Detroit to the British without firing a shot. The city was recaptured in 1813.

Large-scale settlement began with the completion of the Erie Canal in 1825—the population increased 1,000% between 1820 and 1840. However, the path to statehood was blocked by a bitter dispute with Ohio over the area around Toledo that almost resulted in armed conflict. However, in 1835, Michigan agreed to give Toledo to Ohio in exchange for the entire Upper Peninsula. Michigan was admitted to the Union as the 26th state on January 26, 1837.

MICHIGAN'S CITIES

DETROIT (1,027,974) is the nation's seventh-largest city and is center of a metropolitan area with a population 3.5. times its size. In 1701, Antoine de La Mothe, sieur de Cadillac, founded Fort Pontchartrain d'Étroit on the strait that connects Lake Huron to Lake Erie. When the town was founded in 1802, it became Detroit (d'étroit is French for "strait.") The city began to grow with the beginning of steam transportation on the Great Lakes in 1818 and the opening of the Erie Canal in 1825. It's most famous industry began with the opening of automobile production plants by Ransom Eli Olds in 1899 and Henry Ford in 1903. In World War II, the automobile factories converted to defense production, earning the city the nickname "The Arsenal of Democracy." The population of the city is 76% black, due to heavy migration from the South during and after the Depression.

ANN ARBOR (109,582), was incorporated in 1851 on the Huron River west of Detroit. It was named for a grape arbor owned by the wife of an early settler—Ann's Arbor became Ann Arbor. The city is well known today as the home of the University of Michigan, and it produces computer software and electronic equipment.

GRAND RAPIDS (189,126) is located at an area of rapids on the Grand River. The Ottawa Indians lived in the area, and a fur trading post was established on the site in 1826. Grand Rapid's furniture factories sprang up in the 1840s with the development of the lumbering industry. Grand Rapids is the site of the Gerald Ford Presidential Museum.

MICHIGAN'S PEOPLE

With 9,477,545 people, Michigan is the eighth most populated U.S. state. Most live in the Lower Peninsula and 71% reside in urban areas. The state has experienced several waves of immigration since the 1820s. The most recent came between World Wars I and II, when the state's growing industries brought an influx of blacks from the South. The black population jumped from 17,000 in 1910 to 442,000 in 1950. Today, blacks make up 14% of Michigan's population.

MICHIGAN'S ECONOMY

The fertile soil in the southern Lower Peninsula makes Michigan a major agricultural state. The chief sources of income are dairy farming, corn, hogs, soybeans, beef cattle, and beans. Michigan also has many orchards that grow apples, cherries, peaches, plums, and pears.

The Upper Peninsula's economy is based largely on lumbering and mining of iron ore and copper. In recent years, petroleum and natural gas have been produced in the Lower Peninsula.

Michigan is famous as the center of America's automobile industry, and the headquarters of General Motors, Ford, and Chrysler are in Detroit. The state is seventh in industrial output, producing iron and steel, breakfast cereals, processed foods, chemicals, and pharmaceuticals. Michigan has more ports that ship more than one million tons per year than any other state.

Tourism brings nearly $7 billion in annual revenue.

Places to Visit

The Henry Ford Museum and Greenfield Village (313-271-1620) in Dearborn, America's largest indoor-outdoor museum, has more than 40 historical buildings that include the Wright Brothers bicycle shop and Thomas Edison's laboratory. Detroit (313-567-1170) has four major professional sports teams, seveal museums, and lots of entertainment. There are many resorts on the long Lake Michigan shoreline, and Mackinak Island (906-847-3783) is a quaint Victorian village on which no autos are allowed. Tourist information: Michigan Travel Bureau, Box 30226, Lansing, MI 48909 (800-543-2937)

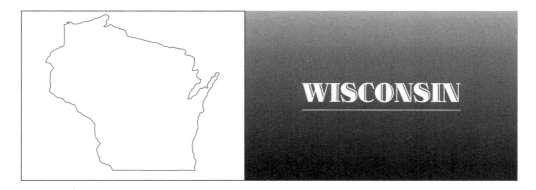

WISCONSIN

Wisconsin, "America's Dairyland," is bordered on the north by Lake Superior and Michigan's Upper Peninsula, on the east by Lake Michigan, on the south by Illinois, and on the west by Iowa and Minnesota. The Menominee River makes up part of the northeast border, while the Mississippi and Saint Croix Rivers make up most of the western border with Minnesota. Wisconsin is roughly rectangular, stretching 320 miles north-south and 295 miles east-west. The state has 675 miles of shoreline on Lakes Michigan and Superior.

Milwaukee, the largest city, is 90 miles north of Chicago on I-43 and 338 miles southeast of Minneapolis on I-90/94. Madison, the state capital, is 77 miles west of Milwaukee on I-94.

THE LAY OF THE LAND

A narrow band of northern Wisconsin is part of the Superior Uplands, an extension of the Canadian Shield. It's ancient, hard rock was cut into a rugged area of moraines, lowlands, and hills. The eastern half of Wisconsin is part of the Western Great Lakes Lowlands, rolling terrain with generally excellent soil deposited by the glaciers. It includes the Kettle Moraine area, one of America's prime examples of glacial carving.

Most of the western part of the state is part of the Till Plains, rolling land that is more elevated than the eastern portion of the state. This area includes the Dells, a scenic gorge on the Wisconsin River. In the far southwestern corner is a small Driftless region. This rugged area was not reached by the glaciers, but it is littered with rocks deposited by glacial streams.

Wisconsin has more than 8,000 lakes, the largest of which is Lake Winnebago.

HOW WISCONSIN BECAME A STATE

The first Europeans to reach the shores of Wisconsin were Étienne Brûlé, who explored Lake Superior in 1621-1623, and Jean Nicolet, who landed at Green Bay in 1634. In 1673, Louis Jolliet and Father Jacques Marquette claimed the area for France and fur traders began to work the area.

The British gained control of Wisconsin after the French and Indian War ended in 1763, but had little impact on the area. Augustin Monet de Langlade founded the first permanent settlement at Green Bay in 1764, and it became a thriving fur trading area. The United States gained control after the Revolutionary War, and in 1816, forts were built Green Bay and Prairie du Chien. The first wave of non-fur traders were prospectors who arrived in the 1820s to mine lead—the state's nickname, the Badger State, comes from the miners who dug into the hillsides like badgers. Indian resistance was strong until Black Hawk was defeated in 1832. Shortly afterwards, the land was opened to public sale. In 1835, the first steamboat arrived.

The result was that the population of the territory grew from 11,700 in 1830 to 305,200 in 1850. Wisconsin became a territory in 1836 (the name was provided by Congress as a derivation of ouisconsin, meaning "grassy place"or "place of the beaver". On May 29, 1848, Wisconsin became the 30th state.

WISCONSIN'S CITIES

MILWAUKEE (628,088), the largest city, was founded as a fur trading post in 1818 by Solomon Laurent Juneau. In 1833, the Indians gave up all rights to the area, which began to grow when the first steamboat arrived in 1835. The name Milwaukee came from the Potawatowi Indian word "mahn-ah-wauk," meaning "gathering place by water." In the 1840s, a large influx of German immigrants arrived, stamping the culture of the new city. Later, large populations of Poles, Italians, and Irish followed. Industries such as tanning, brewing, and iron and steel production grew. Its politically progressive people elected Socialist mayors who governed the city for most of the first half of this century. A revamping of the city's harbor has fueled growth in recent years.

MADISON (191,262) sits on the isthmus between Lakes Mendota and Monona. It was selected as the site of the capital by Judge James Doty in 1836, and was named in honor of James Madison. It began to grow in the 1850s with the arrival of the railroad and the founding of the University of Wisconsin. Madison was a major mustering point for soldiers during the Civil War and became a meat packing center in the 1920s. Today, a mix of government, education, and manufacturing are the basis of the city's economy.

GREEN BAY (96,466) is one of the best known small cities in the United States, because it's the home of the Green Bay Packers of the National Football League. A port on Lake Superior, Green Bay was first visited by Jean Nicolet in 1634. It became a fur trading post and mission that was controlled by the British from 1760 to 1812. Fort Howard was built on the site in 1816, and the city was laid out in 1829. It grew as a lumbering center, and today produces paper, packed meats, cheese, and frozen and canned foods.

WISCONSIN'S PEOPLE

Wisconsin's 5,037,928 live in a state with a progressive political tradition. The election of Robert LaFollette as governor in 1900 ushered in an era that saw Wisconsin become the first state to regulate railroad rates, institute civil service, and develop workmen's compensation and unemployment compensation, among many other significant social programs. Today, Wisconsin is pioneering welfare reform.

WISCONSIN'S ECONOMY

As "America's Dairyland," Wisconsin produces more milk than any other state, 40% of our cheese, and 20% of our butter. The state's farms also produce beef cattle, hogs, turkeys, broiler chickens, corn, hay, potatoes, soybeans, and cranberries.

Wisconsin's first growth was stimulated by lumbering and mining. Today, the state has a moderate lumber industry, but mining consists primarily of abrasive stone, sand and gravel, and lime.

Wisconsin is a major industrial state. Food and food products are a major industry, especially cheese, packed meats, and the beer for which Milwaukee is famous. Wisconsin also leads the nation in paper production, and other industries include industrial machinery, farm machinery, and fabricated metals.

Places to Visit

The Wisconsin Dells (800-223-3557), 15 miles of rock formations cut by the Wisconsin River, are the focus for a tourist area that draws more than 2 million visitors a year. Door County (800-527-3529), a thumb of land that juts into Lake Michigan in northeast Wisconsin, is a paradise for hunters and fishermen. Milwaukee (800-231-0903) has several notable museums as well as its famous German restaurants. Tourist information: Wisconsin Tourism Development, Box 7606, Madison, WI 53707 (800-432-8747/608-266-2161)

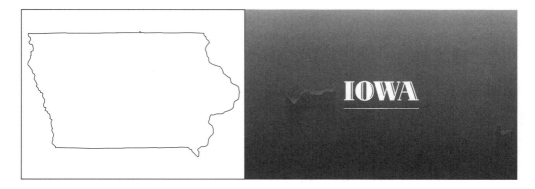

Iowa's reputation as a farm state is justly deserved—90% of the land in the state is devoted to agriculture. Iowa is bordered on the north by Minnesota, on the east by Wisconsin and Illinois, on the south by Missouri, and on the west by Nebraska and South Dakota. The Mississippi River forms Iowa's eastern border and the Missouri makes up most of the western border. Iowa is roughly rectangular, stretching 210 miles north-south and 320 miles east-west.

Des Moines, the largest city and capital, is 345 miles west of Chicago and 137 miles east of Omaha on I-80, and is 194 miles north of Kansas City and 246 miles south of Minneapolis on I-35.

Davenport, on the Mississippi River, is 168 miles east of Des Moines on I-80.

THE LAY OF THE LAND

Iowa is basically prairie, broad and relatively flat grasslands that stretch as far as the eye can see. The northwest corner of the state's part of the Driftless Region does have more rugged land that was not covered by glaciers. Uplands cut by streams make this the most elevated portion of the state. Northeastern Iowa features valleys with steeper sides and bluffs bordering the Mississippi.

However, the north central part of the state, part of the Western Great Lakes Lowlands, is level land that was blessed by the glaciers when they deposited soil that ranks with the most fertile anywhere in the United States. The southern portion of the state is part of the Dissected (eroded) Till Plains, which has more rolling hills and streams. This area is also fertile. In total, Iowa has one-quarter of all the Grade I agricultural soil in the United States.

Iowa is nestled between the two longest of all U.S. rivers, the Mississippi and the Missouri. Most of the state's small lakes are concentrated in the northwest.

HOW IOWA BECAME A STATE

Louis Jolliet and Father Jacques Marquette explored the region in 1673, as did Flemish missionary Louis Hennepin in 1680. In 1682, Robert Cavalier, sieur de La Salle, claimed the entire Mississippi Valley for France. Spain received the area from France in the 1763 Treaty of Paris, and in 1796, it granted territory to Julien Dubuque to mine lead near the Mississippi River city that now bears his name.

Iowa was part of the territory regained by France in 1800, then sold to the United States in the Louisiana Purchase of 1803. Congress passed control of the area to the Territory of Louisiana (1805 -1812), the Territory of Missouri (1812–1821), the Territory of Michigan (1834–1836), and the Wisconsin Territory (1836 - 1838). From 1821 to 1834, the area was totally unorganized. Before 1833, these administrative decisions had little practical impact, because Iowa had been declared Indian Territory and was off-limits to white settlement. However, the Indians began to cede territory after losing the Black Hawk War in 1832. Dubuque, Davenport, and Burlington were founded shortly afterwards, and settlers began to arrive in great numbers. The Iowa Territory, which included parts of Minnesota and the Dakotas, was created in 1838 (the name is from an Indian word meaning "sleepy ones"). On December 28, 1846, Iowa became the 29th U.S. state.

IOWA'S CITIES

DES MOINES (193,187) was the site of a fort built at the junction of the Des Moines and Raccoon Rivers in 1843. Two years later, the area around the fort was opened to settlement. Communities named East Des Moines and Fort Des

Moines soon merged to become Des Moines. The name is from the French "de moyen," meaning "the middle," because Des Moines is halfway between the Missouri and the Mississippi Rivers. The city became the commercial and shipping center of this rich agricultural area. Today, the city's manufacturers also produce printed materials, farm equipment, processed foods, and tires.

CEDAR RAPIDS (108,751) was founded in 1837 at the site of rapids on the Cedar River. The city grew when the railroad arrived in 1859, serving as a gateway for settlers heading west. Today, it is an agricultural shipping and supply center, and manufactures construction machinery, electrical goods, stainless steel, and brass.

DAVENPORT (95,333) is the only Iowa municipality of the Quad Cities—Rock Island, Moline, and East Moline are across the Mississippi River. The city was founded in 1836 by Col. George Davenport, and it developed when a railroad bridge was built across the Mississippi in 1856.

IOWA'S PEOPLE

Because nearly all of the state is suitable for agriculture, Iowa's 2,814,064 people are evenly spread throughout the state. High interest rates and falling agricultural prices produced economic distress that lowered the state's population in the 1980s, but the economy and the population rebounded in the early 1990s. Iowa has a very small minority population.

IOWA'S ECONOMY

Iowa, in the heart of the cornbelt, ranks third among all states in total farm income, income from livestock, and percentage of workers employed in agriculture. The state usually ranks number one in production of corn, and also grows soybeans, hay, alfalfa, oats, clover, flaxseed, sugar beets, and wheat.

Nine out of every ten ears of Iowa corn are fed to the state's livestock. Iowa produces one-quarter of all hogs slaughtered in the U.S., and it ranks fifth in beef cattle. Raising dairy cattle and sheep are also major sources of income.

Given its agricultural riches, it's not surprising that food processing is Iowa's number one industry. The state manufactures livestock feeds, meat, breakfast foods, popcorn, corn oil, corn starch, and white sugar. Farm construction and equipment and home appliances are also important products.

Tourists from outside the state spend about $2.5 billion in Iowa every year.

Places to Visit

Dubuque(800-798-8844), Iowa's oldest community, features elegant Victorian homes nestled against spectacular limestone cliffs on the Mississippi River. From Dubuque, the Great River Road follows the Mississippi southward. The Amana colonies (800-245-5465), southwest of Cedar Rapids, consist of seven villages that were founded as utopian communities in the nineteenth century and today have many craft shops and restored buildings. Outside of Des Moines is the Living History Farms(515-278-2400), a working replica of a nineteenth century agricultural life. Tourist information: Iowa Division of Tourism, 200 E. Grand Ave., Des Moines, IA 50309(515-24-4705)

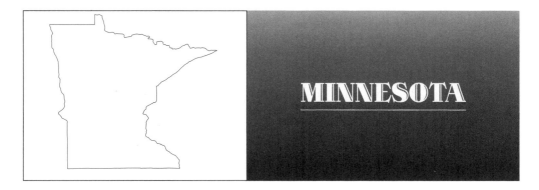

MINNESOTA

I think I've been as cold in Minnesota as I've been anywhere in the U.S.—you've probably heard me report wind chill factors in the state that have reached 80 below zero and lower. But the reward for the hardy residents who endure the bitter winters is that they live in a state with breathtaking natural beauty. Minnesota is bordered on the north by Canada, on the east by Lake Superior and Wisconsin, on the south by Iowa, and on the west by North Dakota and South Dakota. The Red River makes up much of the western border and the Mississippi is part of the southeastern border. This roughly rectangular state stretches 380 miles east -west and 410 miles north-south.

Minneapolis and St. Paul, the twin cities that represent the largest metropolitan area in the state, are 338 miles northwest of Milwaukee and 240 miles southeast of Fargo on I-94, and are 246 miles north of Des Moines on I-35. Duluth is 156 miles north of the Twin Cities on I-35.

THE LAY OF THE LAND

Northern Minnesota's Superior Upland is part of the Canadian Shield, the oldest rock on the North American continent. The rock in this area has been gouged by glaciers into numerous lakes. Most of central Minnesota is part of the Western Great Lake Lowlands. The western part of this area was covered by an ancient lake, Lake Agassiz, and today it is level land with lots of lakes, marshes, or bogs. The eastern section of the Lowlands is better drained.

Most of Southern Minnesota is part of the Dissected Till Plains, glacial deposits of soil and rock that have been eroded (dissected) over time. This land consists of gently rolling hills. In the southeast corner of the state is a Driftless Region, land that wasn't smoothed by glaciers. This area has steep bluffs and narrow valley.

Contrary to the slogan, Minnesota isn't the land of 10,000 lakes—in reality, the state has more than 20,000, of which Red Lake is the largest. Lake Itasca is famous as the source of the Mississippi River, which flows southeastward to the Falls of Saint Anthony in Minneapolis, the head of navigation of this great river.

HOW MINNESOTA BECAME A STATE

French explorers seeking a water passage through the New World to the Pacific Ocean were the first to enter Minnesota, including Médard Chouart, sieur des Groseilliers and Pierre Esprit Radisson. 1679, Daniel Graysolon, sieur Duluth, established a fort on Lake Superior. In 1680, Flemish priest Louis Hennepin discovered and named the Falls of Saint Anthony. French fur traders moved into the area, but in 1763 France ceded the territory east of the Mississippi River to Britain and the area west of the Mississippi to Spain. But it was British fur traders who dominated the entire area, even after Britain lost eastern Minnesota to the new United States in 1783. Finally, in 1805, Zebulon Pike lead an expedition up the Mississippi to enforce U.S. land claims against British traders. In 1815, a law was passed barring foreign fur traders, and in 1819, Fort Saint Anthony (later Fort Snellings) was built on land at the junction of the Mississippi and Minnesota Rivers purchased from the Indians for 60 gallons of whiskey and a few hundred dollars in trade goods. In 1823, the first steamboat reached Fort Snellings, and in 1832, Henry Schoolcraft discovered the source of the Mississippi River.

The door to widespread settlement was opened in 1832 when the U.S. government purchased the triangle of land between the Saint Croix and Mississippi from the Sioux and Chippewa Indians. Farm towns and lumber camps sprang up. During this period, Minnesota was part of the Michigan, Iowa, and Wisconsin territories. Finally, in 1849, when Wisconsin became a state, Congress created the Minnesota Territory (the name comes from two Indian words meaning "sky blue waters"or "cloudy waters"). In 1851, the Sioux ceded most of the western part of the state, and in just ten years, the population jumped from 5,354 to 172,000. On May 11, 1858, Minnesota was admitted to the Union as the 32nd state.

MINNESOTA'S CITIES

MINNEAPOLIS (368,383), one of the Twin Cities, was built on the west bank of the Mississippi River at its junction with the Minnesota River. After the area west of the Mississippi was opened to settlement in 1855, a village was built and given a name that combined the Sioux word for "water" with the Greek word "polis," meaning "city." In 1855, the village of Saint Anthony was built on the east side of the river. When a bridge was built in 1872, the two villages were merged as the city of Minneapolis. With 22 lakes inside the city limits, Minneapolis is one of most beautiful urban areas in the U.S. It is a commercial and financial center for an extensive area, and its leading industries are food and dairy products, printed materials and paper, and manufacturing of machinery.

ST. PAUL (272,238), the state capital, is adjacent to Minneapolis on the east side of the Mississippi River. In 1838, Pierre "Pig's Eye" Parrant founded a settlement that was known as Pig's Eye until 1841, when it was renamed for a local chapel dedicated to Saint Paul. It became the capital of the new Minnesota Territory in 1849, and it became a major transportation center after the arrival of the railroad in 1862. Government and education provide significant employment today.

DULUTH (85,493) is a port on Lake Superior opposite Superior, Wisconsin. In 1679, Daniel Greysolon, sieur Duluth, founded a trading post on the site. In 1817, John Jacob Astor's American Fur Co. built a thriving trading post in the same place. Permanent settlement began in 1852. The railroad arrived in 1870 and 1871, the Duluth Shipping Canal was built. Duluth became the major shipping center for the iron ore, lumber, grain, and coal produced in the interior. The opening of the St. Lawrence Seaway in 1959 gave another boost to the economy.

MINNESOTA'S PEOPLE

A large percentage of Minnesota's 4,517,416 people are of German or Scandinavian descent. The state's black community is small (2% of the population) because it didn't experience the influx of Southern blacks that added to the population of many other Midwestern industrial centers in the 1930s and 1940s. Nearly three-quarters of the state's residents live in urban areas.

MINNESOTA'S ECONOMY

Agriculture is as significant to the economy of Minnesota as it is to its Midwestern neighbors. About half the agricultural income comes from livestock and livestock products, especially milk, beef cattle, and hogs. Minnesota is among the leading producers of oats, spring wheat, hay, and sunflower seed.

Logging and fur trading were the early foundations of the Minnesota economy, but the resources were seriously depleted. A reforesting program has revived the state's logging industry. Mining in Minnesota means iron ore—the state produces 80% of our total national output.

In the last three decades, Minnesota has become a major industrial state, The industrial products are industrial machinery, fabricated metals, dairy processing, canned goods, flour, sugar, printing and printed materials, electronic equipment, and lumber and wood products.

Places to Visit

The Twin Cities of Minneapolis (800-445-7412) and St. Paul (800-627-6101)both have a wealth of cultural institutions, including the Minneapolis Institute of Arts, the James Ford Bell Museum of Natural History, the Minnesota Museum of American Art, and the Science Museum of Minnesota. Shoppers fly from all parts of the country to visit the cavenous Mall of America (612-883-8800) in south of Minneapolis (the structure's so big it houses an amusement park in the center!). Dululth (800-438-5884) is a large port and the gateway to the vast northern Minnesota forests. Tourist information: Minnesota Office of Tourism, 100 Metro Square, 121 7th Place E., St. Paul, MN 55101 (800-766-8687/612-296-5029)

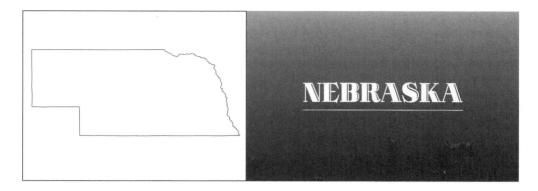

NEBRASKA

Nebraska, one of the West North Central states of the United States, is bounded on the north by South Dakota, on the east by Iowa and Missouri, on the south by Kansas, on the southwest by Colorado, and on the west by Wyoming. The Missouri River forms the eastern boundary. The state is roughly rectangular in shape, and its extreme dimensions are about 205 miles from north to south and about 420 miles from east to west.

Omaha, the largest city, is 540 miles east of Denver and 135 west of Des Moines on I-80, and 190 miles north of Kansas City on I-29. Lincoln, the capital, is 58 miles west of Omaha on I-80.

THE LAY OF THE LAND

The eastern portion of Nebraska is part of the Dissected Till Plains, which were once covered by glaciers. When they melted, they left a deep layer of till (mixed clay and stones). A wind-carried deposit of fine silt (loess) was laid on top of this, and the surface was gradually cut up (dissected) by the action of streams and rivers. The dark, fertile soils of this region form some of the state's richest farmland.

The western four-fifths of Nebraska lies in the Great Plains region. The surface of this region was largely formed as beds of sands, gravel, silts, and muds were deposited by streams flowing from the Rocky Mountains to the west. In the south central area of Nebraska the flat and fertile Loess Plains were formed by the accumulation of windblown silt. To the north of the Platte River are the low Sand Hills, which are ancient sand dunes. In the western Panhandle, erosion has produced isolated buttes and, in the extreme northwest, the picturesque Badlands region.

Nebraska is drained entirely by the Missouri River and its tributaries. The state's principal river, the Platte, is formed by the confluence of the North and South Platte rivers, both of which rise in the Rocky Mountains. Hundreds of small natural lakes are found in the Sand Hills. The state's largest bodies of water are artificial, including Lewis and Clark Lake, Lake C. W. McConaughy, and Harlan County Lake.

HOW NEBRASKA BECAME A STATE

The Spanish explorer Francisco Vásquez de Coronado is believed to have been the first European to see the area that is now Nebraska, in 1541. In 1720 Colonel Pedro de Villasur, a Spanish soldier, led an expedition into Nebraska; he and his party were massacred by Indian tribes. The French controlled the territory from 1700 to 1763, when it was ceded to Spain.

In 1803 Nebraska became a United States possession as a result of the Louisiana Purchase. Between 1804 and 1806 the Lewis and Clark expedition explored a portion of the territory. In 1807 Manuel Lisa (1772-1820), of Spain, established a trading post and became the first permanent white settler in the area. The American Fur Company established (1810) a post in the region at Bellevue. Fort Atkinson was built (1819) on what is now the site of Fort Calhoun, but growth was slow. The Oregon and California trails led through Nebraska; these routes to the West were responsible for the gradual settlement of the region despite the fact that in 1834 the federal government had declared Nebraska part of the Indian Country from which all whites were excluded. The area was successively part of the territories of Indiana, Louisiana, and Missouri. On May 30, 1854, it became the territory of Nebraska. The name comes from an Indain term meaning "flat" or "broad water," a reference to the Platte River. Immigration to the territory increased with the passage of the Pacific Railroad Act and the Free Homestead Act in 1862. On March 1, 1867, Nebraska became the 32nd state.

NEBRASKA'S CITIES

OMAHA (335,795) is located on the Missouri River very near the geographical center of the United States. Mormans heading to Utah spent the winter of 1846-47 on the site, and it became a permanent settlement in 1854. It was named for the Omaha Indians. Omaha became a jumping off place for wagon trains heading west, and it was the eastern terminus the Union Pacific Railroad. Omaha was the site of a world's fair in 1898. The city is a major livestock and meat packing metropolis, and it's also an insurance center.

LINCOLN (191,972) , has a large grain-milling industry and is a manufacturing, government, and education center. The city was founded in a salt basin in 1857, and named for Lancaster, Pennsylvania. In 1867, it was renamed for Abraham Lincoln. It became the capital of Nebraska after a bitter fight between settlers who lived north of the Platte River and favored Omaha and those who lived south of the river and wanted a southern site.

NEBRASKA'S PEOPLE

Nebraska has a population of 1,607,199 people. Most of the population is concentrated in a corridor along the eastern border and in a belt along the Platte and North Platte rivers. Whites made up 93.8% of the population and blacks 3.6%. In 1990 about 66% of all Nebraskans lived in areas defined as urban, and the rest lived in rural areas.

NEBRASKA'S ECONOMY

Farming accounts for 11% of the annual gross state product in Nebraska. The state ranks as one of the top three cattle-producing states in the country. Sheep are raised in most parts of the state, but especially in the Panhandle. Crops make up more than one-third of Nebraska's annual agricultural income. Corn, the most important crop and the dominant feed grain, forms the basis of the state's livestock industry. Other major crops are grain sorghum, grown in the southeastern and south central regions; soybeans, grown in the eastern third of the state; and wheat, grown in the south and the Panhandle.

The mining industry accounts for less than 1% of the annual gross state product in Nebraska. Petroleum makes up about half of the total value of minerals produced; most of the output comes from the southeast and the Panhandle. Other mineral commodities include sand and gravel, stone, clay, cement, lime, and gems.

Manufacturing has become increasingly important. The leading industry is food processing; meat products are particularly important. The other leading industries manufacture industrial machinery and electronic equipment. Among these manufactures are motor-vehicle parts and telephone equipment.

Each year visitors produce more than $1.7 billion for the Nebraska economy. History

Places to Visit

Scotts Bluff National Monument (308-436-4340) is a huge bluff that was such a landmark for pioneers that it was called the "lighthouse of the plains." Not too far away is the equally impressive Chimney Rock National Historic Site (402-471-4754). In Bellevue is the Stratigic Air Command Museum (401-292-2001), a mecca for military aircraft buffs. The Stuhr Museum of the Prairie Pioneer (308-381-5316) in Grand Island includes a 60 building "Railroad Town." Tourist information: Nebraska Division of Travel and Tourism, Box 94666, Lincoln, NE 68509 (800-228-4307)

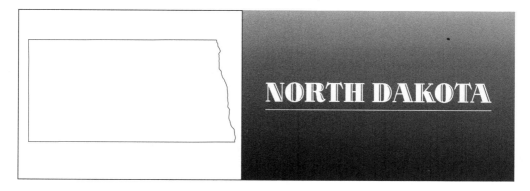

NORTH DAKOTA

North Dakota, one of the nation's great agricultural states, is bounded on the north by the Canadian provinces of Saskatchewan and Manitoba, on the east by Minnesota, on the south by South Dakota, and on the west by Montana. The Red River of the North forms most of the eastern boundary. The state is roughly rectangular in shape, and its extreme dimensions are about 210 miles from north to south and about 360 miles from east to west.

Fargo, the largest city, is 240 miles northwest of Minneapolis on I-94 and is 248 miles north of Sioux Falls on I-29. Bismarck is 191 miles west of Fargo on I-94.

THE LAY OF THE LAND

The eastern part of North Dakota lies in the Western Great Lakes Lowland region. This is an area of plains containing many glacial features such as moraines and flat plains that were formerly the beds of glacial lakes. The bed of the ancient Lake Agassiz, along the eastern border, contains the state's richest soils. Scattered marshes and small lakes are found throughout the Western Great Lakes Lowland.

To the west of this escarpment lies the Great Plains, known in North Dakota as the Missouri Plateau. Many flat-topped buttes stand as high as about 180 m (about 600 ft) above the plains, and a strip of Badlands, which are spectacular formations produced by the erosion of soft sedimentary rocks, is found in the southwest.

Western North Dakota is drained by the Missouri River and its tributaries, while the eastern plains are drained by the north-flowing Red River and its tributaries. The north central part of the state is drained by the Souris River, and the southeastern part by the James River, a tributary of the Missouri River. Numerous small natural lakes are in the glaciated part of the state; the largest of these is Devils Lake. The largest body of water in the state is Lake Sakakawea, formed behind Garrison Dam on the Missouri River.

HOW NORTH DAKOTA BECAME A STATE

The first European known to have been in the area was the French-Canadian explorer Pierre Gaultier de Varennes, sieur de La Vérendrye, who visited a Mandan village near what is now Bismarck in 1738. In the ensuing period, fur traders from his posts in Canada dealt with the Indians on the Red River of the North as far south as Grand Forks. In the 1790s the Canadian North West Company and Hudson's Bay Company built trading posts on the Red River in the northeast corner of the state.

North Dakota became a United States possession as part of the Louisiana Purchase of 1803, but its boundary with Canada was not agreed on until 1818. White settlement began in 1812, when people from the Selkirk Settlement at Winnipeg founded a colony at Pembina. A community of Indians and matis (persons of mixed Indian and white ancestry) grew up around the fur-trading posts. Matis staffed the trains of carts carrying furs and merchandise between Winnipeg and Saint Paul, Minnesota. After 1859, steamboats on the river sped such goods between the two cities.

The Dakota Territory, which included North and South Dakota, as well as Wyoming and Montana, was created in 1861. The name comes from the Indian tribal name, Dakota, meaning "allies". When warfare broke out between the Sioux and white settlers in neighboring Minnesota the following year, many Indians sought refuge in the Dakota Territory, but most were eventually confined to the area west of the Missouri River. In the 1870s railroad links with St. Paul brought settlers from the East, among them many Norwegian and German immigrants. The bonanza farm craze of 1875-90 attracted so many settlers that North Dakota was admitted to the Union as the 39th state on November 12, 1889.

NORTH DAKOTA'S CITIES

FARGO (74,111) is at the head of navigation on the Red River, opposite Moorhead, Minnesota;. Founded in 1875, it was a center for workers on the Northern Pacific Railway. The city is named for American express shipper William G. Fargo. It is the commercial, financial, cultural, and medical center of a rich grain, sugar-beet, and livestock producing area. Farm equipment, processed food, and fabricated metal are manufactured. North Dakota State University (1890) is here.

BISMARCK (49,256), was founded in 1872 as a camp to protect Northern Pacific railway workers. It was named for Prince Otto von Bismarck of Germany. The city began to grow when gold was discovered nearby, and it was named capital of the new state in 1889. Today, it is the center of a rich agricultural and mining region, and it produces farm machinery and steel.

NORTH DAKOTA'S PEOPLE

North Dakota has just 634,935 people, making it the fourth least populated state. It also has one of the lowest population densities—just 9.2 people per square mile. Whites made up 94.6% of the population and blacks 0.6%. North Dakota is one of the nation's least urbanized states; in 1990 about 53% of all state residents lived in areas defined as urban and the rest lived in rural areas. The largest cities were Fargo; Grand Forks; Bismarck, the capital; Minot; and Dickinson.

NORTH DAKOTA' ECONOMY

Since its settlement by non-Indians in the mid-19th century, North Dakota has had an economy dominated by agriculture. Wheat, barley, hay, sunflowers, and sugar beets are leading crops. Wheat is grown in all areas, but is especially important in the north. Other crops include potatoes, rye, and flax. North Dakota leads the nation in the production of durum and other spring wheat, as well as barley, flax, and sunflowers. Irrigation agriculture is increasing in importance, especially in the semiarid west along the Missouri River. Livestock accounts for about 31% of the annual farm income. Beef cattle are most important and are raised primarily in the west; hogs, dairy cattle, and sheep are raised in the southeast.

Petroleum, which makes up about two-thirds of the annual mineral value, is found primarily in the west. Lignite, the second most important mineral, is strip-mined, primarily in the west central part of the state. Natural gas is produced in the southwest and the northwest. Other mineral products include sand and gravel, clays, and lime.

The leading industries in North Dakota are the manufacture of industrial machinery, and food processing, followed by printing and publishing. About one-sixth of the manufacturing labor force is employed in producing industrial machinery, particularly farm equipment. Among the principal food products are flour, cereals, butter, cheese, and sugar (processed from sugar beets). Other important manufactures are transportation equipment and electronic items. North Dakota also has several oil refineries. Although manufacturing has grown in importance, North Dakota still has one of the smallest manufacturing outputs of any state.

Each year several million visitors produce in excess of $790 million for the North Dakota economy.

Places to Visit

Theodore Roosevelt National Park (701-623-4466)is divided in two parts by 50 miles of spectacular badlands. You can get a great view of the badlands from the Painted Canyon Overlook and Visitors Center (701-623-4466). Rugby , the geographical center of North America, is the site of the Geographical Center Historical Museum (701-776-6414). Bonanzaville U.S.A. (701-282-2822) north of Fargo is a replica of a pioneer village as well as a museum. Tourist information: North Dakota Tourism Department, 604 E. Blvd., Bismarck, ND 58505 (800-435-5663)

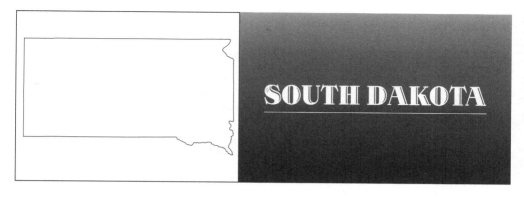

SOUTH DAKOTA

South Dakota is bounded on the north by North Dakota, on the east by Minnesota and Iowa, on the south by Nebraska, and on the west by Wyoming and Montana. The Missouri River forms part of the southeastern boundary. The state is roughly rectangular in shape, and its extreme dimensions are about 245 miles from north to south and about 379 miles from east to west.

Sioux Fall is 248 miles south of Fargo and 318 miles north of Kansas City on I-29, and 347 miles east of Rapid City on I-90.

THE LAY OF THE LAND

The eastern third of South Dakota lies within the Central Lowland region of the midwestern U.S. It is an area of rolling glacial plains, with numerous lakes and ponds. A higher zone, lying between the James and Big Sioux rivers, is covered with loess, a wind-deposited silt. The dark, fertile loess soil makes this area the state's most productive agricultural region..

Central and western South Dakota are part of the Great Plains. The portion of the Great Plains lying east of the Missouri River was smoothed by glaciers and resembles the rolling plains of the Central Lowland. In the portion not affected by glaciers, to the west, deep valleys of the tributaries of the Missouri River interrupt the plains, and flat-topped buttes may rise as much as 500 feet above them. In the extreme west are the Black Hills, granite peaks that rise almost 4,000 feet above the surrounding plains. Encircling the granite core of the Black Hills are sharp ridges formed of upturned sedimentary rock.

The northeastern part of South Dakota abounds in glacial lakes, the largest being Lake Traverse and Big Stone Lake. The state's largest lakes, however, are artificial. The most important of these are Lake Oahe, Lake Francis Case, and Lewis and Clark Lake, all of which are formed behind dams on the Missouri River.

HOW SOUTH DAKOTA BECAME A STATE

The first authenticated European exploration of the region comprising present-day South Dakota was made by Franácois (1715-94) and Louis Joseph (1717-61) de La Vérendrye, sons of the French Canadian explorer Pierre Gaultier de Verennes, sieur de La Vérendrye, in 1743. A lead marker establishing the region as a French possession was planted on the site of present-day Fort Pierre. Toward the close of the 18th century the region was visited by fur trappers and traders operating on the Missouri River. In 1803 territory now occupied by the Dakotas became part of the United States as a result of the Louisiana Purchase. The Lewis and Clark expedition passed through the region in 1804 and in 1806.

The first permanent settlement in South Dakota was established in 1817 opposite the site of modern Pierre. In 1832, Fort Pierre Chouteau was constructed by the American Fur Company of the German-American merchant John Jacob Astor, which also operated steamboats on the Missouri River. In 1855 the company sold the fort to the federal government. In 1849 the region east of the Missouri River became part of the territory of Minnesota, and in 1854 the region west of the river became part of the territory of Nebraska. In 1861 the entire region, including present-day Wyoming, Montana, and a part of eastern Idaho, was established as the territory of Dakota, with Yankton as the capital. The name comes from the Indian tribal name Dakota, meaning "allies". Settlement of the region progressed slowly until 1874, when gold was discovered in the Black Hills, in the great Sioux Indian reservation, and large numbers of whites began to flock to the region, hoping to strike it rich. The federal government halted the settlers and attempted to keep them out of the Black Hills until an agreement could be reached with the Sioux.

In 1875, after the refusal of the Indians to cede their land, the government made no further attempts to stop the gold seekers. In 1876 other gold lodes were discovered, including the famous Homestake Lode at the Homestake Mine, near Lead, in the southwestern part of the state. The Great Dakota Boom, a period of rapid settlement, occurred between 1879 and 1886. A statehood movement was started, and in 1889 the Dakotas were separated. On November 2 of that year South Dakota was admitted to the Union as the 39th state.

SOUTH DAKOTA'S CITIES

SIOUX FALLS (100,814) is located at the site of falls on the Big Sioux River (hence its name); incorporated 1883. It is a commercial, manufacturing, and data-processing center situated in a farm region producing corn and soybeans.. Sioux Falls was founded in 1856, but was abandoned during the Sioux Indian uprising of 1862. It was resettled after the establishment of Fort Dakota here in 1865.

RAPID CITY (54,523) is located on Rapid Creek at the base of the Black Hills. It was founded in 1876 when gold was discovered in the area and grew as a mining center. Today it is the commercial center for a ranching, grain-growing, and lumbering region. It is also the gateway to Mount Rushmore National park.

PIERRE (12,909) is the second smallest capital city in the U.S. It was founded on the Mississippi as Fort Pierre, built opposite an Indian fort. Fur trading was a major source of income before the railroad arrived in 1880. It became the capital in 1889.

SOUTH DAKOTA'S PEOPLE

South Dakota has 715,392 inhabitants, the sixth least of any state. Like its northern neighbor, its population density is low—9.37 people per square miles. Whites made up 91.6% of the population and blacks 0.5%. In 1990, South Dakota had some 50,500 American Indians; the Sioux formed the largest Indian group in the state. About 50% of all people in South Dakota lived in areas defined as rural, and the rest lived in urban areas. The major area of population concentration was in the east.

SOUTH DAKOTA'S ECONOMY

Since the area's early settlement in the mid-19th century, South Dakota's economy has been based on cultivating the fertile soils in the east and ranching on the abundant grazing lands of the west.About 59% of the state's yearly farm income is derived from sales of livestock and livestock products. Beef cattle, hogs, milk, and sheep and lambs account for almost all the total. Cattle and sheep are raised primarily in the west. The main crops are corn, wheat, and hay (alfalfa). South Dakota leads the nation in the production of oats and rye. Most of the crops are grown in the eastern half of the state.

Although forests cover only about 3% of South Dakota's land area, forestry is a significant industry in the state. One of the nation's largest gold mines, the Homestake, is located in Lead; sand and gravel deposits are worked in all parts of the state. . Of several uranium deposits in the west, the most important is at Edgemont.

The leading industry is the manufacture of food products. Other major manufactures include industrial machinery, lumber and wood products, electronic goods, metal items, precision instruments, and transportation equipment. Printing and publishing are also important.

Each year several million visitors spend at least $380 million in South Dakota.

Places to Visit

Nature, with a little assist from man, provided many of South Dakota's attractions. In the 1.3 million acre Black Hills National Forest is famous Mount Rushmore (605-574-2523), on which are carved the portraits of Presidents Washington, Jefferson, Lincoln, and Teddy Roosevelt. Jewel Cave National Park (605-673-2288) contains the world's fourth largest cave, and Wind Cave National Park (605-745-4600) has the world's sixth longest cavern. The rust, gold, and pink rock of Badlands National Park (605-433-5361) make it seem like the surface of another planet. The infamous gold-mining town of Deadwood (800-999-1876) has been revitalized through the legalization of casino gambling. South Dakota Department of Tourism, Capital Lake Plaza, Pierre, SD 57501 (800-843-1930)

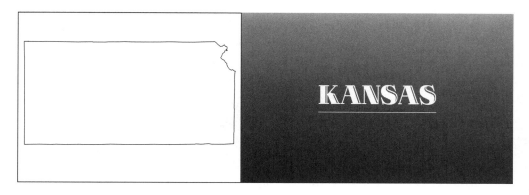

KANSAS

Kansas, is bordered on the north by Nebraska, on the east by Missouri, on the south by Oklahoma, and on the west by Colorado. The Missouri River forms the northeastern boundary. The state is situated at the geographical center of the 48 contiguous states. The state is approximately rectangular in shape, and its extreme dimensions are about 205 miles from north to south and 410 miles from east to west.

Wichita is 190 miles southwest of Kansas City and 157 miles north of Oklahoma City on I-35. Topeka is 61 miles west of Kansas City on I-70.

THE LAY OF THE LAND

The terrain of Kansas is flat to gently rolling. Elevations increase with general uniformity from east to west. The highest and flattest region, the Great Plains, occupies the western half of the state. Topographic relief here is provided by stream valleys. Some of the underlying rocks in the region are soluble, and they have collapsed or subsided as they weathered near the surface, forming depressions that dot the surface. Soils of this region have a higher sand content than those in other parts of the state. The Osage Plains in the east is a gently rolling region with some low hills. The Flint Hills in the western part of this region are composed of an erosion-resistant flinty limestone. The Dissected Till Plains form the northeastern part of the state. Glaciers deposited a mixture of silt and stones (drift), which was eventually eroded (dissected) by streams that formed on the surface. Rich prairie soils are found in most of the eastern part of the state.

The state's only natural lakes are small, intermittently dry ponds in the surface depressions of the Great Plains. The major bodies of water are artificial. These include the following: Waconda (Glen Elder), Tuttle Creek, Cheney, Milford, and Kanopolis lakes.

HOW KANSAS BECAME A STATE

Beginning about 10,000 years ago, five different prehistoric cultures were established in the area of present-day Kansas. They were the predecessors of historic Plains tribes: the Wichita, Pawnee, Kansa, Osage, and Plains Apache. These natives were hunters who also farmed and were joined on the Plains by the nomadic Cheyenne, Arapaho, Comanche, and Kiowa about 1800. Despite some temporary alliances, inter-tribal warfare continued into the 19th century.

The Spanish, under Francisco Coronado, came in 1541 seeking Quivira, a fabled land of gold. Other Spaniards came occasionally until 1601, and the French traded for furs from 1682 to 1739. The first American explorers to enter the area, in 1804, were Meriwether Lewis and William Clark. The Santa Fe Trail was opened in 1821 and remained a trade artery for 50 years, and the Oregon-California road, in northeastern Kansas, was a major route of westward migrants. After 1830 thousands of eastern Indians, including the Shawnee and Potawatomi, were removed to Kansas. Traders, missionaries, and troops at Forts Leavenworth, Scott, and Riley made up the white population of Kansas before its opening as a territory.

Kansas comes from an Indian word meaning "people of the south wind". Kansas Territory was opened for settlement on May 30, 1854, with its western border at the Rocky Mountains. Disputes arose at once over whether Kansas would enter the Union as a free or slave state, and they continued until 1858. Because of some violent incidents between pro slavery and anti-slavery settlers, the territory became known as "Bleeding Kansas"; the name was also applied to the conflict itself. For several years the territory had two governments; one extralegally, and fraudulent elections were common. Finally, in 1859, a constitution acceptable to the U.S. Congress was written, and Kansas entered the Union as a free state in 1861.

KANSAS'S CITIES

WICHITA (304,011) is at the junction of the Arkansas and Little Arkansas rivers. It was founded about 1867 as a trading post and is named for the Wichita Indians. The Chisholm Trail, a major cattle-driving route from Texas, passed through Wichita, and after the arrival of the railroad in 1872 the city boomed as a cattle-shipping point. Petroleum was discovered in the region in 1915, and aircraft manufacturing began in 1920. Today, as the largest city in the state, it is a major transportation center.

TOPEKA (119,887) was founded in 1854, and was a stop on the Oregon and Santa Fe trails. It became a division point on the Atchison, Topeka, and Santa Fe Railroad , and was named the capital of the state in 1861. Topeka is the center of a major wheat and corn growing region, and produces processed foods, printed materials, and rubber and plastic products.

KANSAS'S PEOPLE

Kansas has 2,477,574 inhabitants, an increase of 4.8% over 1980. The average population density in 1990 was 30 people per square mile. The eastern part of the state, however, had a considerably higher population density than the west, where, in most areas, there were fewer than 10 per square mile. Whites made up 90.1% of the population and blacks 5.8% . In 1990 about 69% of all Kansas residents lived in areas defined as urban, and the rest lived in rural areas.

KANSAS'S ECONOMY

Kansas is a leading wheat-producing state in the U.S. and is a major producer of corn, alfalfa, and sorghum grains. Winter wheat, which is grown in virtually every county in Kansas, is grazed by cattle during the fall and spring and allowed to grow and ripen during the summer months. Other crops include soybeans, rye, oats, barley, and sunflowers. About 5.7 million cattle graze on Kansas farms, and the prairies of the Flint Hills in eastern Kansas support large-scale ranching. Kansas is also a leading hog-raising state; corn and sorghum crops furnish feed. Dairy and poultry farming are also important.

Petroleum is found in most areas, but especially in the central part of the state. Most natural gas is produced in the southwest. Helium is extracted from the natural gas obtained in several locations. Other important minerals are salt recovered from brine wells and rock-salt mines, coal, and clay.

The state's leading manufacture is transportation equipment, primarily aircraft. The next most important manufactures are industrial machinery and processed foods. Kansas is a major flour-milling state, and meat packing has grown in importance since the 1960s. Other manufacturing includes printed materials, chemicals, rubber and plastic products, fabricated metals, and electronic equipment. Oil refining is also a significant industry.

Each year visitors to Kansas produce more than $2 billion for the Kansas economy.

Places to Visit

You may be surprised to learn that the Kansas Cosmosphere and Space Center (800) 397-0330) in Hutchison has the largest collection of space memorabilia outside of the Smithsonian. Nearby, in western Kansas, are several sites rich in frontier history, such as the Sante Fe Trail Center (316-285-2054) and Dodge City's Boot Hill Museum (316)227-8188). Abilene, hometown of Dwight Eisenhower, is home of the Eisenhower Center (913-263-4751), which includes a museum, library, and the former president's boyhood home. Tourist information: Kansas Travel & Tourism Division, 700 S.W. Harrison St., Topeka, KS 66603 (800-252-6727/913-296-2009)

MISSOURI

Anybody who's read Mark Twain knows where Missouri is or at least knows that the state touches the Mississippi River that carried Huck Finn and Jim southward on their great adventure. Missouri is bordered on the north by Iowa; on the east by Illinois, Kentucky, and Tennessee; on the south by Arkansas; and on the west by Oklahoma, Kansas, and Nebraska. The Mississippi River forms most of the eastern boundary, and the Missouri River forms the northwestern boundary. The state is roughly trapizoidal in shape, and its extreme dimensions are about 285 miles from north to south and about 305 miles from east to west.

Kansas City, the largest city, is 605 east of Denver and 256 miles west of St. Louis on I-70, and is 194 miles south of Des Moines on I-35. St. Louis, the state's other major metropolitan area, is 299 miles southwest of Chicago and 311 miles north of Memphis on I-55, and is 258 miles west of Louisville on I-70.

THE LAY OF THE LAND

The land surface of Missouri is more diverse than that of any other midwestern state. The largest physical region is the Ozark Plateau (sometimes called the Ozark Mountains, or Ozarks), which occupies most of the southern part of the state. The limestone bedrock of the Ozark Plateau dissolves easily, and many extensive caverns have been formed. From these underground caverns flow thousands of springs. Missouri ranks second to Idaho in the U.S. in its number of large springs. The soils of the Ozark Plateau are generally thin and stony. Along the western border of the state is the Osage Plains, a gently sloping region.

North of the Missouri River is the Dissected Till Plains region. The glaciers of the Ice Age deposited a deep layer of rich drift (silt and sand) hundreds of thousands of years ago. The surface was gradually eroded by streams to form a hilly terrain. The extreme southeastern portion of the state is a part of the Mississippi Alluvial Plain; it consists of low, flat, poorly drained land.

No large natural lakes are found in Missouri. The Lake of the Ozarks, impounded on the Osage River, is the largest of several reservoirs.

HOW MISSOURI BECAME A STATE

The founding of Sainte Genevieve in 1735 by the French marked the first European settlement in the Missouri region, which was then a part of the French territory of Louisiana. The second European settlement in the state was St. Louis, established as a trading post in 1764, a year after the cession of Louisiana to Spain by the French. The colonization of the region was greatly accelerated by the Ordinance of 1787, which excluded slavery from the Northwest Territory of the United States. The Spanish also encouraged immigration by offering liberal bounties to settlers. Louisiana was returned to France in 1800 and sold to the United States three years later. In 1812 the region became the territory of Missouri.

After 1815, immigration increased rapidly. In 1816 the first steamboat reached St. Louis. In 1818 the territorial legislature applied to the U.S. Congress for permission to prepare a state constitution. Missouri was admitted to the Union as a slave state on August 10, 1821, under the terms of the Missouri Compromise. This state's moniker comes from the Missouri Indians, whose name means "place of the large canoes" or "big and muddy".

MISSOURI'S CITIES

SAINT LOUIS (396,685) is in eastern Missouri on the Mississippi River, just south of its confluence with the Missouri River. The city is a leading railroad center and is one of the nation's busiest inland ports, with ship connections to the Upper Mississippi, to Chicago and the Great Lakes, to the Ohio River system, and to the Gulf of Mexico. 1764, Pierre Laclède Liguest (1724-78), a French merchant, selected the site of St. Louis for a trading post. Construction of a village, named for Louis IX of France, began the following year. From its founding St. Louis was both a market and an outfitting point for fur traders and explorers of the American West. It was transferred to the Spanish (1770), returned to France during the reign of Napoleon the First and, following the Louisiana Purchase of 1803, became part of the United States. St. Louis continued to grow quickly after the war, and by 1900 it was a major manufacturing center. In 1904 a world's fair and the Olympic Games were held in the city.

KANSAS CITY (435,146) is located at the junction of the Kansas and Missouri Rivers in the western part of state, adjacent to Kansas City, Kansas. The city was founded by French fur trapper Francois Chouteau in 1821, and was later known as Westport Landing. It became the City of Kansas in 1850 (and Kansas City in 1889) . Kansas City was famous as a gateway to the Santa Fe and Oregon Trails, and growth increased when the railroad to St. Louis was completed in 1865. Today, it is one of America's leading transportation centers, with many grain elevators and warehouses.

JEFFERSON CITY (35,481) the captial, was visited by Lewis and Clark in 1804, but not settled until seventeen years later. The city was laid out by Daniel Morgan Boone, son of the famous frontiersman. The state government is the focus of the economy.

MISSOURI' S PEOPLE

Missouri has 5,233,899 inhabitants. The average population density is 28 persons 73 people per square mile. The population density is lightest in the rugged central portion of the Ozark Plateau and in the rolling farmlands of the north. Whites make up 87.7% of the population and blacks 10.7%; about 69% of Missouri's residents lived in areas defined as urban, and the rest live in rural areas.

MISSOURI'S ECONOMY

Farming accounts for about 2% of the annual gross state product in Missouri. The state has some 107,000 farms (only Texas has more). Sales of livestock and livestock products make up about 52% of Missouri's yearly farm income. Missouri is a leading state in the raising of dairy cows and beef cattle. Dairy farming is most important in the southwest. Sheep production is concentrated in the northeast. Soybeans are the leading crop and are grown in the fertile northern part of the state and in the southeastern lowlands. Corn and hay, the next most important crops, are grown throughout the state.

The most important mineral is lead, of which Missouri is the principal national supplier. Other major mineral products are cement, stone (including limestone, marble, and granite), zinc, coal, sand and gravel, and barite. Lead, zinc, and barite are all mined in the vicinity of the St. Francois Mountains. Bituminous coal deposits, largely exploited by strip-mining, are widespread; the largest deposits are worked in the west central and north central parts of the state. Smaller amounts of copper, silver, and petroleum are also produced.

The state's leading manufactures make transport equipment, fabricated metals, printed materials and processed foods. Missouri is a major producer of automobiles and aerospace equipment, with production concentrated in the St. Louis area. Important food products include dairy items, flour, and beer. Among the chemical products manufactured in Missouri are fertilizers, insecticides, and pharmaceuticals.

Each year more than 40 million visitors produce over $6 billion for the Missouri economy.

Places to Visit

The building of the soaring Gateway Arch symbolized the revitalization of St. Louis (800-888-3861), a city that has played a key role in U.S. history since it's founding on the Mississippi River in 1764. St. Louis has shorefront riverboats, museums, and one of America's best zoos. The tiny Ozark Mountain town of Branson (417)334-4136) has become a country music mecca with over 61,000 theater seats and numerous other attractions. Lake of the Ozarks (800-325-0213), the state's largest lake, has 1,300 miles of shoreline. Kansas City (800-767-7700) has more fountains than Rome, Italy and more miles of boulevards than Paris, France. Tourist information: Missouri Division of Tourism,Truman State Office Building, Box 1055, Jefferson City, MO 65102 (800-877-1234/314-751-4133)

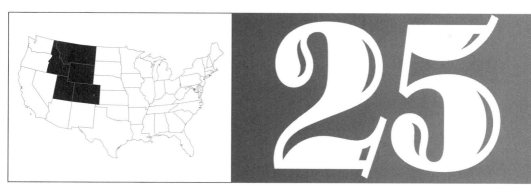

THE MOUNTAIN STATES

Comparing The Mountain States

STATE National rankings in parenthesis	POPULATION (1993 U.S. Bureau of the Census)	SQUARE MILES (U.S. Geological Survey)	PER CAPITA INCOME (U.S. Bureau of Labor Statistics)
Colorado	3,565,959 (26)	104,100 (8)	$14,821 (17)
Utah	1,829,582 (34)	84,904 (13)	$11,029 (46)
Idaho	1,099,096 (42)	83,574 (14)	$11,457 (41)
Montana	839,422 (44)	147,046 (4)	$11,213 (43)
Wyoming	470,242 (50)	97,818 (9)	$12,311 (36)

Towering peaks soaring upward from the plains is an unforgettable image common to the entire region we call the Mountain States. The distinguishing characteristic of these state is their rugged and often spectacularly beautiful terrain. In the Mountain States are such famous sites as Yellowstone National Park, Glacier National Park, Hells Canyon, Pike's Peak, and the Great Salt Lake. Much of the land in these states is inaccessible, which limits the population but preserves much of their natural beauty.

A second characteristic of these states is their mineral wealth. Almost all experienced gold rushes in the late 1880s that brought waves of settlers to the most sparsely populated of American regions. Over 200 minerals are found in the region, including coal, petroleum, natural gas, gold, and silver, and mining is still important to every state's economy.

The plains and plateaus of the Mountain States tend to be dry, and farming is difficult without irrigation. But grasses grow freely, so cattle and sheep ranching have been major sources of income for the last century and a half. The lone cowboy sitting on his horse watching cattle graze in the valley below is an image that would fit every state.

Despite that image, however, the culture of these states is anything but male dominated. The Mountain States were among the first to give women the vote, the first to elect a woman governor, and the first to send a woman to Congress. Fair-minded, hard-working, and independent are the traits of the people who survive in these rugged lands.

EARLY EXPLORATION OF THE MOUNTAIN STATES

Although the Spanish claimed most of the land in this area, they never settled it. The most famous expedition into these states was that led by Lewis and Clark, who were ordered by President Jefferson in 1804 to explore the lands acquired by the U.S. in the Louisiana Purchase. However, by far the most important explorations of this vast territory were made by trappers and traders who survived the often harsh climate and often hostile Indians to establish trading posts and forge trails into the wilderness that would later be pathways for prospectors and settlers moving westward from the Mississippi River.

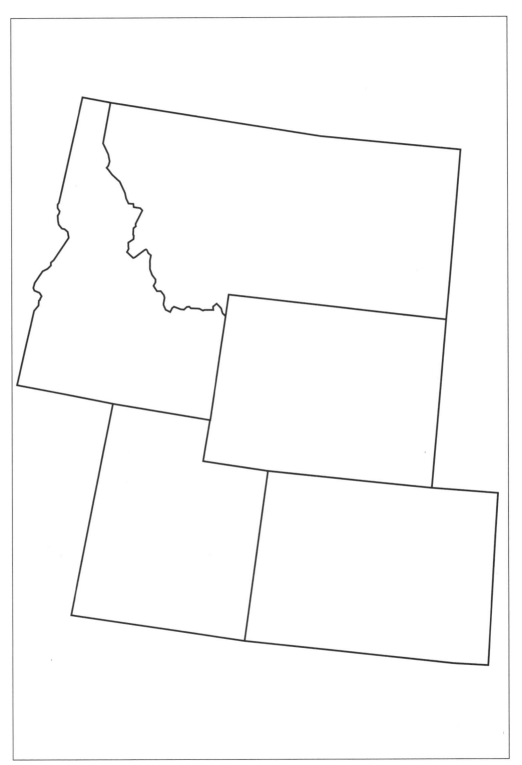

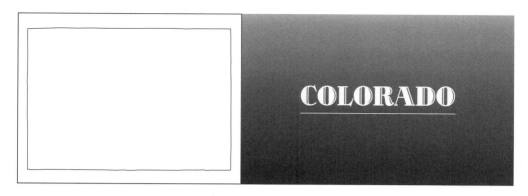

COLORADO

Anyone who has seen the Rocky Mountains rise dramatically out of Colorado's Great Plains can understand the nearly mystical appeal the state has had to so many Americans fleeing more polluted and congested areas. Colorado is bordered on the north by Wyoming and Nebraska, on the east by Nebraska and Kansas, on the south by Oklahoma and New Mexico, and on the west by Utah. Colorado is nearly a perfect rectangle, stretching 385 miles east-west and 275 miles north-south. Its southwestern point is the Four Corners, where you can lie down and touch the states of Arizona, New Mexico, Utah, and Colorado.

Denver, the capital and largest city, is 605 miles west of Kansas City on I-70, and 439 miles north of Albuquerque and 100 miles south of Cheyenne on I-25. Colorado Springs is 67 miles south of Denver on I-25, and Grand Junction is 246 miles west of Denver on I-70.

THE LAY OF THE LAND

Colorado has the highest mean elevation of any state. The eastern two-fifths are part of the Great Plains known as the High Plains, because their elevations are as much as a mile above sea level. This land is generally flat and rises gently from the Kansas-Nebraska border to the foothills of the Rockies. Much of the area is semi-arid, but fertile, and irrigation has turned it into productive farmland.

The Southern Rocky Mountains begin at the Wyoming Basin in the north and cut a huge swath through the center of the state before entering New Mexico. This majestic, rugged area contains more than 50 "fourteeners," peaks over 14,000 feet, with Mount Elbert (14,433 ft.) being the highest. Through the Rockies runs the Continental Divide, the separation between waters that flow eastward and those that flow westward.

On the western side of the Rockies is the Colorado Plateau. This upland area has been cut by wind and water erosion into numerous canyons and mesas. Mesa Verde, in the southwestern corner, was the home of an important pre-Columbian civilization.

Colorado has many natural lakes and more than 1,900 artificial lakes used to store water. The Rocky Mountains are the source of the Arkansas, South Platte, Rio Grande, and Colorado Rivers.

HOW COLORADO BECAME A STATE

Spanish explorers from New Mexico ventured into the state as early as 1694, but neither the Spanish nor the Mexicans ever established any settlements. The U.S. acquired eastern Colorado in 1803 as part of the Louisiana Purchase. In 1806, Zebulon Pike led an expedition of 22 men into the state to seek the source of the Arkansas River, and he saw many natural wonders including the mountain that bears his name, Pike's Peak. Many fur trappers moved into the mountains, and in 1833, William Bent built Fort Bent, the state's first permanent settlement.

In 1848, Mexico ceded western Colorado to the United States. The state's area was divided up between the territories of Utah, New Mexico, Kansas and Nebraska in 1858, when gold was discovered in Cherry Creek. The tens of thousands of fortune seekers who headed west under the slogan "Pike's Peak or Bust" led to the settlement of numerous mining camps and commercial centers. Some of the new settlers sought a separate identity for the region. In 1859, they voted to create the "Territory of Jefferson" and adopted a constitution. Congress declared that the formation of the Territory of Jefferson was illegal, but it did create the Territory of Colorado in 1861 (the name comes from the Spanish for "color red," which was first applied to the Colorado River).

When the gold rush petered out, many settlers returned to their homes. But the arrival of the railroad at Denver in

1870 brought farmers and ranchers, and the discovery of silver in the San Juan Mountains stimulated a second rush. Colorado was finally admitted to the Union as the 38th state on August 1, 1886. Because that year marked the Centennial of the U.S., Colorado is called the Centennial State.

COLORADO'S CITIES

DENVER (467,610), the largest city and capital, is called the Mile High City because it sits at the base of the Rocky Mountains a mile above sea level. Two communities, Aurora and Saint Charles, were founded in the 1850s to supply prospectors heading to the gold fields on Cherry Creek. The name of Saint Charles was changed to Denver City in honor of Governor. James Denver. In 1860, the two communities were combined as Denver, and the city became the capital of the Territory of Colorado in 1867. Denver is the financial and commercial center for a manufacturing, agricultural and tourist area. The city has the nation's largest sheep market, a large cattle market, and it is the home of a branch of the U.S. Mint.

COLORADO SPRINGS (214,821) was founded as a resort at the foot of Pike's Peak in 1871 by Gen. William Palmer, builder of the Denver and Rio Grande Railroad. The city grew rapidly after gold was discovered at Cripple Creek in 1882. Today, Colorado Springs has significant defense and high-technology manufacturing. It's the home of the U.S. Air Force Academy and nearby is the Army's Fort Carson.

GRAND JUNCTION (29,034) was founded in 1871 at the junction of the Gunnison and Colorado Rivers. As the largest Colorado city west of the Rockies, it is the commercial center for a large farming and dairy region. Uranium, vanadium, and oil share are also mined in the area.

COLORADO'S PEOPLE

The vast majority of Colorado's 3,565,959 people live in the eastern portion of the state. The state has experienced significant, steady growth over the last three decades, although Denver's population has declined. Almost 13% of Colorado's residents are Hispanic, the fifth highest percentage of any state.

COLORADO'S ECONOMY

Mining built the state of Colorado, and is still a significant industry. Petroleum, coal, and natural gas are the most lucrative products, and silver, gold, vanadium, lead, copper, zinc, moldybdenum, and uranium are also important. The amount of oil trapped in shale in Colorado is estimated to be 5 times the entire known world reserves, but to date there is no profitable way to extract most of this petroleum.

Colorado's two most important agricultural products are beef cattle and winter wheat. Other sources of revenue for the state's farms are sheep, corn, hay, milk, potatoes, and vegetables.

Colorado is also home to a diverse light manufacturing economy. The leading products are scientific instruments, high-tech equipment, electronic equipment, food, printed items, aerospace equipment, and fabricated metals.

Fourteen million tourists come to Colorado annually, many to the famous Rocky Mountain ski resorts such as Aspen and Vail. These tourists spend almost $6 billion.

Places to Visit

Colorado is dominated by the Rocky Mountains, which include such famous ski resorts as Aspen (303-925-5656), Vail (303)476-1000), and Steamboat Springs (800-922-2722). Colorado Springs (800-368-4748) is nestled at the foot of the Rocky Mountains, and is home to the Air Force Academy and the U.S. Olympic Center. Nearby are Pikes Peak (719-636-1602) and Royal Gorge (719-275-7507), site of the world's highest suspension bridge. Mesa Verde National Park (303-529-4465) features cliff-dwellings of the ancient Anasazi Indians. Tourist information: Colorado Tourism Board, 1625 Broadway, Denver, CO 80202 (800-433-2656/303-592-5510)

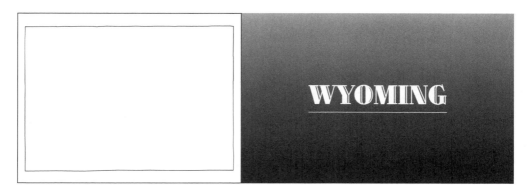

WYOMING

Wyoming is the least-populated of the 50 U.S. states, but such spectacular natural attractions such as Yellowstone Park make it one of the most visited. Wyoming is bordered on the north by Montana, on the east by South Dakota and Nebraska, on the south by Colorado and Utah, and on the west by Utah, Idaho and Montana. Like Colorado, Wyoming is nearly a perfect rectangle, stretching 365 miles east-west and 276 miles north-south.

Casper is 281 miles northwest of Denver and 281 miles southeast of Billings, Montana on I-25. Cheyenne, the capital, is 180 miles south of Casper on I-25

THE LAY OF THE LAND

The Rocky Mountains fan out south and southeast from the northwest corner of the state. Rugged ranges are punctuated with arid basins that include rock towers like the spectacular Devils Tower, the first national monument. The Wind River Range includes Gannett Peak (13,804 ft.), Wyoming's highest point. The Continental Divide crosses the state from the northwest corner the south-central border. Yellowstone National Park, America's first national park, is located primarily in the extreme northwest corner of Wyoming.

The eastern quarter of Wyoming is part of the Great Plains, which stretch southward all the way to Texas. This largely treeless land rises from the east to the foothills of the Rocky Mountains, and its high plateaus are home to large cattle ranches.

Wyoming has many natural lakes, the largest of which are Yellowstone, Jackson, Fremont, and Shoshone. Its rivers have many waterfalls, including the Lower Falls of the Yellowstone River, which are twice the height of Niagara Falls.

HOW WYOMING BECAME A STATE

The first Europeans to cross the area may have been French brothers Francois and Louis Joseph de la Vérendrye in 1542. John Colter, who had traveled with Lewis and Clark, wintered in the state in 1807-1808 and was the first to see the hot springs and geysers of Yellowstone National Park. Wilson Price led a group of fur traders into the state in 1811, and in 1824 a group of trappers from the Rocky Mountain Fur Company may have been the first to cross the South Pass of the Rocky Mountains, blazing a route that was soon known as the Oregon Trail.

In 1834, Fort Laramie was established as a trading post near the South Pass, and James Bridger established Fort Bridger in 1843. In a decade, more than 150,000 people passed through Wyoming, including gold prospectors headed to California and Mormons led by Brigham Young headed for Utah. In 1845, the Republic of Texas gave up claims to part of Wyoming when it joined the Union. In 1846, the British gave up a claim to part of northwestern Wyoming, and Mexico ceded southwest Wyoming in 1848. The Union Pacific Railroad reached the area in 1867, bringing many settlers. Wyoming had been part of the Dakota Territory, but its growth led Congress to make it a territory of its own in 1868 (the name comes from the Wyoming Valley in Pennsylvania, which in turn came from an Indian word meaning "large prairie place.") In 1870, a scientific expedition catalogued the wonders of Yellowstone, and on March 1, 1872, it became the very first national park. On July 10, 1890, Wyoming became the 44th state.

WYOMING'S CITIES

CHEYENNE (50,008), the largest city and capital, was founded in 1867 at a division point on the Union Pacific

Railroad. It was named for the Cheyenne Indians. In 1869, it became the capital of the new Wyoming Territory, and in the 1870s its growth was fueled by the discovery of gold in the Black Hills. Cheyenne was also the distribution point for hundreds of thousands of cattle brought northward from Texas to be raised by Wyoming ranchers. Cheyenne's annual Frontier Days, a festival and rodeo, have been held continuously since 1897.

CASPER (46,742) was the built on the site where the Oregon Trial crossed the North Platte River. A Mormon ferry began operations in 1847, a bridge was built in 1859, and Fort Casper was erected in 1861. The Fort and city were named for Caspar Collins, who had been killed by the Indians(the change in spelling from Caspar to Casper was a clerk's error). Today, Casper has a significant petroleum producing and refining industry and is in the center of a large ranching area.

LARAMIE (26,687), located in the southeastern corner of the state, was the site of a Fort built in 1834 and named for a French fur trapper, Jacque La Ramie. A town was built when the railroad arrived in 1868. Laramie produces cement and wood products, and it is a commercial center for a ranching and timber-producing region.

WYOMING'S PEOPLE

Wyoming's 470,242 people are primarily the descendants of the original settlers of an area that symbolizes the ruggedness of the Old West. However, the cowboy image contrasts with the state's long history of giving equal rights to women. Women were granted the right to vote when a Territorial government was formed in 1869, and in 1925, Nellie Tayloe Ross became the first woman governor of any U.S. state.

WYOMING'S ECONOMY

Ranching and mining have been the backbone of the state's economy since its earliest days. Today, livestock, especially cattle and sheep, provide over two-thirds of Wyoming's agricultural income. The most important crops are hay, sugar beets, barley, and wheat.

Wyoming has the largest U.S. coal reserves and significant amounts of petroleum and natural gas. Clays and sodium carbonate are also major sources of revenue.

Manufacturing plays a very small role in Wyoming's economy, employing less than 10,000 people. Petroleum and oil products, chemicals, and wood products are its major manufactured goods.

Yellowstone and Grand Teton National Parks attracted huge numbers of summer visitors who spend almost $2 billion in Wyoming.

Places to Visit

Yellowstone National Park (307-344-7381) has been one of America's treasures since 1872. Just south is the also scenic Grand Teton National Park (307-543-2851). Wyoming has several ski areas, including Jackson Hole (307-733-2292). Cheyenne is the home of the annual Frontier Days and the Frontier Days Old West Museum (307-778-7290). tourist information: Wyoming Division of Tourism, I-25 at College Dr., Cheyenne, WY 82002 (307-777-7777)

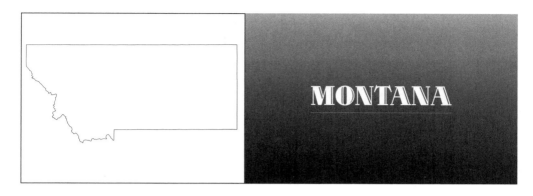

Montana's sweeping plains and towering mountains were portrayed in loving detail in A.B. Guthrie's novel, "The Big Sky," which provided the nation's fourth largest state with it's nickname, Big Sky Country. Montana is bordered on the north by Canada, on the east by North Dakota and South Dakota, on the south by Wyoming, and on the south and west by Idaho. The state, which is shaped like a rectangle with the southwest corner cut off, stretches 535 miles east-west and 275 miles north-south.

Billings, the largest city, is 539 miles east of Spokane and 420 miles west of Bismarck, North Dakota in I-94 and 456 miles north of Cheyenne, Wyoming on I-25. Great Falls is 561 miles north of Salt Lake City on I-15.

THE LAY OF THE LAND

The eastern two-thirds of Montana are the northernmost part of the Great Plains. This high, gently rolling plateau is largely covered with grasslands ideal for grazing. The landscape includes isolated mountains, and in the south is an area of badlands, land dotted with oddly-shaped rock masses and barren hills.

The western third of the state contains the northernmost of the American Rocky Mountains. In the far north, the Bitterroot Mountains contain some of the most remote and inaccessible regions of our entire country, and in the northwestern portion of the state is Glacier National Park, which has more than 50 glaciers and more than 200 lakes. The Continental Divide cuts through the northwestern part of the state and makes up part of the southwestern border.

Glaciers carved countless lakes out of the Montana landscape. Flathead Lake in the Rockies is the largest natural lake west of the Mississippi. Numerous fast-flowing rivers originate in the Rockies.

HOW MONTANA BECAME A STATE

French trappers and traders worked the area in the eighteenth century, and the U.S. acquired eastern Montana in the Louisiana Purchase of 1803. In 1805-1806, Lewis and Clark explored the area on their way to the Pacific and back. Trappers followed: in 1807, Manual Lisa established a trading post on the Bitterroot River, in 1808 a post was established by the British on the Kootenai River ,and in 1809, David Thompson of the North West Fur Company established Salish House near Thompson Falls. In 1828, the American Fur Company built Fort Union on the Missouri River, and soon paddle steamers were heading upriver with settlers and traders. Missionaries also came to the state, establishing St. Mary's Mission in the Bitterroot Valley in 1842 and St. Ignatious on Flathead Lake in 1854.

Gold was discovered at Helena, Bannack, and Virginia City beginning in 1862, and with a flood of prospectors came lawlessness. Congress joined Montana with Idaho in creating the Idaho Territory in 1863, then made the state a separate territory in 1864. The name Montana is Spanish for "mountainous." In the 1860s and early 1870s, Indian tribes angered by intrusions into their prime hunting grounds waged war against the settlers. General George Custer was hunting the Sioux when he and his force of 200 men were ambushed and massacred on June 25, 1876 at Little Big Horn. However, the Indians were defeated by the arrival of the railroad in 1883. Montana applied for statehood and was admitted to the Union on November 8, 1889 as the 41st state.

MONTANA'S CITIES

BILLINGS (81,151) was built on the Yellowstone River in 1882 as a home for employees working on the

Northern Pacific Railroad. it was named for Frederick Billings, a president of the Northern Pacific. It was an early trading and shipping center. Today, it is the commercial center for a cattle, wheat, sugar beets, and petroleum producing region.

GREAT FALLS (55,097) was built at the junction of the Missouri and Sun Rivers near large falls on the Missouri River. The city was settled in 1882 and began to grow when the railroad arrived in 1887. In 1891, the first hydroelectric dam was built to harness the energy of the falls, and the availability of power led to the development of copper smelting and refining. Today, the city produces flour and refined petroleum.

HELENA (24,561) was founded in the colorfully named Prickly Pear Valley when gold was discovered in the equally colorfully named Last Chance Gulch. The city was named after Helena, Minnesota. It became the capital of the Montana Territory in 1878, and the arrival of the railroad in 1887 helped it survive after the gold rush was over.

MONTANA'S PEOPLE

Montana's 839,422 people includes 47,527 Native Americans, 6% of the population. Only .3% of the state's residents are black, the lowest percentage of any state. Montana, like its southern neighbor, Wyoming, has a history of equality for women. In 1917, Montana elected Jeannette Rankin as the first woman U.S. Representative.

MONTANA'S ECONOMY

Montana has been and still is cattle country. Hogs, sheep, lambs, and dairy products also provide significant agricultural income. The major crops are wheat, hay, barley, and sugar beets.

Montana's gold and silver discoveries gave it the nickname "The Treasure State." In the 1890s, copper was discovered in Butte, which was called the "richest hill on earth," and the copper barons ran the entire state. Today, coal, petroleum, and natural gas are the main sources of mineral revenue, and lead, zinc, manganese and precious stones are also mined.

For a long time, Montana actively discouraged manufacturing, but today has a small industrial base consisting of lumber and wood products, petroleum and coal products, processed foods, and chemicals.

Yellowstone and Glacier National Parks are major tourist attractions, and both professional and amateur fossil hunters flock to a state that is the richest source of dinosaur skeletons. Tourism brings Montana annual revenues of over $1 billion.

Places to Visit

Among Montana's vast open spaces is the breathtaking Glacier National Park (406-888-5441), which has glaciers, peaks, waterfalls, and lots of wildlife. Flathead Lake is the largest freshwater lake west of the Mississippi. Little Bighorn Battlefield National Monument (406-638-2621) preserves the site of Custer's last stand. In Great Falls, the C.M. Russell Museum (406-727-8787) has a collection of works by the cowboy artist. tourist information: Travel Montana, 1424 9th Ave., Helena, MT 59620 (800-541-1447/406-444-2654)

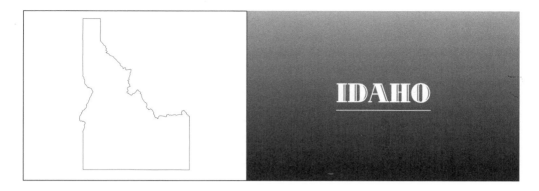

Idaho is a state with an unusual L-shape that resulted when it was put together with pieces left over from the creation of other states. It is bordered on the north by Canada, on the east by Montana and Wyoming, on the south by Utah, and on the west by Oregon and Washington. The northern panhandle is only 45 miles wide at the Canadian border but the state stretches 300 miles east-west in the south. Idaho spans 500 miles north-to-south. The Snake River forms part of the Washington and Oregon borders.

Boise, the largest city and capital, is 349 northwest of Salt Lake City and 494 miles southwest of Seattle on I-84. Pocatello is 237 miles east of Boise on I-84.

THE LAY OF THE LAND

The northern two-thirds of Idaho is covered by more than 20 ranges of the Rocky Mountains. This rugged, heavily forested land is one of the most inaccessible areas in the United States. The Seven Devils Range in western Idaho make up one wall of the famous Hells Canyon on the Snake River, America's deepest gorge.

Most of the southern third of the state is part of the Columbia Plateau, which is a broad plain built by lava flows. The soil is fertile in river valleys and where irrigation is possible. In the southeast corner is a section of the Great Basin, which consists of long, parallel ranges separated by narrow valleys.

The Snake River is Idaho's major river system. The Middle Forks of the Salmon and Clearwater Rivers were named as two of the eight wild and scenic rivers in the system created in the Wild and Scene Rivers Act of 1968. The state has numerous lakes, the largest of which are Lake Pend Orielle and Coeur d'Alene.

HOW IDAHO BECAME A STATE

Lewis and Clark crossed Idaho on their way to and from the Pacific in 1805-1806. Fur traders drifted north into the territory, and a trading post was built on Lake Pend Orielle 1809. Additional posts were built by the American Missouri Fur Company in 1810 and John Jacob Astor in 1811. The British seized control during the War of 1812, but in 1818, the British and American governments agreed to allow joint exploration of what was part of the huge Oregon country. In 1834, Fort Hall was built near present day Pocatello, and Fort Boise was built later that same year.

In 1848, Idaho was included with Oregon and Washington in the Oregon Territory, then was divided into parts attached to both states when Oregon and Washington were divided in 1853. Oregon became a state in 1859 and all of Idaho was made part of the Washington Territory.

Gold was discovered on Orofino Creek in 1860, bringing tens of thousands of settlers into the area. The Mormons also established the first permanent agricultural settlement at Franklin the same year. Because residents of Washington feared these settlers would grab political control, they persuaded Congress to create the Territory of Idaho in 1863, which included Wyoming, Montana, and parts of the Dakotas and Nebraska. The origin of the name is a mystery—it may be an invented Indian word meaning "gem of the mountains" or the Kiowa Indian term for Comanches. Montana became separate in 1864 and by the time the Wyoming Territory was created in 1868, Idaho had its present boundaries.

The 1870s and 1880s brought additional mineral discoveries, the railroad, and Mormon settlers moving north from Utah. Idaho was finally admitted to the Union as the 43rd state on July 3, 1890.

IDAHO'S CITIES

BOISE (125,738), the largest city and capital, began as Fort Boise on the Boise River in 1834. The name comes from the French "boise," meaning "wooded," and was originally applied to the River. After the settlers were killed by the Indians in Wards Massacre in 1854, the fort was closed. After the discovery of gold in the area, Boise was laid out in 1863 and became the capital in 1884. Today, it is the commercial capital of southern Idaho.

POCATELLO (46,080) was built as a railroad center on the Portneuf River in southwestern Idaho in 1882. It was named for a Bannock Indian leader. The city is the center of a large irrigated farm area that produces the famous Idaho potatoes as well as grain and dairy products. The largest industries are fertilizer and electronic equipment.

IDAHO FALLS (43,929) was built in 1860 on the site of falls on the Snake River and was given its present name in 1891. The city produces processed foods, vitamins, and toiletries.

IDAHO'S PEOPLE

Idaho's 1,099,906 people include one of the smallest minority populations of any state. Mormons make up the largest religious group, and most of the population lives in the southern third of the state. The opening of new recreation areas has helped make Idaho one of the fastest growing states in the early 1990s.

IDAHO'S ECONOMY

To a lot of Americans, Idaho means "potato," and the state does produce two-thirds of our country's spuds. Idaho's farms also sell wheat, hay, milk, beef cattle, barley, and sugar beets. Idaho is one of the few states with large tracts of farmable land that has not yet been developed.

Gold brought a rush of settlers to Idaho, and today precious metals are still important. Idaho leads the U.S. in silver production, and antimony, vanadium, garnets, and phosphates are also mined. Idaho also has a significant timber industry.

Manufacturing came late to Idaho, but is now helping to fuel its population growth. The major products of the state are food, lumber and wood products, fertilizers, industrial machinery, paper, and electronic equipment.

Sun Valley in the Sawtooth Mountains is one of America's most famous ski resorts, and the state's rivers are meccas for whitewater rafters. Tourism brings $1.5 billion in annual revenues to the state.

Places To Visit

Sun Valley (800-634-3347), which opened in 1935, was the first famous U.S. ski resort. The 212 foot high Shishone Falls are the highlight of the spectacular Snake River Canyon area of South Central Idaho (800-255-8946). U.S. astronauts trained for lunar landings in the eerie Craters of the Moon National Monument (208-527-3257). Hell's Canyon (208-743-2363) is, at 5,500 feet, America's deepest canyon. Tourist information: Idaho Travel Council 700 W. State St., Boise, ID 83720 (800-635-7820/208-334-2470)

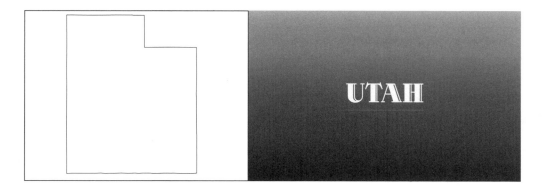

The variety of terrain found in the state have justifiably given rise to the phrase, "The Different Worlds of Utah." Utah is bordered on the north by Idaho and Wyoming, on the east by Colorado, on the south by New Mexico and Arizona, and on the east by Nevada. The state is rectangular with a "hat" on the northwest corner. It stretches 350 miles north-south and 275 miles east-west. Its southeastern most point is part of the Four Corners, where you can lie down and touch Utah, Colorado, Arizona, and New Mexico.

Salt Lake City, the largest city and capital, is 558 miles east of Reno and 434 miles west of Cheyenne on I-80, and is 415 miles northeast of Las Vegas on I-15. Provo is 43 miles south of Salt Lake City and Ogden is 34 miles north of Salt Lake City on I-15.

THE LAY OF THE LAND

The Rocky Mountains jut into the state from the northeast, with the Wasatch Mountains running north-south and the Uinta Mountains running east-west. The Rockies taper off to the Colorado Plateau, which covers the south central and eastern Utah. Included in these undulating plateaus are rock formations in different colors and the spectacular gorges of Canyonland, which includes the wildly shaped rock formations of Bryce Canyon.

The western third of Utah is part of the Great Basin. This area, which includes some of the flattest land in the United States, is largely desert broken up by isolated mountains and barren ridges.

Utah's most famous natural feature is the Great Salt Lake in the northwestern part of the state. This lake is a remnant of the prehistoric Lake Bonneville, which covered an area 350 miles by 150 miles during the period 50,000 to 25,000 years ago. Because the land lifted higher than the streams that followed into it, Lake Bonneville gradually evaporated, leaving the Bonneville Salt Flats, absolutely level ground on which almost every automobile speed record has been set. The Great Salt Lake has no outlet, so evaporation builds up salt content more than twice that of the oceans.

HOW UTAH BECAME A STATE

While Spanish explorers may have ventured into the state as early as 1540, the first written description was provided by two Franciscan fathers who led a party across the state towards California in 1776. . Beginning about 1820, a number of trappers came into the area looking for skins. Among them were James Bridger, who sighted the Great Salt Lake in 1824 and Jedediah Smith, who crossed the state north-to-south and east-to-west in 1826-1827. Capt. Benjamin Bonneville led a Federal party through northern Utah in 1833-34, and John C. Fremont wrote a scientific report on his travels in 1843-45.

The history of Utah changed dramatically in 1847, when Brigham Young led a large group of members of the Church of Jesus Christ of the Latter-day Saints, better known as Mormons, who were fleeing persecution to search for a "land nobody else wanted." Young and the Mormons found such a place on the shores of the Great Salt Lake. These very industrious people set up irrigation systems that allowed them to become self-supporting, and many more Mormons followed.

In 1848, the U.S. was ceded Utah by Mexico in the Treaty of Guadeloupe Hidelgo. But Congress was slow getting around to forming a government for the area, the Mormons took it upon themselves to form the "State of Desert" in 1849 and applied for admission to the Union. Desert means "land of the honeybees," the reason Utah is known as the Beehive State.

Congress was reluctant to act quickly and instead created the Territory of Utah as part of the Compromise of 1850, and Brigham Young was named the first governor. The name comes from the Shoshone Ute tribe, which in turn comes from an Indian word meaning "top of the mountains." Non-Mormons began moving into the state, and conflicts arose that resulted in a Mormon attack on a non-Mormon immigrant party that killed many people. Young was removed as governor and U.S. troops were sent in to keep order.

What followed was nearly 40 years of conflict between the Mormons and the Federal government. One major sticking point was the Mormon's acceptance of polygamy, although only about 15% of men had more than one wife. In 1862, Congress made polygamy illegal; in 1874, it substituted Federal courts for Mormon courts that refused to convict polygamists; and in 1882, it took the vote away from Mormons based on their approval of polygamy. Finally, in 1890, the Church gave in. Mormon President Wilfred Woodruff issued a manifesto prohibiting polygamy. Congress in turn passed legislation allowing Utah's admission to the Union on January 1, 1896 as the 45th state.

UTAH'S CITIES

SALT LAKE CITY (159,231) when the Mormons arrived on July 24, 1847. Brigham Young laid out a city of 10 acre plots radiating from a Temple Square. In 1856, the city that was originally called Great Salt Lake City became the capital of the territory. In 1862, a fort was built on the site so Federal troops could prevent violence between Mormons and non-Mormons. Economic growth was further stimulated with the arrival of the railroad in 1869. Today, Salt Lake City is the transportation and commercial center for the state, and has a modern and diverse economy. Temple Square and the headquarters of the Mormon Church are popular tourist attractions.

PROVO (86,835) was founded as Fort Utah by Mormons in 1849, and was named for French-Canadian fur trapper Etienne Provost in 1850. Its growth was stimulated by the arrival of the railroad in the 1870s. Provo is the home of Brigham Young University, is a gateway to Utah's ski country, and has a diverse economy that produces computer hardware and software, clothing, electronic equipment, and printed materials.

UTAH'S PEOPLE

Of Utah's 1,859,582 people, about 69% are members of the Mormon Church. Because Mormon families tend to have more children, Utah has America's third highest birth rate and the largest percentage of its population age 18 and under. Utah's population is overwhelmingly urban (87%) and is concentrated in the northwestern portion of the state.

UTAH'S ECONOMY

The Mormons who settled Utah built an elaborate irrigation system to sustain agriculture. Today, those areas still produce wheat, hay, and fruits and vegetables. With arable land at a premium in the state, it's understandable that 70% of the farm income comes from beef cattle, dairy farming, and sheep.

Like the other Rocky Mountain states, Utah has considerable mineral wealth. Copper, gold, silver, coal, petroleum, and salt are among the state's 210 mineable minerals.

Manufacturing has been important to Utah's economy since 1940. Among the leading industries are metal fabricating, machinery production, grain milling, meat packing, defense-related electronics, and production of campers and mobile homes.

Utah's ski areas and other natural wonders brings in $3 billion in annual tourism revenues.

Places to Visit

In Bryce Canyon National Park (801-834-5322), millions of years of erosion have carved fabulous stone formations in gigantic natural bowls. Zion National Park (801-772-3256) features towering cliffs and spring-fed vegetation. Ski Utah (801-534-1779) offers information about conditions at such famous ski resorts as Sundance, Solitude, and Snowbird. Salt Lake City (801-521-2822) features Temple Square, which contains sites central to Mormonism, and the Great Salt Lake. Tourist information: Utah Travel Council, Council Hall, Capital Hill, Salt Lake City, UT 84114 (801-538-1030)

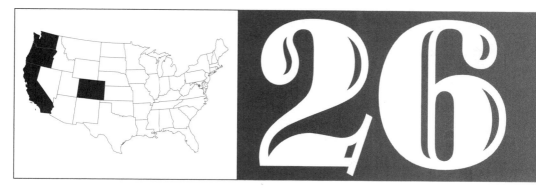

THE PACIFIC STATES

Comparing The Pacific States

STATE	POPULATION	SQUARE MILES	PER CAPITA INCOME
National rankings in parenthesis	(1993 U.S. Bureau of the Census)	(U.S. Geological Survey)	(U.S. Bureau of Labor Statistics)
California	31,210,750 (1)	163,707 (3)	$16,409 (8)
Washington	5,255,276 (15)	71,302 (18)	$14,923 (16)
Oregon	3,031,867 (29)	98,386 (9)	$13,418 (26)
Hawaii	1,171,592 (40)	10,932 (43)	$15,770 (11)
Alaska	599,151 (48)	656,424 (1)	$17,610 (5)

Five U.S. States border the Pacific Ocean the vast body of water that covers one-third of the earth's surface. California, Oregon, and Washington all of which are part of the contiguous 48 States, the Pacific States share a coastline that could not be more different from our nation's other coastline. The broad Atlantic Coastal Plain with its many harbors, barrier islands, and sandy beaches was formed by sediment washed down from the eroding Adirondack mountains over millions of years. The Pacific Coast, on the other hand, is the site of the recent and violent collision between the North American and Pacific Plates. This collision forced up three lines of mountains; first the Rockies, then the Cascade-Sierra Nevada ranges, then the Coastal Mountains that hug the shore. These mountains were formidable barriers to settlement from the more eastern areas, while the long trip around the tip of South America made the sea journey a long and expensive venture. That's why the Pacific States were settled so much later than those on the east coast.

EXPLORATION OF THE PACIFIC STATES

The eventual spurs to exploration and settlement were the natural riches of this vast area—first the fur bearing animals that lived in the forests and on the coast, then the timber and mineral riches inland. Behind the miners and the loggers came farmers who found ways to use water from the mountains to irrigate semi-arid but fertile valley land.

The sailors of many nations, including Spain, Great Britain, Russia, the Netherlands, and the United States, poked their ships into the harbors on the Pacific Coast between the late 1500s and the beginning of the nineteenth century. While governments may have argued over conflicting claims to the territory, there were so few Europeans actually in these states that no armed conflicts were possible. Trappers, traders, pirates, and missionaries carved out their places in this wilderness until the mid-1800s, when the U.S. acquired clear title to the area and the widespread settlement began.

Today, the mild climate and great natural beauty have made the three Pacific States among the most desirable places to live. Constant influx of new residents give these states a forward-looking mentality that has placed them on the cutting edge of technology and social change. The only dark spot on the horizon stems from the gigantic collision that formed these states—areas on the edge of plates are prone to devastating earthquakes and volcanic eruptions.

Alaska and Hawaii share only their proximity to the Pacific Ocean with California , Oregon, and Washington. Their cultures, histories and physical geography are unlike any of the other states. And they are so totally different that they serve as an excellent contrast to each other.

I have had the good fortune to travel extensively in both states in my role as ambassador for "Good Morning America." I have found both to be exotic and fascinating, and I can't wait until my duties take me back for additional explorations.

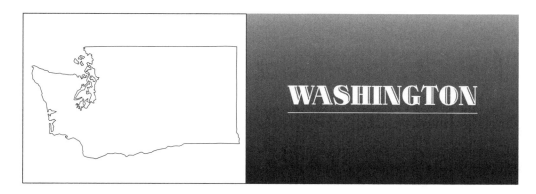

WASHINGTON

Snow-capped peaks and vast forests that include some of our last virgin timber are the most outstanding features of Washington, the northernmost of the Pacific states. Washington is bordered on the north by Canada, on the east by Idaho, on the south by Oregon, and on the west by the Pacific Ocean. The Columbia River makes up most of the southern border, and the Strait of Juan de Fuca, Haro Strait, and the Straits of Georgia separate the state from the Canadian Vancouver Island. Washington is roughly rectangular, stretching 345 miles east-west and 235 miles north-south. The state has 157 miles of coastline.

Seattle, the largest city, is 174 miles north of Portland and 143 miles south of Vancouver, British Columbia, on I-5. Spokane is 282 miles east of Seattle and 310 miles northwest of Helena, Montana on I-90.

THE LAY OF THE LAND

Puget Sound is a wedge of water that cuts 80 miles into the northwest corner of Washington. On the Sound's inlets are many communities, including the major ports of Seattle, Tacoma, and Bremerton. A band of flat land surrounds the Sound.

South of Puget Sound are the Coastal Ranges. The northernmost range is the spectacular Olympic Mountains, which have snow-capped peaks and more than 60 glaciers. This area of virgin forest is a national park. South of the Olympic Peninsula is an area of high hills or low mountains dotted with small lakes and fertile valleys.

The backbone of the state, running through the center north to south, is the towering Cascade Range. This range averages 8,000 feet above sea level and serves as a natural barrier to weather systems and ground travelers alike. The Cascades feature many volcanic mountains, including Mount Ranier(14,410 feet, the highest in the state) and Mount Saint Helens, which caused great devastation when it last erupted in 1980. Many peaks are snow-covered and the northern part of the range has many glaciers.

Most of central and southern Washington on the eastern side of the Cascades is part of the Columbia plateau, which was created by vast lava flows. Much of this land is flat and arid, but rivers such as the Columbia and the Snake have cut deep trenches. In the northeastern corner is part of the Rocky Mountains. Ranges that don't exceed 5,000 feet are interspersed with valleys, and much of the land is forested.

Washington has almost 1,000 natural lakes, most of which are in the Cascades and the western portion of the state. The largest is 50 mile long Lake Chelan in the eastern Cascades.

HOW WASHINGTON BECAME A STATE

In 1775, Bruno Heceta and Juan Francisco de la Bodega y Quadra became the first known Europeans to land on Washington soil, and they claimed the area for Spain. However, the British and the Americans had the most interest in the area. In 1792, Captain Robert Gray discovered the mouth of the Columbia River and Captain George Vancouver explored Puget Sound.

In 1805, Lewis & Clark reached the most distant point on their incredible expedition, navigating the treacherous Snake and Columbia Rivers to reach the coast. The British North West Company established a trading post near Spokane in 1810, and John Jacob Astor established a second in 1811. Although the British and the Americans agreed to a ten year period of joint occupancy of the area in 1818, tensions grew almost to the brink of war. In 1845, James Polk campaigned to the slogan "54'40" or fight," a pledge to set the border far into what is now Canada. But in 1846, in the Oregon Treaty, the U.S. and Britain agreed to set the border at 49 degrees north latitude, the same parallel used to set the eastern part of the U.S.-Canadian border. In 1848, Congress created the Territory of Oregon, which included Washington, Oregon and part of Idaho. In 1853, Congress created the Washington Territory.

At the time, the population of the state was only 4,000 people. But the arrival of the Northern Pacific Railroad in Spokane in 1881 opened the way for increased settlement. Washington was admitted to the Union on November 11, 1889, as the 42nd state.

WASHINGTON'S CITIES

SEATTLE (516,259) is built on a series of hills between Puget Sound and Lake Washington. The city was founded in 1852 and named for Seattle, chief of a local Indian tribe. A sawmill was built in 1853, but the town grew slowly until the first transcontinental railroad arrived in 1884. Seattle became one of the gateways to the Alaska and Klondike gold rushes, and growth was further stimulated by the completion of the Panama Canal in 1914. Today, Seattle is the home to Boeing, the nation's largest aircraft manufacturer, as well as to computer software and biotechnology firms. The city hosted a World's Fair in 1962.

SPOKANE (177,196) is on the Spokane River in eastern Washington. Trading posts were established in the area as early as 1810. A sawmill was built on the site in 1971, and the city that grew around it was named for the River, which had been named for the Spokane Indians. The railroad arrived in 1881 because Spokane was located at one of the few places where the Rocky Mountains could be easily crossed. The city became the center of the "Inland Empire," a rich farming and logging region.

OLYMPIA (33,840) is located at the tip of Puget Sound, and the majestic summit of Mount Ranier can be seen from the city. It was founded in the 1840s as Smithfield, but in 1850 was renamed for the Olympic Mountains. It became the capital of the Washington Territory in 1853. Olympia is a fishing port (the Olympia oyster is the most famous catch) and produces wood products, processed foods, metal and paper containers, and mobile homes.

WASHINGTON'S PEOPLE

Half of Washington's 5,255,276 people live in the Puget Sound area. About three-quarters of the states' residents live in urban areas. Washington has 77,627 Native Americans, but blacks make up only 3% of the population. Washington is one of the ten fastest growing states.

WASHINGTON'S ECONOMY

Washington's forest wealth fueled the state's early growth, and it still provides 10% of the nation's lumber. Wood provides the raw materials for a large wood products industry.

Agriculture is the second foundation of the economy. Washington grows more apples than any other area of the world, and it's also number one in peas, hops, cherries, and rhubarb. Other important crops include plums, grapes, pears, blackberries, cattle, and sheep.

Washington's dams produce almost a third of America's hydroelectric power,, which stimulated the growth of industries such as aluminum smelting and refining and electrochemicals that depend on plentiful, cheap electricity. Boeing is the state's largest single employer. Recently, Washington has attracted an increasing number of high-tech electronic companies.

Tourism brings annual revenues of nearly $5 billion.

Places to Visit

Seattle's (206-461-5840) hills and scenic views make it one of America's most beautiful cities. The 14,411 ft. Mt. Ranier (360-569-2211) is surrounded by a stunning national park. A vivid contrast is the still-steaming crater which can be seen from the Mt. St. Helens National Volcanic Monument (360-274-2100) The towering Olympic Mountains dominate the rugged terrain of the Olympic Peninsula (800-942-4042) Tourist information: Washington Tourism Development Division, Box 42500,Olympia, WA 98504 (800-544-1800/360-586-2088)

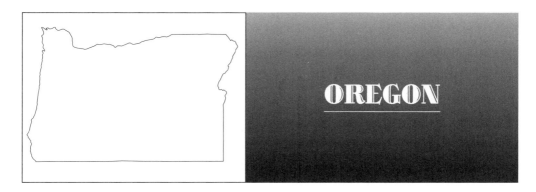

OREGON

Oregon, like its Pacific Coast neighbor Washington, has been blessed with great natural beauty. It is bordered on the north by Washington, on the east by Idaho, on the south by Nevada and California, and on the west by the Pacific Ocean. The Columbia River makes up about three-quarters of the northern border. Oregon is roughly rectangular, stretching 295 miles north-south and 376 miles east-west. It is the tenth largest state, covering about 50% more area than Washington. Oregon has 296 miles of coastline.

Portland, the largest city, is 174 miles south of Seattle and 636 miles north of San Francisco on I-5, and is 428 miles northwest of Boise on I-84. Eugene is 109 miles south of Portland on I-5.

THE LAY OF THE LAND

Arching and folding of rock and frequent lava flows created the Coastal Range that give Oregon a rugged Pacific coastline. This heavily forested area is not as high as the Coastal Ranges to the north, consisting primarily of hills interrupted by narrow valleys.

The broad Willamette Valley, which contains the state's most fertile soil, separates the Coastal Range from the Cascade Range for 150 miles south of the Columbia River. Most of the state's major cities are located along the Willamette River. A band of low mountain ranges connects the Coastal and Cascade Ranges south of the Valley.

The Cascade Range, a wall of mountains averaging 6,000 feet high and 50 to 100 miles separates the western third from the eastern two-thirds of Oregon. Crown jewel of the Cascades is the volcanic Mt. Hood (11,245 ft.). The slopes of the mountains are heavily forested.

Most of the eastern two-thirds of Oregon is part of the Columbia Plateau. Gently rolling hills in the west give way to flatter, drier land cut by narrow valleys. In southeastern Oregon is a small section of the Basin and Range region, while the Wallowala Mountains cut across the northeastern corner.

Oregon is home to one of America's most famous lakes, Crater Lake, which fills the crater of the extinct volcano, Mount Mazama. The state is fortunate to have numerous natural lakes, as well as man-made lakes formed by dams.

HOW OREGON BECAME A STATE

For more than two centuries, the Europeans who sailed up the Pacific Coast were searching not for lands to settle, but for the elusive Northwest Passage that would connect the Atlantic and Pacific Oceans. The first were the Spanish in the 1540s, and in the eighteenth century came British sailors, including Captain James Cook. The first British traders arrived in 1785, and in 1792, Robert Gray sailed up the mouth of a river he named Columbia, after his ship. These traders soon learned that they could trade beads, nails, and other trinkets to the Indians in exchange for sea otter and beaver pelts, then sail across the Pacific to China to trade the furs for a fortune in tea, silk, and spices.

The land route was pioneered by Lewis and Clark, who spent the winter of 1805-1806 at the mouth of the Columbia River. John Jacob Astor established the Pacific Fur Company in a town he called Astoria in 1811. In 1818, the Americans and British agreed to jointly develop the territory. But in the early 1840s, missionaries began arriving from the east, and they were followed in 1843 by the first large group to follow what became known as the Oregon Trail. Most settled in the Wilmette Valley, and they soon formed a government with the hope of joining the Union. Tension with the British almost forced a war until the two countries agreed to establish the U.S. - Canadian border at the 49th parallel in 1846. Congress combined Oregon, Washington, and part of Idaho into the Oregon Territory in 1848 (the name is thought to be derived from the Indian name for the Columbia River), but split off Washington into a separate territory in 1853. Oregon joined the Union as the 33rd state on February 14, 1859.

OREGON'S CITIES

PORTLAND (437,319) is a major deepwater port on the Willamette River near its junction with the Columbia River. The city was founded in 1845 and named for Portland, Maine, the birthplace of some of the settlers. It became a major supply center and transfer point for fortune seekers heading to the California gold rush in the 1850s and the Alaska and Klondike gold rushes in the 1890s. Portland was the site of the Lewis & Clark Centennial Exposition, a 1905 World's Fair.

EUGENE (112,669) is the center of a rich logging and agricultural region. The city was founded in 1840 and named for Eugene F. Skinner, one of the first settlers. The railroad arrived in 1871, and Eugene became a trade and transportation center. In recent years, it has attracted high-tech research companies and has developed a significant tourist industry.

SALEM (107,786) was founded on the Willamette River in 1840 by Jason Lee, a Methodist minister. The city was laid out in 1848, and in 1851 it became the capital of the Oregon Territory. The name Salem is an Anglicization of the Hebrew word "shalom," which means "peace," and it probably was used because the Calapooya Indians had called the site Chemekeke, which means "place of rest."

OREGON'S PEOPLE

Most of Oregon's 3,031,867 people live along I-5 in the Willamette Valley, where the state's four major metropolitan areas are located. Oregon has traditionally had a very small minority population, but that has changed over the last decade as more Asians and Hispanics have moved to the area. The growth rate of the state was slowed by the recession in the 1980s, but has begun to climb in the 1990s.

OREGON'S ECONOMY

If you love the sound of chain saws cutting through wood, Oregon is the place for you. The state has the nation's top timber reserves and it is first in timber production, supplying 20% of our lumber and one-sixth of our plywood. The production of lumber, wood products, and paper accounts for one-third of Oregon's manufacturing employment. Oregon also helps our balance of payments by exporting timber to Japan.

The early settlers came to the Willamette Valley to farm, and Oregon is still a major agricultural state. Crops account for two-thirds of farm income, especially vegetables and fruits, wheat, potatoes, barley, and oats. The major livestock are cattle, sheep, poultry, and hogs.

In addition to wood and paper products, the leading manufactured goods are processed foods, electronic equipment, industrial machinery, and precision instruments.

Tourism annually brings nearly $3 billion to Oregon.

Places to Visit

Crater Lake National Park (503-594-2211) surrounds America's deepest lake, formed 6,800 years ago when volcanic Mount Mazama decapitated itself. Astoria (503-325-6311), at the outlet of the Columbia River, was where Lewis and Clark reached the Pacific Ocean and is the oldest city west of the Mississippi. Majestic Mount Hood is the focal point of the 1.1 million acre Mount Hood National Forest (503-666-0700). Oregon's wine country (503-228-8403) is nestled in the valley between the Coast and Cascade Ranges. Tourist information: Oregon Tourism Division, 595 Cottage St., NE, Salem, OR 97310 (800-547-7842)

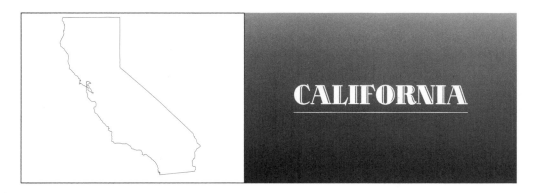

CALIFORNIA

One out of every nine Americans lives in California, which has been a mecca for fortune seekers since the gold rush began in 1848. California is bordered on the north by Oregon, on the east by Arizona and Nevada, on the south by Mexico, and on the west by the Pacific Ocean. The state, which is shaped roughly like a crescent moon, has 840 miles of coastline and stretches 770 miles north-south and 350 miles east-west. If California were magically moved to the east coast, it would stretch from Cape Cod to Charleston, South Carolina. Off the coast are two groups of islands, the Channel Islands (which include Santa Catalina) and the Farallon Islands.

Los Angeles is 369 miles west of Phoenix on I-10 and 380 miles south of San Francisco and 124 miles north of San Diego on I-405 and I-5. San Francisco is 95 miles west of Sacramento and 228 miles west of Reno on I-80, and 636 miles south of Portland, Oregon in I-5. San Diego is 409 miles west of Tucson on I-8.

THE LAY OF THE LAND

A chain of Coastal Mountains runs the entire length of California's Pacific Coast. In the far north are the Klamath Mountains, which have peaks that range up to 7,000 feet. A series of north-south ranges stretch southward, broken up by depressions such as the Salinas Valley and the San Francisco area. South of San Francisco are the Transverse Ranges, which run east-west instead of north-south. The Los Angeles Basin, a huge depression, separates the Transverse Ranges from the north-south Peninsula Ranges that continue southward to the Mexican border. This entire coast region is undercut with numerous major faults, such as the San Andreas fault, which generated more than 6,000 earthquakes between 1980 -1991.

Inland from the Coastal Ranges is the California Valley, a broad area with deep fertile soil that stretches nearly 500 miles southward from Redding to north of Los Angeles. The Valley is California's major agricultural area.

The lower end of the Cascade Range thrusts into north central California. This range includes such major peaks as Mt. Shasta (14,162 ft.), an extinct volcano, and the active volcano Lassen Peak. South of the Cascades are the rugged Sierra Nevada, a towering range with six peaks over 14,000 feet, including Mount Whitney (14,494 ft.), the state's highest. Streams running down from the Sierra Nevada provide the water for most of Southern California.

The northeast corner of the state contains a small portion of the Great Basin. A much larger area of the Great Basins covers the southeastern section of California. Isolated mountain ranges are interspersed with broad deserts. The Mojave Desert alone covers one-fifth of California's area. The most famous part of this Basin is a depression known as Death Valley, which is the lowest point in the United States at 282 feet below sea level. In the southeastern corner is the Imperial Valley, where irrigation has turned desert land into the state's second most important agricultural area.

Northern California has numerous lakes, including Lake Tahoe on the Nevada border. Southern California has few. One exception is the brackish Salton Sea, the state's largest lake.

HOW CALIFORNIA BECAME A STATE

In 1539, the Spanish governor of Mexico sent navigator Juan Rodríguez Cabrillo up the Pacific Coast to support Coronado's inland explorations. In 1542, Cabrillo reached the coast of present-day California. For the next sixty years Spanish ships visited the coast, as did English and Dutch pirates that preyed on the Spanish. In 1579, Sir Francis Drake landed on the coast to repair his ship and claimed the land he called Nova Albion ("New England" in Latin) for England. The Spanish named the area after the mythical paradise California that was the setting of a in a hugely popular Spanish romance novel called "Las Serges de Esplandian," written in 1510.

However, the first settlement was a mission established at San Diego in 1769 by Gaspardo Portolá and Father Junípero Serra. Portolá went on to establish a second mission at Monterey. The Spanish soon had a network of 21 missions, four presidios (forts), and three pueblos (villages). Control passed to Mexico in 1825, which issued grants for more than 800 huge ranches. Livestock, not crops, was the major industry during a period glamorized by the Disney TV series "Zorro."

During the 1830s, American trappers worked the state, and in 1841, the first wagon train of settlers arrived in the San Juaquin Valley. The U.S. government made several unsuccessful attempts to purchase California. After some local settlers launched the Bear Flag Revolt in June, 1846, a U.S. military force consisting of John Fremont's California Battalion, Gen. Stephen Kearny's Army of the West, and a Regiment of New York Volunteers led by Jonathan Stevenson defeated the Mexicans and took control of California.

On January 24, 1848, just ten days before the signing of the Treaty of Guadalupe Hidelgo, in which Mexico formally granted California to the U.S., a man named James Marshall found gold while working at a sawmill owned by his John Sutter. This discovery launched a gold rush that brought an astounding 250,000 settlers to the state in just four years. Anxious to become a state with a government that could stem the violence that accompanied the gold rush, Californians adopted a constitution and elected a governor in 1849 before their application for statehood had been approved.

Congressional approval was held up by Southerners reluctant to add another free state. Finally, in a series of measures called the Compromise of 1850, southerners agreed to the admission of California in exchange for enactment of the Fugitive Slave Law and a bill that opened the New Mexico and Utah territories to both slaveholders and nonslaveholders. California joined the Union as the 31st state on September 9, 1850.

CALIFORNIA'S CITIES

LOS ANGELES (3,485,398) is California's largest city and the second largest in the U.S. It was founded in 1781 at the direction of the Spanish governor, Felipe de Neve, who named it El Pueblo de Nuestra Señora la Reina de Los Angeles de Poricuncula, or "The Village of Our Lady the Queen of Angels of Poricuncula." The site passed to Mexico in 1821 and the U.S. in 1846, but it grew slowly until the arrival of the Southern Pacific Railroad in 1876. There was a land boom in the 1880s as irrigation proved valuable in growing citrus fruits and other produce. In the early twentieth century, Los Angeles became the center of the motion picture industry, and today it's a television, radio, and music-recording center as well. The city has always attracted people who want to better their lot in life, and a recent influx of Asians and Mexicans has produced a population that's 38% foreign-born.

SAN FRANCISCO (723,959) is at the northern tip of a peninsula at the head of San Francisco Bay. In 1776, Juan Bautista de Anza built a presidio on the site to guard San Francisco Bay. In 1790, the Mission San Francisco de Assisi was built nearby. In the 1830s, a settlement called Yerba Buena Cove was established in what is now the northeastern section of the city. When the U.S. acquired California in 1846, Yerba Buena was renamed San Francisco. Two years later came the discovery of gold at Sutter's Mill. San Francisco became an important port and supply point for the people pouring into the state. The city had a population of 340,000 when it was destroyed by the devastating earthquake of April 18, 1906. The city was quickly rebuilt, and today it is one of the world's leading finance and international trade centers. One visit, and you'll know why Tony Bennett left his heart there.

SAN DIEGO (1,110,549), California's second largest city, is located on San Diego Bay near the Mexican border. That bay was discovered by the Portuguese explorer Juan Rodríguez Cabrilla in 1542. In 1602, it was named San Diego by Sebastian Vizcaino, either for San Diego de Alcala (St. Didacus) or for his ship, the San Diego. In 1769, California's first mission and first presidio were established here. By the 1800s, the city had a small trade in furs and hides. After the U.S. acquired California in 1846, San Diego was a whaling center. It's rapid growth stems from the arrival of the railroad in 1889. Today, San Diego is a major shipbuilding center, the home of the Pacific fishing fleet, and the site of a Naval shipyard and training base.

SACRAMENTO (369,365), the state capital, was founded in 1839, when a Swiss-American named John Sutter was granted land by the Mexican government to build a community he named New Helvetia. Fort Sutter was built in 1844, and in 1848, gold was discovered nearby at Sutter's Mill. The city, renamed for the Sacramento River, grew rapidly and became the capital in 1859. Sacramento was the first terminus of the first trans-continental rail line in 1869. It's economy received another boost in 1963 when a deepwater channel to San Francisco was completed.

CALIFORNIA'S PEOPLE

California's 31,210,750 people represent America's most ethnically diverse population as well as its largest. More than 7,000,000 Californians are Hispanic (mostly Mexican), about one-quarter of the population. The state also has 731,000 residents from the Philippines, 704,000 Chinese, 362,000 Japanese, 280,000 Vietnamese, and 236,000 Native Americans, the country's second largest concentration. About 93% of Californians live in urban areas.

CALIFORNIA'S ECONOMY

California is America's leading agriculture state, generating 11% of the total national farm income. The state's farms produce more than 200 different crops, and its production of 50 of these leads the nation. The major crops are grapes, vegetables and fruits, hay, cotton, and nuts. The state also raises beef cattle, sheep, poultry, and hogs.

Oil was discovered in the Los Angeles basin in the 1920s, and today California still produces 12% of America's oil and significant amounts of natural gas. Other leading mineral products are boron, tungsten, and asbestos. California trails only Oregon and Washington in annual timber production.

California also leads the nation in manufacturing income. Major industries include aircraft and other transportation equipment, electronics, industrial machinery, scientific instruments, processed foods, printed items, and fabricated metals. The state is the center of the entertainment industry and is a major world finance and international trade mecca.

Finally, California annually attracts 40 million tourists who spend $53 billion, by far the nation's highest total.

Places to Visit

From its steep hills to its diverse neighborhoods, San Francisco (415-974-6900) is one of America's most distinctive cities. San Diego (619-276-8200) offers an almost perfect climate and the atmosphere of old Spanish California along with its famous Zoo. The sprawling Los Angeles area (213-624-7300) has beaches, movie stars, and Disneyland. Yosemite National Park (415-556-0560) contains Yosemite Falls, the tallest in North America. Tourist information: California Division of Tourism, 801 K St., Sacramento, CA 95814 (800-862-2543/916-322-1397)

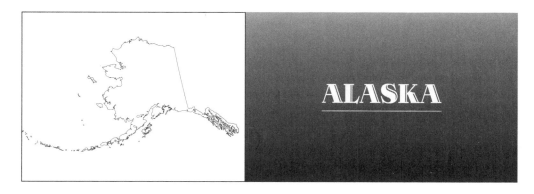

ALASKA

Alaska, the land of the midnight sun and our northernmost state, covers a vast area as large as Texas and 2 Californias. It is bordered on the north by the Arctic Ocean, on the east by the Yukon Territory and British Columbia, on the south by the Pacific Ocean, and on the west by the Bering Strait, the Bering Sea, and the Arctic Ocean. Alaska has two major island chains, the Alexander Archipelago and the Aleutian Islands, which stretch all the way to the Eastern Hemisphere—the easternmost point in the United States is in far western Alaska! The state stretches 1,100 miles north-south and 2,000 miles east-west. Alaska has 5,580 miles of shoreline on the Pacific Ocean and 1060 miles on the Arctic Ocean.

THE LAY OF THE LAND

Southern Alaska is part of the Pacific Mountain chain. This chain of towering mountain ranges was thrown up by the massive impact of two of the earth's plates, and consequently is an extremely active earthquake and volcano area—Alaska has registered more than 10,000 quakes in the last decade. These mountain ranges include Mt. McKinley (20,320 ft.), the highest point in the U.S, and Mt. Wrangell (14,006), the highest active volcano. The panhandle area is cut by spectacular fjords and covered by huge glaciers.

Inland is the vast Central Highland and Basin Region, sometimes known as the Yukon Plateau. In the west, the land is low and frequently floods, while the east has more mountain ranges and steep valleys. Running across the entire state from east to west is the little-explored Brooks Range, which has elevations up to 10,000 feet. From the Brooks Range, the Arctic Lowlands gradually slope to the Arctic Ocean—it's on this North Slope that huge reserves of oil were discovered in 1968. Much of this flat region is tundra.

The major river in Alaska is the Yukon, which begins in Canada before crossing the state to its outlet in the Bering Sea. Alaska has countless lakes, many of which are unexplored.

HOW ALASKA BECAME A STATE

Alaska was populated by three main cultures before the arrival of Europeans. The Intuit, or Eskimos, hunted whales, seal, and polar bears in the frigid north. The Aleuts were a maritime culture on the Aleutian Islands. Finally, several Indian tribes inhabited southern Alaska.

Russian explorer Vitus Bering was the first European to explore Alaska in 1741. After he returned with a load of rich furs, many Russian fur traders poured into the area. Gregory Shelekhou established a trading post on Kodiak Island in 1784 and Genasin Pribilof landed on the Sea Islands in 1788. In 1799, the Russian-American Fur Trading Company declared a monopoly over the entire area. Under the direction of Chief Manager Aleksandr Berenov, the Russians established a network of 24 settlements.

However, by the 1850s, the fur trade was declining, and Russia began to worry that Canada might invade Alaska. Russian explored the idea of selling Alaska to the United States. The Civil War postponed the negotiations, but in 1867, Secretary of State William Seward arranged to purchase the state for $7.2 million, an agreement that many called "Seward's Folly."

There were no detractors when gold was found in the Klondike in 1896 and in Nome in 1898. Settlers poured into the state. In 1906, Alaska was given one delegate to Congress and in 1912, it became a Territory. However, it wasn't until Alaska's strategic value became evident when Japan invaded the Aleutian Islands in World War II that the U.S. government had serious interest in the state. The Cold War also led to a military build-up in the state. On January 3, 1959, Alaska was admitted to the Union as the 49th state.

ALASKA'S CITIES

ANCHORAGE (226,338), the largest city, is a port on Cook Inlet. It was founded in 1914-1915 as the main supply point for the building of the Alaska Railroad. In World War II, it became the headquarters of the U.S. Alaska Defense Command. Today, it is an air freight hub and a gateway to the rest of Alaska. The city has rebuilt after a 1964 earthquake that was the strongest ever recorded in North America.

JUNEAU (26,751), the state capital, is located on the Alaskan Panhandle. It was founded in 1880, when Joseph Juneau and Richard Harris discovered gold in the area. This gold mining town became the capital of the Alaska Territory in 1900 and capital of the new state in 1959. In 1974, Alaskans voted to move the capital to a more central location, but they haven't gotten around to implementing the move in more than two decades. At 2594 square miles, Juneau is one of the largest cities in area in the U.S. It is a fishing, lumbering, and tourist center.

FAIRBANKS (30,843) is on the Chino River in central Alaska. It was founded in 1902 when gold was discovered in the region, and was named for U.S. Vice President Charles W. Fairbanks. The city began to grow when the Alaska Railroad arrived in 1924 and the Alaska Highway was completed in 194. Recently, it was the headquarters for the building of the Alaska Pipeline. Fairbanks is the administrative and commercial center for interior Alaska, and a departure point for tourists visiting nearby Mt. McKinley.

ALASKA'S PEOPLE

With just 599,591 people, Alaska is by far the most sparsely populated U.S. state. Its residents include 44,401 Eskimos, 31,245 Native Americans, and 10,052 Aleuts. Most of the population lives in the southern region of the state.

ALASKA'S ECONOMY

Despite its vast area, Alaska has very little land that is suited for agriculture. The state only has slightly more than 500 farms, which produce greenhouse plants, dairy products, and potatoes. The state is not self-sufficient in food, and food prices tend to be America's highest.

What Alaska does have is trees. Even though only one-third of the state is forested, Alaska is so huge that those forests contain 22.8 million acres of commercial timber. The state also has a major fishing industry, with salmon, shellfish, halibut, herring, and flounder the major catches.

The economic profile of Alaska changed dramatically when huge oil reserves were discovered on the North Slope in 1968. That discovery led to the building of the Trans-Alaska Pipeline, which stretches more than 1,000 miles to the ice free port of Valdez. Oil has made Alaska rich, but the Exxon Valdez oil spill was vivid proof that a steep price might have to be paid.

Because of its climate and its distance from major markets, Alaska has a very small manufacturing base. The major industries are fish processing, and lumber and wood products.

Places to Visit

Glacier Bay National Park (907-697-2230) and the Kenai Fjords National Park (907-224-3175) offer glaciers and wildlife most commonly seen from cruise ships. Juneau (907-586-2201) is a charming gold-rush town crammed against the mountains. Anchorage, set between mountains and the sea, is the only major city where you can regularly scheduled sled dog races. In the Interior, the Denali National Park and Preserve (907-683-2686) has six million acres of tiaga and tundra as well as Mt. McKinley, the tallest North American peak. Tourist information: Alaska Division of Tourism, Box 110801, Juneau, AK 99811 (907-465-2010)

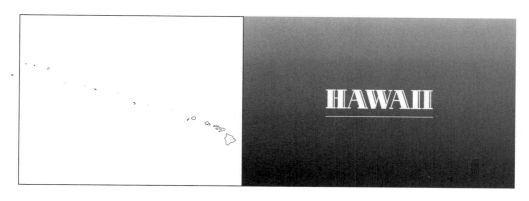

HAWAII

Hawaii is unusual among the 50 states, because it isn't bordered by anything. The state consists of 8 islands and 124 islets, reefs, and shoals located in the geographic center of the Pacific Ocean, 2,557 miles west of Los Angeles. Hawaii is the southernmost of all states, with a latitude about the same as Mexico City. The state stretches 1450 miles in an arc from east to west, and has 750 miles of coastline on the Pacific. The entire state consists of the tops of gigantic shelf volcanoes that erupted from the ocean floor.

Honolulu, the largest urban area, is 3,859 miles southeast of Tokyo, 5,505 miles northwest of Melbourne, Australia, 5,077 miles southeast of Beijing, and 4,969 miles from New York City.

THE LAY OF THE LAND

All of the 8 Hawaiian islands are the tops of one or more shield volcanoes. The biggest island, Hawaii, consists of five volcanoes, Mauna Kea, Mauna Lua, Hualalai, Kohalu, and Kilauea. Mauna Kea and Mauna Lua are still active. Between the mountains are plateaus on which sugar cane, fruits and nuts are grown.

Oahu consists of two volcanoes eroded to mountain ranges, with a central plateau in between. Honolulu is located on a narrow plain on the coast. Maui consists of two shied volcanoes connected by an isthmus. Kauai, which was formed from one volcano, has some of the most spectacular scenery in the Islands, including its dramatic Sea Cliffs, inland canyons, and Mt. Waialeale, the wettest spot on earth.

Molokui is a plateau with a mountainous east coast. Lanai consists of one volcano with a central plateau that contains the largest pineapple plantation in the world. Niihau, which is privately owned and closed to tourists, is a plateau. Kaboolawe, the smallest island, is uninhabited and has often been used as a naval gunnery range.

Because Hawaii's lave-based soil is so porous, most of the precipitation is quickly absorbed. None of the islands have any significant rivers or lakes.

HOW HAWAII BECAME A STATE

Hawaii was settled in the more than one thousand years ago by Polynesian peoples who took to the ocean to find a new home. They developed a civilization based on fishing, growing a root called taro, and raising pigs, chickens, and dogs.

That civilization was shattered when Captain James Cook landed on Kauai in January, 1778. He named the territory he found the Sandwich Islands, after his patron John Montagu, the fourth Earl of Sandwich. Cook was killed by the islanders on a return trip in 1779, but word of his discovery spread. After 1790, western traders and missionaries arrived in droves, spreading contagious disease and inducing alcoholism that devastated the local population. Between 1790 and 1810, the islands were consolidated into one government headed by King Kamehameha I, whose descendants ruled until 1872. The Hawaiian royalty worked to defend their culture against the westerners.

The first sugar plantation was planted on Kauai in 1835, and soon the descendants of the missionaries began acquiring land and building large plantations. They imported large numbers of Chinese and Japanese to provide labor. Discontent arose on the islands in the late nineteenth century when the McKinley Tariff levied a duty on sugar, crippling the Hawaiian economy. A U.S. government group called the Committee on Safety seized control of the islands in January 17, 1893, and petitioned the U.S. to annex the territory. But President Cleveland refused to approve the acquisition, and the Republic of Hawaii was created on July 4, 1894.

However, President McKinley was of a different mind, and Hawaii was annexed on August 12, 1898. The island was dominated by five companies that controlled the growing and processing of agricultural products. Hawaii's strategic importance was emphasized in World War II, and after the war the local residents wanted to become a state. However, some members of Congress were suspicious of Hawaii's large Japanese population, and the state wasn't admitted until August 21, 1959, when it became the 50th state.

HAWAII'S CITIES

HONOLULU (365,272), on the island of Oahu, is officially not a city—Hawaii doesn't have any incorporated cities or city governments. So what we're going to be talking about is a "census designated place"—albeit, one of the most beautiful cities in America. Honolulu was founded when Captain William Brown sailed into the harbor in 1794. European and American missionaries and traders soon followed, and Honolulu (which means "sheltered harbor") became the residence of the Hawaiian royal family. Honolulu covers the area from Makapuu Pond to Pearl Harbor, and includes the craters of two volcanoes, Diamond Head and the Punchbowl. The city is forever famous for the devastating air strike launched by the Japanese on ships moored in Pearl Harbor on December 7, 1941. Honolulu is a tourist and military center, as well as the administrative center for Hawaii.

HILO (37,808), is a seaport on the big island of Hawaii. It is the commercial center for an agricultural economy that produces sugar cane, fruits, orchids, and macadamia nuts. It is also a gateway to Hawaii Volcano National Park, which includes the island's two active volcanoes.

HAWAII'S PEOPLE

Hawaii's 1,171,592 people are by far the most ethnically diverse of any U.S. state. The population is 33% white, 2% black, 22% Japanese, 15% Filipino, 13% native Hawaiian, 6% Chinese, 2% Korean, and 1% Samoan. Intermarriage has produced more than 60 different racial mixes. Almost 90% of the population live in urban areas. Over the last decade, Hawaii has experienced a high rate of population growth.

HAWAII'S ECONOMY

In 1778, Hawaii went from a subsistence economy to one based on trade. After the founding of the first sugar plantation in 1835, the economy became focused on agriculture. Today, Hawaii's main products are sugar cane, pineapples, papayas, flowers, and macadamia nuts. It is also the only U.S. state in which coffee is grown.

There is very little manufacturing on the Islands. The few industries process the state's agricultural yield, producing raw sugar and canned fruits and juices.

Tourism is by far the most important element of the economy, generating $10 billion in annual revenues. Hawaii's exotic locales and excellent climate have also made it a favorite spot for filming movies and television programs.

Places to Visit

Honolulu, on the island of Oahu, offers the sun-drenched Waikiki beach as well as the intense human drama of Pearl Harbor recalled by the USS Arizona Memorial. The big island of Hawaii is features Hawaii Volcanoes National Park, that focuses on the still-active Kilauea. Maui has perfect beaches and sophisticated resorts. Kauai has some rugged coastline and some of the wettest areas of the 50 states. Tourist information: Hawaii Visitors Bureau, 2270 Kalakaua Ave., Honolulu, HI (808-923-1811)

The Other United States

American Samoa is a group of seven islands in the southern Pacific Ocean about 2,300 miles southwest of Honolulu, Hawaii and about the same distance north of New Zealand. Western Samoa, an independent nation, is to the west of American Samoa, while New Zealand's Cook Islands are to the east. Tutuila is the largest island of the group, covering an area of about 55 sq. mi. The other islands are Aunu'u, the three islands of the Manua'a Group, Rose, and Swain. All of the islands are remnants of volcanoes except Rose, which is uninhabited and used largely for naval target practice. Pago Pago, on Tutuila is the capital of American Samoa and has one of the finest harbors in the South Pacific. The total area of American Samoa is about 84 square miles, about 17 square miles larger than Washington, D.C. The population of the seven islands is 52,860.

According to native tradition the Samoan Islands were the original home of the Polynesian race, from which colonists peopled the other Polynesian islands of the Pacific. However, most historians believe that the Samoan Islands were first colonized by peoples from Southeast Asia. A later group of colonists kicked out the original Samoans, who then began to colonize the more easterly islands of Polynesia.

The islands were discovered in 1722 by Jacob Roggeveen a Dutch navigator. In 1768 Louis Antoine de Bougainville, a French explorer, named the group the Navigators Islands. During the 19th century Germany, Great Britain, and the United States established commercial posts on the islands. In 1878 the United States annexed Pago Pago for use as a naval coaling station. Beginning in 1888, the U.S., Great Britain, and Germany squabbled over control of the islands. Eleven years later, the three finally signed a treaty that gave the U.S. sovereignty over the islands west of longitude 171. . The chiefs of Tutuila and Aunuu ceded these islands to the United States in 1900, and the Manua group was ceded in 1904. Swains Island was annexed by the United States in 1925 and added to American Samoa. The Territory was administered by the Department of the Navy until 1951.

American Samoa is now administered by the U.S. Department of the Interior and is under the executive authority of a governor, who has been popularly elected since 1977. Samoans are U.S. nationals, and a significant number of them have moved to the continental U.S. and Hawaii. American Samoa has a non-voting delegate to the U.S. House of Representatives.

Agriculture is the basis of the economy, with the most important crops being taro, coconuts, bananas, oranges, pineapples, papayas, breadfruit, and yams. Samoans can fish and make grass mats and other handicrafts for export. But American Samoa isn't self-sufficient—it depends largely on U.S. grants.

Guam, is the largest and southernmost of the Mariana Islands, in the western North Pacific Ocean. This island is 3,700 miles west of Hawaii, 1,200 miles east of the Philippines, and about 1,200 miles north of New Guinea. Guam is 30 miles long and approximately 8 miles wide, and its total area is 209 square miles. It's strategic location makes it an important U.S. military base. The capital is Agana, which is located on the western coast north of Apra Harbor.

The northern and southern halves of Guam are like two different islands pasted together. The northern half is a plateau built of coral, while the southern portion is made up of volcanic hills. Apra Harbor, on the western coast, is the only port. If I was a weatherman on Guam, I'd be out of business—the temperature hovers around 80 degrees centigrade all year.

Guam was discovered in 1521 by the Portuguese navigator Ferdinand Magellan, who, claimed it for Spain, the country for which he was sailing. In 1898, Spain ceded the island to the U.S. in the Treaty of Paris that ended the Spanish-American War. The Japanese captured Guam in December 1941, during World War II., but is was recaptured by American forces between July 20 and August 10, 1944 in one of the bloodiest and most memorable invasion of the Pacific War. In 1950, the people of Guam became U.S. citizens. Up to 90% of the buildings on the island were damaged by Typhoon Omar in August, 1992, but most have been rebuilt.

The original natives of Guam, who are called the Chamorro, have a primarily Micronesian origin and they speak their own language. But English is the official language and is taught in the schools. The Chamorro make up a little less than half the population of Guam. The territory is administered by a legislature and an elected governor, and Guam sends one nonvoting delegate to the U.S. House of Representatives.

Executive power is vested in a governor, who is popularly elected to a 4-year term.

U.S. Naval, Army, and Air Force installations are the basis of the economy of Guam. The main crops are vegetables, citrus and tropical fruits, coconuts, and sugarcane. There is a very small manufacturing industry that produces some apparel, cement and plastic items.. Guam has become an increasingly popular tourist destination.

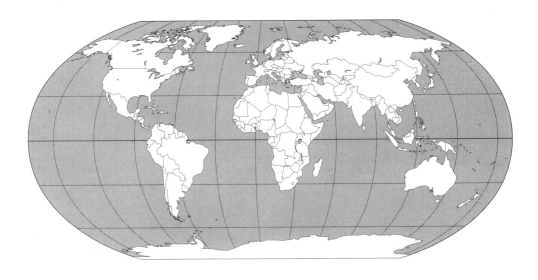

Epilogue

Charting Our Geographic Future

Just as surely as we do not understand all the mysteries of the earth's past, we also cannot speak with absolute certainty about its future. We are not yet able to predict precisely, for example, when natural calamities like earthquakes and volcanoes will occur. We do not know exactly when we will finally exhaust the earth's supply of precious resources, like oil. And, there is no widely accepted belief within the scientific community about when—or even if—the earth will begin to experience dramatic and devastating climatic changes, like global warming or the greenhouse effect. But, there are some things we do know—our planet is in a constant state of change, and the more knowledgeable we are, the better equipped we will be to survive.

The earth continues to move. Geological plates are shifting, oceans are expanding and shrinking, continents are gradually moving, volcanic activity is ongoing. I am not suggesting that we live in a constant state of panic, but I believe it is important for us to gain the greatest possible understanding of these natural forces, so we can be prepared for whatever the future may hold. If "the big one" ever does hit California and the far western United States, it will not only reshape geographic boundaries, it will have a profound life-and-death effect on tens of millions of people. The economic impact would be unthinkable. Such an event would have worldwide importance.

Also, as we look to the future, common sense should tell us that we need to treat our environment with greater care. You don't have to read this book to understand that the natural elements exist in a very delicate balance, and that balance has been significantly disturbed by human behavior and human inventions. Air quality, water quality, the atmosphere's protective ozone layer, and more have been seriously affected by our reckless disregard for the consequences of rapid industrialization and gluttonous human consumption of natural resources. It is amazing, but true, that it has taken mankind only about one century to consume a natural oil supply which was 250 million years in the making. Obviously, we cannot continue that rate of consumption. If individuals and nations adopt a "kinder, gentler" way of treating the planet, and a more careful use of its resources, our future *can* be a healthy one.

A sound knowledge of geography is essential in understanding and appreciating who we are and how we got here; but it is just as important in helping us determine where we are going and what our quality of life will be in the future. The world is rapidly shrinking as technology continues to advance, particularly in the field of mass communication. Boundaries will be redrawn, cultures will merge—or clash, the balance of economic and military power will shift, political systems will change, and population will grow at a staggering rate. A better understanding of our natural environment will greatly enhance our ability to survive and prosper during the next century of change.

INDEX